Methods in Molecular Biology™

AF774190

Series Editor
John M. Walker
School of Life Sciences
University of Hertfordshire
Hatfield, Hertfordshire, AL10 9AB, UK

For other titles published in this series, go to
www.springer.com/series/7651

METHODS IN MOLECULAR BIOLOGY™

Nucleic Acid and Peptide Aptamers

Methods and Protocols

Edited by

Günter Mayer

University of Bonn, Germany

Humana Press

Editor
Günter Mayer
University of Bonn
Bonn, Germany
gmayer@uni-bonn.de

Series Editor
John M. Walker
University of Hertfordshire
Hatfield, Herts.
UK

ISSN 1064-3745 ISSN 1940-6029 (eISSN)
ISBN 978-1-62703-893-5 ISBN 978-1-59745-557-2 (eBook)
DOI 10.1007/978-1-59745-557-2

Cover illustration: Derived from Figure 2 in Chapter 22

Printed on acid-free paper

springer.com

Preface

After the deciphering of the human genome and those of other organisms, the investigation of the function of gene products and their orchestral interplay is now one of the most important challenges in the life sciences. In this regard, specific ligands are required that allow the sensitive detection and functional assignment of gene products, favourably in the native context, which can be a model organism or at least cultured cells. Darwinian-like evolutionary methods, which enable the identification of such ligands, are described in this volume of the Humana Press Methods in Molecular Biology Series, entitled *Nucleic Acid and Peptide Aptamers*. The identified active compounds according to the protocols described in *Nucleic Acid and Peptide Aptamers* harbour information about both their active conformation and the blueprint for their own synthesis. This feature allows the simultaneous screening of up to 10^{16} different molecules in one test tube by the application of appropriate selection schemes and the rapid synthetic access to adapt the ligands for certain purposes. Selection procedures can be performed solely in vitro, allowing the most convenient control of the selection process and thus retaining control of the characteristics of the identified ligands. Target molecules can be either small compounds (or metabolites), proteins, nucleic acids or even complex targets such as living cells.

The present protocol collection covers methods related to the two major classes of molecules employed for in vitro selection procedures: Nucleic acids and peptides/proteins. The 22 chapters of *Nucleic Acid and Peptide Aptamers* highlight important methodologies in the field of evolutionary molecular biology approaches. The collection allows researchers not only to identify ligands for their target molecules but also describes protocols for the application of these ligands in certain research issues. These ligands, unless they are of nucleic acid or of peptidic nature, can act as potent inhibitors and enable the functional investigation and/or the detection of the target molecule. This volume is meant to support students, postdoctoral fellows, and senior scientist in their efforts to investigate biomolecules by using specific nucleic acid and/or peptide aptamers and offers guidelines for their identification and application.

Chapter 1 (by Ellington) describes the synthesis of nucleic acid libraries and methods to investigating their diversity. In the following Chapters 2–6 protocols for the application of different in vitro selection methods targeting distinct molecules are illustrated in detail. These protocols cover the modification of proteins with biotin, enabling access to streptavidin–biotin chemistry for the separation step during the selection process (by Höver and Mayer) and protocols for the identification of aptamers by capillary electrophoresis (by Mosing and Bowser), a method that circumvents any protein modification prior selection. Chapters 4–6 describe the selection of aptamers targeting small molecules (by Piganeau), complex targets (by Franciscis) and ribonucleic acids (by Toulmé). Protocols for the characterization of aptamers by state-of-the-art methods can be found in the Chapters 7 (by Werner and Hahn), describing the application of fluorescence correlation spectroscopy for determining the dissociation constant of an aptamer-target interaction. On the subject of structural investigations Chapters 8 and 9, by Wakeman and Winkler and Batey et al., respectively, give protocols for the application of in-line probing of RNA structures and the growth and analysis of crystals to determine the secondary and tertiary structure of RNA molecules by X-ray crystallography. Applications of aptamers are highlighted in the Chapters 10–14. These chapters give insight how aptamers can be adapted to different assay formats, covering locked nucleic acids (by Erdmann), the application of aptamers for diagnostic purposes (by Gronewold and by Lu et al.), the use of aptamers to control gene expression (by Weigand and Suess) and as molecular probes for the identification of small molecule inhibitors of protein function with aptamer inherited properties (by Yamazaki and Famulok). The nucleic acid aptamer part is then closed by a Chapter 15 by Tavitian et al. describing protocols allowing the in vivo imaging of aptamers.

The peptide aptamer part commences with protocols that describe different methods for the identification of peptide aptamers. These chapters also include detailed explanations of the construction of suitable peptide libraries for the selection process. Chapter 16 by Arndt et al. describes the use of phage display and complementation assays for the identification of peptides that interfere with protein-protein interactions. Chapter 17 by Takahashi and Roberts introduce the mRNA display methodology and strategies based on the well-known and wide spread used yeast "two hybrid" system for the specific enrichment of peptide aptamers and ligand-regulated peptide aptamers can be found in Chapter 18 (by Miller). Lopez-Ochoa et al. give details for the high-throughput identification and characterization of peptide aptamers in Chapter 19. A recently described variant of peptide aptamers, namely Microbodies, which are embedded in a certain three-dimensional, highly stable scaffold, is introduced in Chapter 20 (by Blind). Chapters 21 and 22 illustrate how peptide aptamers can be used to identify small molecules (by Colas) and for the application of peptides as drug carriers (by Beck-Sickinger).

The protocols given herein represent a state-of-the-art collection of methodologies for the isolation, characterization and application of both peptide and nucleic acid aptamers and will allow researchers to apply these compounds to address distinct research issues.

Günter Mayer, PhD.

Contents

Contributors

KATJA M. ARNDT • *Institute of Biology III, Albert-Ludwigs-University of Freiburg, Freiburg, Germany*
JAN BARCISZEWSKI • *Institute of Bioorganic Chemistry of the Polish Academy of Sciences, Poznan, Poland*
CARINE BARDOU • *Aptanomics S.A., Lyon, France; Imaxio, Lyon, France*
ROBERT T. BATEY • *Department of Chemistry and Biochemistry, University of Colorado, Boulder, CO, USA*
ANNETTE G. BECK-SICKINGER • *Institute of Biochemistry, Faculty of Biosciences, Pharmacy and Psychology, Leipzig University, Leipzig, Germany*
MARC BICKLE • *Aptanomics S.A., Lyon, France*
MICHAEL BLIND • *NascaCell Technologies AG, Munich, Germany*
RAPHAËL BOISGARD • *Inserm U803, Laboratoire d'Imagerie Moléculaire Expérimentale, Service hospitalier Frédéric joliot, Intitut d'Imagerie Biomédicale, CEA, Orsay, France*
CHRISTOPHE BORIE • *Aptanomics S.A., Lyon, France*
MICHAEL T. BOWSER • *University of Minnesota, Minneapolis, MN, USA*
ANTHONY BUGAUT • *Department of Chemistry, University of Cambridge, Cambridge, UK*
LAURA CERCHIA • *Istituto per l'Endocrinologia e Oncologia Sperimentale "G. Salvatore," Naples, Italy*
PIERRE COLAS • *Aptanomics S.A., Lyon, France*
MATIN DANESCHDAR • *Institut für Biochemie und Organische Chemie, TU Darmstadt, Darmstadt, Germany*
ERIC DAUSSE • *INSERM, Institut Européen de Chimie et Biologie, Pessac, France; Université Victor Segalen, Bordeaux, France*
VITTORIO DE FRANCISCIS • *Istituto per l'Endocrinologia e Oncologia Sperimentale "G. Salvatore," Naples, Italy*
FREDERIC DOLLE • *Groupoe de Radiochimie, Laboratoire d'Imagerie Moléculaire Expérimentale, Service hospitalier Frédéric joliot, Intitut d'Imagerie Biomédicale, CEA, Orsay, France*
FREDERIC DUCONGE • *Inserm U803, Laboratoire d'Imagerie Moléculaire Expérimentale, Service hospitalier Frédéric joliot, Intitut d'Imagerie Biomédicale, CEA, Orsay, France*
ANDREA L. EDWARDS • *Department of Chemistry and Biochemistry, University of Colorado, Boulder, CO, USA*
ANDREW D. ELLINGTON • *Department of Chemistry and Biochemistry, Institute for Cell and Molecular Biology, University of Texas at Austin, Austin, TX, USA*

VOLKER A. ERDMANN • *Institute for Chemistry and Biochemistry, Free University Berlin, Berlin, Germany*
MICHAEL FAMULOK • *Life and Medical Sciences, Program Unit Chemical Biology and Medicinal Chemistry, University of Bonn, Bonn, Germany*
ANNETTE GAIDA • *Institute of Biology III, Albert-Ludwigs-University of Freiburg, Freiburg, Germany*
ANDREW D. GARST • *Department of Chemistry and Biochemistry, University of Colorado, Boulder, CO, USA*
PALOMA H. GIANGRANDE • *Department of Internal Medicine, Division of Cardiology, University of Iowa, Iowa City, IA, USA*
THOMAS M. A. GRONEWOLD • *Biosensor GmbH, Bonn, Germany*
URS B. HAGEMANN • *Institute of Biology III, Albert-Ludwigs-University of Freiburg, Freiburg, Germany*
ULRICH HAHN • *Department of Chemistry, Institute for Biochemistry and Molecular Biology, University of Hamburg, Hamburg, Germany*
BRADLEY HALL • *Department of Chemistry and Biochemistry, Institute for Cell and Molecular Biology, University of Texas at Austin, Austin, TX, USA*
LINDA HANLEY-BOWDOIN • *Department of Molecular and Structural Biochemistry, North Carolina State University, Raleigh, NC, USA*
THOMAS HÖVER • *Life and Medical Sciences, Program Unit Chemical Biology and Medicinal Chemistry, University of Bonn, Bonn, Germany*
TROELS KOCH • *Santaris Pharma, Horsholm, Denmark*
HARALD KOLMAR • *Institut für Biochemie und Organische Chemie, TU Darmstadt, Darmstadt, Germany*
JENS KURRECK • *Institute for Chemistry and Biochemistry, Free University Berlin, Berlin, Germany; Institute of Industrial Genetics, University of Stuttgart, Stuttgart, Germany*
ISABELLE LEBARS • *CNRS – Université Bordeaux, Institut Européen de Chimie et Biologie, Pessac, France*
JUEWEN LIU • *Department of Chemistry, University of Illinois Urbana-Champaign, Urbana, IL, USA*
LUISA LOPEZ-OCHOA • *Department of Molecular and Structural Biochemistry, North Carolina State University, Raleigh, NC, USA*
YI LU • *Department of Chemistry, University of Illinois Urbana-Champaign, Urbana, IL, USA*
DINAH MATTAY • *Institute of Biology III, Albert-Ludwigs-University of Freiburg, Freiburg, Germany*
GÜNTER MAYER • *Life and Medical Sciences, Program Unit Chemical Biology and Medicinal Chemistry, University of Bonn, Bonn, Germany*
DEBAPRIYA MAZUMDAR • *Department of Chemistry, University of Illinois Urbana-Champaign, Urbana, IL, USA*
JAMES O. MCNAMARA • *Department of Internal Medicine, Division of Cardiology, University of Iowa, Iowa City, IA, USA*
MICHAEL MEDGAARD • *Santaris Pharma, Horsholm, Denmark*
RUSSELL A. MILLER • *Clinical Research Building, University of Pennsylvania, Philadelphia, PA, USA*

Renee K. Mosing • *University of Minnesota, Minneapolis, MN, USA*
Kristian M. Müller • *Institute of Biology III, Albert-Ludwigs-University of Freiburg, Freiburg, Germany*
Tara E. Nash • *Department of Molecular and Structural Biochemistry, North Carolina State University, Raleigh, NC, USA*
Ines Neundorf • *Institute of Biochemistry, Faculty of Biosciences, Pharmacy and Psychology, Leipzig University, Leipzig, Germany*
Shawn K. Piasecki • *Department of Chemistry and Biochemistry, Institute for Cell and Molecular Biology, University of Texas at Austin, Austin, TX, USA*
Nicolas Piganeau • *Department of Chemistry, Institute for Biochemistry and Molecular Biology, University of Hamburg, Hamburg, Germany*
Jorge Ramirez-Prado • *Department of Plant Pathology, North-Carolina State University, Raleigh, NC, USA*
Christina Räuber • *Institute of Biology III, Albert-Ludwigs-University of Freiburg, Freiburg, Germany*
Bernard Rayner • *INSERM, Institut Européen de Chimie et Biologie, Pessac, France; Université Victor Segalen, Bordeaux, France*
Robert Rennert • *Institute of Biochemistry, Faculty of Biosciences, Pharmacy and Psychology, Leipzig University, Leipzig, Germany*
Richard W. Roberts • *Departments of Chemistry, Chemical Engineering, and Biology, University of Southern California, Los Angeles, CA, USA*
Brian B. Rudkin • *Differentiation and Cell Cycle Group, Laboratoire de Biologie Moléculaire de la Cellule, UMR 5239 CNRS/ENS Lyon, Univerité Lyon 1, Lyon, France*
Hans-Ulrich Schmoldt • *NascaCell Technologies AG, Munich, Germany*
Beatrix Suess • *Institut für Molekulare Biowissenschaften, Johann-Wolfgang-Goethe-Universität Frankfurt, Frankfurt, Germany*
Terry T. Takahashi • *Department of Chemistry, University of Southern California, Los Angeles, CA, USA*
Bertrand Tavitian • *Inserm U803, Laboratoire d' Imagerie Moléculaire Expérimentale, Service hospitalier Frédéric joliot, Intitut d' Imagerie Biomédicale, CEA, Orsay, France*
Jean-Jacques Toulmé • *INSERM, Institut Européen de Chimie et Biologie, Pessac, France; Université Victor Segalen, Bordeaux, France*
Catherine A. Wakeman • *Department of Biochemistry, The University of Texas Southwestern Medical Center, Dallas, TX, USA*
Marguerite Watrin • *INSERM, Institut Européen de Chimie et Biologie, Pessac, France; Université Victor Segalen, Bordeaux, France*
Julia E. Weigand • *Institut für Molekulare Biowissenschaften, Johann-Wolfgang-Goethe-Universität Frankfurt, Frankfurt, Germany*
Arne Werner • *Department of Chemistry, Institute for Biochemistry and Molecular Biology, University of Hamburg, Hamburg, Germany*
Wade C. Winkler • *Department of Biochemistry, The University of Texas Southwestern Medical Center, Dallas, TX, USA*
Satoko Yamazaki • *Life and Medical Sciences, Program Unit Chemical Biology and Medicinal Chemistry, University of Bonn, Bonn, Germany*

Part I

Nucleic Acid Aptamers

Chapter 1

Nucleic Acid Pool Preparation and Characterization

Shawn K. Piasecki, Bradley Hall, and Andrew D. Ellington

Abstract

Random sequence nucleic acid pools can be used in a variety of applications, including the selection of functional nucleic acids such as protein binding sites, aptamers, and ribozymes. While the design, synthesis, and purification of pools is relatively straightforward, keeping track of the size and complexity of a nucleic acid pool can sometimes task even an experienced researcher. The following protocol takes the reader through the steps necessary for the preparation of a pool of known complexity.

Key words: Nucleic acid pool, random sequence, complexity, primer extension assay, SELEX, in vitro selection.

1. Introduction

Random sequence nucleic acid pools are used in a variety of applications, but most especially during the selection of nucleic acid aptamers *(1–4)*. The design, synthesis, and purification of a pool are critical to the success of in vitro selection experiments, and pool preparation can take upwards of a month.

Most random sequence pools for aptamer selections consist of a central random region flanked by primer binding sites for amplification and transcription (*see* **Fig. 1.1**; **Note 1**). When designing a new pool, a number of variables must be considered. First, the length of the pool will determine both relative functionality of individual members of the pool and the number of different nucleic acid sequences that can be made. Longer pools can form more complex structures that may show better binding to target molecules. However, the overall yield of longer pools falls off quickly, especially above 100 random sequence positions. In

Günter Mayer (ed.), *Nucleic Acid and Peptide Aptamers: Methods and Protocols, vol. 535*

DOI 10.1007/978-1-59745-557-2_1 Springerprotocols.com

```
            T7 Promoter
5'-GATAATACGACTCACTATAGGCGCTCCGACCTTAGTCTCTG-3'
41.62F                  5'-GGCGCTCCGACCTTAGTCTCTG-N62-GAACCGTGTAGCACAGCAGA-3'
                        N62 pool                       3'-CTTGGCACATCGTGTCGTCT-5'
                                                       20.62R
```

Fig. 1.1. The N62 pool consists of a 'forward' primer (41.62F) binding site, a central random region (of 62 random nucleotides in this case), and a 'reverse' primer (20.62R) binding site. Forward primers are often designed with a T7 RNA polymerase promoter (*underlined*) in order to transcribe the DNA pool into RNA.

addition, it can be difficult to readily identify sequence and structural motifs selected from longer random sequence pools. Therefore, we typically make pools that contain at most 70 random sequence positions, and frequently as few as 30. For most pools, it is suggested either that they be synthesized by researchers familiar with DNA synthesis, or that a single random sequence oligonucleotide be ordered from a supplier who is known to be capable of synthesizing oligonucleotides of up to 150 residues with little error. Since pools should be synthesized so as to appropriately balance the differential reactivities of phosphoramidites *(5)*, it is strongly suggested that the sequence compositions of pools from commercial sources be determined by cloning and sequencing several members of the pool prior to amplification (*see* **Note 2**). If the composition of a pool is significantly skewed, the pool may not be suitable for many applications.

Additional modifications may be necessary depending on what chemistry is desired for the pool that will be used during selection. A RNA polymerase promoter can be introduced into one of the constant regions in order to transcribe the DNA pool into RNA or modified RNA; a biotinylated primer can be used for amplification in order to separate one DNA strand from the other and thereby generate a single-stranded DNA pool.

In this report, we will assume that a given pool (N62, *see* **Fig. 1.1**) has been designed and ordered, and that the pool must be amplified, that its complexity must be determined, and that it will eventually be transcribed into RNA.

2. Materials

2.1. Pool Purification of a Newly Synthesized Pool via Polyacrylamide Gel Electrophoresis (PAGE)

1. TBE (Tris/borate/EDTA) electrophoresis buffer (10 ×): 890 mM Tris base, 890 mM boric acid, 20 mM EDTA (pH 8.0). Stable at room temperature.
2. 8% Acrylamide: Mix 50 mL 10 × TBE, 100 mL 40% acrylamide/bis-acrylamide solution (19:1), 210 g urea and water up to 500 mL (*see* **Note 3**). Filter through 0.45 μm PES

membrane. Unpolymerized acrylamide is a neurotoxin, avoid exposure to bare skin.

3. N,N,N,N'-Tetramethyl-ethylenediamine (TEMED, EMD Chemicals Inc., Gibbstown, NJ). Store at 4°C in a flammables refrigerator.
4. Ammonium persulfate (APS): Prepare a 10% w/v solution in water. It is a very strong oxidizing agent and a radical initiator, avoid exposure to skin. Stable at 4°C for 6 months or –80°C for long-term storage.
5. 2X Denaturing dye: 7 M urea in 1 × TBE, 0.1% bromophenol blue. Stable at room temperature for 2 months or –20°C for long-term storage.
6. Model V16 Whatman/Biometra vertical PAGE apparatus, 170 × 150 × 1.5 mm gel size (Labrepco, Horsham, PA).
7. 500 V Power source. Can be purchased from most large lab supply warehouses.
8. K6F TLC plate 20 × 20 cm, 250 um, 60 A pore size, with fluorescent indicator (Whatman, Florham Park, NJ).
9. 13 mL 16.8 × 95 mm PP Sarstedt tubes (VWR, West Chester, PA).
10. Ultrafree-MC (Durapore 0.45 um) spin filter (Millipore, Billerica, MA).

2.2. Labeling and Purifying Primers for Extension Assay

1. T4 polynucleotide kinase buffer, T4 polynucleotide kinase (at 10,000 U/mL) (New England BioLabs, Ipswich, MA).
2. γ^{32}P-ATP (ICN Biomedical Inc, Aurora, OH).
3. Centri-Sep Spin Columns (Princeton Separations, Freehold, NJ).
4. Phenol/chloroform/isoamyl alcohol (25:24:1) saturated with 10 mM Tris–HCl, pH 8.0, 1 mM EDTA. Can be purchased from any major chemical supplier. Store at 4°C in a flammables refrigerator.
5. 99% Chloroform from any major chemical supplier. Store at room temperature in a flammables cabinet.
6. Ethanol precipitation: 3 M NaOAc, glycogen (DNase free), and 95% ethyl alcohol. Can be stored at room temperature, ethanol should be stored in a flammables cabinet.

2.3. Primer Extension Assay

1. 10X PCR buffer including 50 mM $MgCl_2$, 4 mM dNTPs, γ^{32}P-labeled primer at 4 pmol/μL, ssDNA template at 100 pmol, Taq polymerase (5 U/μL). Can be purchased from any supplier. Components are stable at –20°C for extended periods, and polymerase should not undergo freeze thaw cycles.
2. Materials from **Sect. 2.1** to PAGE purify.

3. Gel drying paper.
4. Access to phosphorimaging equipment.
5. Ultrafree-MC (Durapore 0.45 μm) spin filter (Millipore, Billerica, MA).

2.4. Polymerase Chain Reaction (PCR) of Pool ssDNA

1. 10X PCR buffer including 50 mM $MgCl_2$, 4 mM dNTPs, 200 μM forward and reverse primers, and Taq polymerase (5 U/μL). Can be purchased from any supplier. Components are stable at −20°C for extended periods, and polymerase should not undergo freeze thaw cycles.
2. 3.8% w/v 3:1 NuSieve agarose (Cambrex Corporation, East Rutherford, NJ), (*see* **Note 4**).
3. 6X Orange dye: 60% glycerol, 0.025–0.1% w/v Orange G, (*see* **Note 4**).
4. 100 base pair DNA ladder (Invitrogen, Carlsbad, CA). Can be stored at room temperature.
5. 100 ng/μL 200 bp quantitation standard (GenSura, San Diego, CA).
6. Ultrafree-MC (Durapore 0.45 μm) spin filter (Millipore, Billerica, MA).
7. Materials from **Sect. 2.2** for ethanol precipitation.

2.5. Transcription and Purification to New RNA Pool

1. AmpliScribe T7 High Yield Transcription Kit, (Epicentre, Madison, WI).
2. Materials from **Sect. 2.1** for PAGE purification of the new RNA pool.
3. Materials from **Sect. 2.2** for ethanol precipitation.

3. Methods

3.1. Pool and Primer Purification via Polyacrylamide Gel Electrophoresis (PAGE)

Once a pool has been designed, a crude oligonucleotide corresponding to the design should be synthesized, purified, and amplified. The deprotected oligonucleotide from a 1 μmol synthesis can typically be purified on two to three 17 × 15 × 1.5 mm 8% denaturing polyacrylamide gels.

3.1.1. Pool Purification

1. The 1 umol synthesis should be resuspended in 500 μL diH_2O.
2. Clean a large 17 cm and small 15 cm glass per gel being run (*see* **Note 5**).
3. Prepare an 8% denaturing polyacrylamide gel (*see* **Note 6**).
4. While the gel is polymerizing, add 500 μL 2 × denaturing dye to your crude oligo and mix well.

5. Heat denature for 5 min at 90°C and cool to room temperature for 10 min.
6. Once the acrylamide gel has polymerized, remove the comb and wash under tap water to remove any unpolymerized acrylamide.
7. Set up the gel rig according to the manufacturer's instructions using 1 × TBE buffer. The urea will begin to diffuse from the gel into the buffer, so it is necessary to blow out the wells immediately prior to loading.
8. Load between 30 and 50 μL of the sample per well.
9. Run the gel at a relatively low voltage (350 V), which will likely take between 1 and 2 h.
10. Once the bromophenol blue dye has run to the bottom of the gel, electrophoresis is stopped and the gel is wrapped in Saran Wrap and placed onto a large TLC plate containing an embedded fluorophore. This allows UV visualization of the DNA bands as dark shadows on the TLC plate.
11. Cut out the nucleic acid band (*see* **Note 7**).
12. The gel slice (top 1 cm of the bands) is further chopped up with a razorblade to facilitate elution of the DNA oligonucleotide.
13. The PAGE separated ssDNA can be eluted by combining the chopped gel chunks and 10 mL 0.3 M NaOAc in two 13 mL Sarstedt tubes.
14. Elution should proceed overnight with rotation in a 37°C incubator.
15. Transfer the eluate (leaving behind the gel chunks) into a 50 mL high speed conical tube for ethanol precipitation.
16. Precipitate the purified oligonucleotides with ethanol: add 10 μL glycogen and 2.5 volumes of 95% ethanol (25 mL).
17. Vortex and incubate at –80°C for 30 min or –20°C for several hours.
18. Centrifuge the precipitate at 13,000 *g* for 1 h.
19. Decant the supernatant and wash the pellet with 95% chilled ethanol
20. Centrifuge for an additional 5 min, remove the residual supernatant by pipetting.
21. Dry the pellet at room temperature for 30 min or in a speed-vac to remove any contaminating ethanol.
22. Once dry, resuspend the pellet containing the pool in 500 μL sterile diH_2O
23. To remove any insoluble material, purify the resuspended oligonucleotide by spinning through an Ultrafree-MC filter.

This unamplified, single-stranded pool is the stock for the production of multiple, double-stranded libraries. The concentration of the pool should be determined via UV spectrometry on a NanoDrop (or other spectrophotometer). The OD at 260 nm (A_{260}) is multiplied by a conversion factor of 33 ng/µL per A_{260} unit (AU); this conversion factor is an average for ssDNA oligonucleotides *(6)*.

A recent synthesis of the N62 pool yielded 1,321 ng/µL purified ssDNA. The molar concentration was calculated by dividing the concentration (ng/µL) by the length of the synthesized pool (104 bases) times the average molecular weight of a ssDNA base (0.330 ng/pmol); for the N62 pool this corresponded to 38.5 pmol/µL, or 38.5 µM. The final concentration is frequently found to be between 30 and 70 µM. The total complexity of the synthesis is calculated by multiplying the concentration by the resuspension volume: 38.5 pmol/µL * 500 uL = 19.3 nmol or 1.2E16 total sequences. Keep in mind this synthetic complexity is not the same as the usable pool complexity, which will be further calculated below. In order to keep track of individual syntheses, the pool should be explicitly labeled with its name, type of nucleic acid (ssDNA), who synthesized it, date of synthesis, concentration, and resuspension volume.

3.1.2. Primer Purification and Labeling

DNA primers will be used not only to amplify the pool, but to characterize its complexity. Primers are routinely ordered from commercial sources like IDT, Invitrogen and others. Depending on the purity, they can be resuspended and used immediately or purified via PAGE gel using the protocol described for DNA pools above. When calculating the primer concentration, the exact primer extinction coefficient (calculated using online resources such as SciTools OligoAnalyzer on the IDT website) should be used. Primers are typically diluted to either 400 or 200 µM prior to use.

Once the ssDNA pool has been purified, the pool complexity should be calculated. In order to assess pool complexity an extension assay is performed with radiolabeled primers. The labeling reaction is as follows:

2 µL	10 × T4 polynucleotide kinase (PNK) buffer
1 µL	T4 PNK (10,000 U/mL)
0.5 µL	γ^{32}P ATP (about 20 pmol, 100 µCi)
4 µL	Reverse primer at 20 µM (80 pmol total)
12.5 µL	diH_2O

1. Per 20 μL kinase reaction, combine:
2. Incubate the reaction at 37°C for 1 h. To effectively manage time, it is helpful to set up the Centri-Sep columns for the subsequent purification step (*see* **Note 8**).
3. Heat inactivate the kinase at 70°C for 10 min.
4. Adding the 20 μL kinase reaction directly onto the middle of the Centri-Sep gel bed, being careful not to touch the pipette tip to the gel. If this happens, the column must be re-made.
5. Place the column into its corresponding collection container, maintaining proper orientation, and centrifuge at 450 *g* for 2 min.

The labeled oligonucleotide is then further purified to remove contaminating enzymes and small organics via a phenol/chloroform extraction.

6. Add equal volumes of the Centri-Sep flow-through sample (20 μL) and phenol/chloroform/isoamyl alcohol (25:24:1) are mixed.
7. Vortexed and then centrifuge at 13,000 *g* to separate the organic layer from the aqueous layer (containing radiolabeled primer).
8. Extract the aqueous layer again with 20 μL 99% chloroform to remove any trace phenol.
9. Ethanol-precipitate the final aqueous layer by adding 1:10th volume of 3 M NaOAc (2 μL), 3 μL glycogen and 2.5 volumes of ethanol. Follow the protocol described in **Sect. 3.1.1** (**Steps 17–21**). The drying step not only removes contaminating ethanol, but also trace chloroform.
10. The resulting pellet is resuspended in 20 μL diH_2O (*see* **Note 9**).

3.2. Primer Extension Assay

Some fraction of the DNA pool will have been damaged during the repeated acid deprotection steps during chemical synthesis (since the pool is synthesized in a 3'–5' direction, the 3' end of the pool is generally much more damaged than the 5' end, as it has seen many more acid wash steps). In order to determine what fraction of the pool is functional, we routinely carry out primer extension reactions, which mimic the end-to-end DNA synthesis that occurs during PCR amplification of the pool.

For the extension assay, 10 pmol of primer is incubated with 100 pmol of ssDNA template (10-fold molar excess of pool). The DNA template is in excess so that all of the primer will hybridize and potentially be extended. A 'no template' reaction should be run in parallel, to ensure that there are no artifactual bands due to the primer alone. The assay can be performed with either reverse transcriptase or a DNA polymerase, which will treat lesions in the template DNA differently. The polymerase that provides the most complete extension of the synthetic pool should be used for the scale-up reactions, as well (below, we have used Taq polymerase, which routinely extends upwards of 30% of the pool to full length).

1. Set up two 30 μL reactions, differing in the addition of template:

3 μL	10X PCR buffer
1 μL	50 mM $MgCl_2$
1.5 μL	4 mM dNTPs
2.5 μL	γ^{32}P-labeled reverse primer at 4 pmol/μL (*see* **Note 9**)
X μL	100 pmol ssDNA template "+" or water "-" (*see* **Fig. 1.2**)
21.5-X μL	diH_2O
0.5 μL	Taq polymerase (5 U/μL)

2. Run one cycle of PCR with a 30 min extension step.

95°C	3 min
50°C	1 min (*see* **Note 10**)
72°C	30 min

3. Add an equal volume (30 μL) of 2 × denaturing dye and separate on a 0.75 mm 8% denaturing polyacrylamide gel similar to **Sect. 3.1.1** (**Steps 2–9**). The reaction will fit into one well of a 10-well comb. The gel can be run at 450 V instead of 350 V.
4. Once the bromophenol blue has run to the bottom of the gel, dry the gel under vacuum on drying paper at 75°C for about 1 h.
5. Cool to room temperature before removing vaccum.
6. The dried gel can be analyzed following exposure to a Phosphorimager screen.

The extension efficiency of the pool is calculated by dividing the radioactive signal corresponding to the fully extended primer (top band) by the total signal of fully and partially extended sequences (*see* **Fig. 1.2**). The extension efficiency in this case was 12.3%; the typical range for a pool of this size is between 10 and 30%.

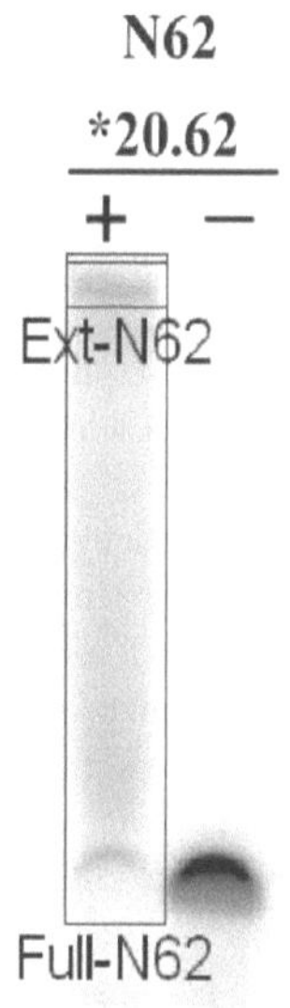

Fig. 1.2. The extension assay of the N62 pool. *20.62 represents the reverse primer used. The template addition is labeled "+" and "–". The extension efficiency is calculated by dividing the signal within the small *grey box* (fully extended primer) by the signal in the *large grey box* (all extended sequences).

These experiments should provide the information necessary to calculate the complexity of the pool. The total number of different fully extendable sequences from our synthesis: 1.2E16 total sequences * 12.3% = 1.5E15 extendable sequences in 500 μL total volume. This number is calculated prior to pool amplification because amplification is merely increasing the overall number of each unique sequence, but is not significantly increasing the complexity of the pool.

3.3. Polymerase Chain Reaction (PCR) of Pool DNA

The nascent pool will be amplified into a library that contains multiple copies of each individual sequence. Prior to amplification, the researcher needs to decide both on the total complexity of the desired pool, and the number of copies of the pool. Many selections start with between 1E13 and 1E15 different sequences. If the total number of extendable sequences from the original synthesis (above) is less than the number of sequences desired for a selection experiment, then the pool should be resynthesized. In the current example, it will be assumed that 1E14 different N62 sequences are desired for a selection experiment. To amplify 1E14 sequences of the 1.5E15 total extendable sequences will thus require 33 μL of the 500 μL total volume.

Having decided on the amount of input DNA that will be used, the volume of the PCR required to amplify the single-stranded DNA pool into a double-stranded DNA library must also be calculated. In general, it is assumed that the total amount of DNA that can be made in a given volume of a PCR is relatively constant, and that the variable is therefore how many copies of the original pool are desired in the final library. If ten copies of the pool are desired, a smaller overall reaction volume will be required than if 100 copies of the pool are desired.

As a working hypothesis, it is generally useful to assume that PCR of single-stranded DNA oligonucleotides of 100–200 base-pairs in length will yield approximately 1 ug of double-stranded DNA product in a 100 μL reaction. Since the average molecular weight of members of the N62 pool is 74,400 g/mol (based on the average molecular weight per base pair of dsDNA (*see* **Note 11**)), this means that there would ultimately be on the order of 8E12 total double-stranded DNA molecules in each 100 μL reaction (alternately, it is often easy to assume that a 100 μL reaction will yield 1E13 sequences). Thus, to make ten copies of the pool each 100 μL reaction would be seeded with 8E11 single-stranded DNAs, which in turn implies a total reaction volume of 12.5 mL of PCR seeded with 33 μL of the N62 ssDNA pool.

The large-scale PCR can be performed in a variety of ways, including as a batch reaction incubated in water baths of different temperatures. However, with the advent of high-throughput thermal cyclers, it is generally easiest to perform even such large reactions in multiple 96-well PCR plates.

When dealing with large PCR reactions, preliminary optimizations will prevent wasted time and money. In order to determine whether the large-scale PCR reaction will work, it is necessary to first attempt amplifications on a smaller scale. In theory, since only ten copies of the pool are being made, only ca. 3–4 thermal cycles of amplification (yielding 2^3–2^4 progeny) will be needed. However, in practice since amplification capacity (nucleotides and polymerases) will be going towards the production of both fully extended and partially extended templates during the early rounds of amplification, it is generally useful to empirically determine how many thermal cycles are required for full amplification of a given number of DNA molecules. The optimum number of thermal cycles can be determined using a cycle course reaction. Optimization of primer annealing temperature, $MgCl_2$, dNTP, and primer concentrations can also be performed at this juncture. PCR efficiency (whether a doubling of molecules or a smaller exponent is observed) can be experimentally verified (for methods, *see* ref. *(7)*).

The cycle course is set up so as to mimic the eventual large-scale reaction. The pool is first extended with one primer (the reverse primer) to generate full-length templates, and then the other primer (the forward primer) is added and a PCR is initiated.

1. A 100 μL reaction can be set up as follows:

10 μL	10X PCR buffer
3 μL	50 mM $MgCl_2$ (*see* **Note 12**)
5 μL	4 mM dNTPs
2.5 μL	20 nM reverse primer
1 μL	1:4 dilution of ssDNA template (*see* **Note 13**)
75.5 μL	diH_2O
0.5 μL	Taq polymerase (5 U/μL)

2. Initial extension carried out at:

95°C	3 min
50°C	1 min (*see* **Note 10**)
72°C	30 min

1. 2.5 μL of the Forward primer (20 nM) is added, and the reaction is run for additional thermal cycles consisting of:

92°C	2 min
50°C	1 min (*see* **Note 10**)
72°C	3 min

4. The amplification reaction is monitored by taking 5 μL samples near the end of the extension steps of cycles **4, 6, 8, 10, 12,** and **14** (*see* **Note 14**).
5. The samples are readily compared by electrophoresis on a 3.8% agarose gel (*see* **Fig. 1.3**). For estimating size and quantity of PCR products, a 100 base-pair DNA ladder along with 100, 50, and 25 ng quantitation standards should be included on the gel.

The number of thermal cycles that yield a stained band with density similar to the 50 ng quantitation standard should also be used for the large-scale PCR (in our example, the band is ca. 120 bp long).

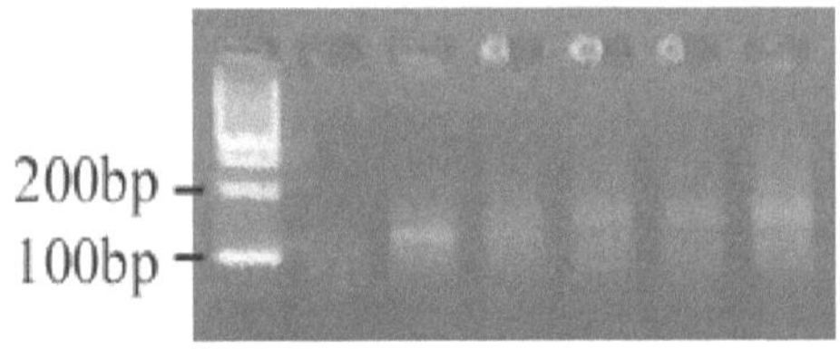

Fig. 1.3. Cycle course PCR of N62 pool. Lanes from *left* to *right*: 100 bp DNA ladder, cycles 4, 6, 8, 10, 12, 14 (respectively).

Care should be taken not to over-amplify a pool and thereby accumulate artifacts of spurious size. For example, it can be seen that larger amplification products begin to arise after cycle 6, and eventually predominate (*see* **Fig. 1.3**). These larger products may be members of the single-stranded DNA pool that fold back on themselves, are extended, and are then amplified by a single primer.

6. Once the optimalreaction conditions have been determined, the large-scale PCR is mixed and 100 μL reactions are aliquoted into wells of a 96-well PCR plate and cycled on a PCR machine.
7. The resultant 100 μL reactions are combined and ethanol-precipitated in a 50 mL conical centrifuge tube with 1:10 volume of 3 M NaOAc, 10 μL glycogen and 2.5 volumes of 100% ethanol similar to **Section 3.1.1** (**Steps 17–21**).
8. The pellet is resuspended in 300 μL diH_2O.
9. Once resuspended, there may be undissolved salts or other insoluble materials, so it is frequently useful to perform a final clean-up of the DNA sample either via a phenol/chloroform extraction and/or by running it through a Ultrafree-MC filter (as explained previously).

Based on the assumption that each of the 125 PCR reactions yielded 1 μg of product, 125 μg of N62 pool product should have been recovered after ethanol precipitation and sample clean-up. To verify this amplification yield several dilutions of the dsDNA pool should be separated via agarose gel electrophoresis and compared with several dilutions of a quantitation standard (*see* **Fig. 1.4**). For this example, we estimate that a 1:10 dilution contains about 10 ng/ul, and thus that we recovered about 25% of the theoretical 125 μg. This value is perfectly reasonable, given that we tend to limit amplification in order to limit the accumulation of amplification artifacts. Also, note that the complexity of the amplified pool is roughly the same as that of the synthetic DNA that served as starting material; there are just more overall copies of each individual sequence.

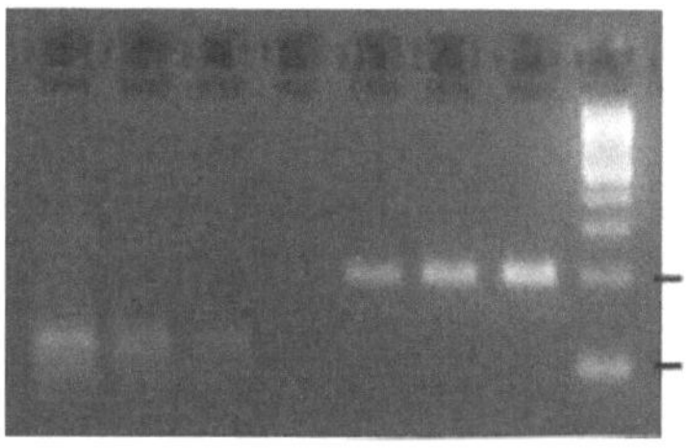

Fig. 1.4. Verification of N62 pool amplification. Lanes from *left* to *right*: N62 dilution series from stock (1:10, 1:25, 1:50), blank, quantitation standards (50, 100, 200 ng), 100 bp ladder.

3.4. Transcription and Purification of a RNA Pool

The double-stranded DNA pool can be used directly, or single-stranded DNA can be purified from it. The N62 pool in this example contains a promoter for T7 RNA polymerase, and thus can also be transcribed into RNA.

Based on the PCR yield determined above, one copy (1E14 sequences or 12.5 μg) of the final, double-stranded N62 DNA pool would be contained in 125 uL (12.5 μg divided by 10 ng/μL * dilution factor of 10; *see* **Note 15**). Per the manufacturer's instructions, the Epicentre High Yield T7 kit can be used to produce about 100 μg of RNA from 500 ng starting dsDNA template in a 20 μL reaction. Each reaction can therefore potentially yield roughly 450 RNA molecules per each DNA molecule, although actual yields are often much less.

1. The in vitro transcription protocol should be followed per the manufacturer's instructions (*see* **Note 16**).
2. Incubate the reaction at 37°C for between 4 and 16 h.
3. Add 50 μL RNase free DNase I and incubate at 37°C for 30 min.
4. Add 550 μL 2X denaturing dye.
5. Incubate at 70°C for 5 min.
6. Purify on a 1.5 mm 8% denaturing polyacrylamide gel similar to **Section 3.1.1 (Steps 2–21)** with the following changes: it is acceptable to load up to 100 μL into each well. Instead of a smear, there should be a very concise band upon separation. The gel chunks should be eluted in 13 mL 1X TE pH 7.5 to prevent nuclease degradation. In **Step 16**, 1:10 elution volume of 3 M NaOAc must also be added during ethanol precipitation.
7. Resuspend in 200 μL diH_2O.

To calculate the complexity of the RNA pool, first calculate the concentration (in μM) by multiplying the A_{260} times 40 ng/μL *(6)* divided by the pool length (102 for the N62 pool) and 0.345 ng/pmol (the average molecular weight of a nucleoside monophosphate). In a recent 500 μL transcription of the N62 pool, the final concentration of purified RNA was found to be 61.1 μM. This would have corresponded to 73 copies of the original DNA pool (61.1 pmol * 200 μL divided by 166 pmol/1E14 sequences). The RNA pool can be aliquoted and the stocks stored at –80°C for up to a year.

4. Notes

1. Much of the information contained within is further expanded upon in Pollard et al. *(7)*, although this method contains more up-to-date examples of many of the procedures described therein.
2. It is frequently useful to clone and sequence the initial synthetic pool, as well as the amplified pool. In this way, it should be possible to determine whether the initial pool used for selections is skewed in terms of sequence or any other parameter, such as secondary structural elements *(8)*. Moreover, the 'Round 0' sequences will frequently prove to be useful negative controls for other functional sequences derived from the pool.
3. All solutions and reactions should be made with deionized water (diH_2O) that has a resistance of 18.2 MΩ-cm and contain relatively few organics or pyrogens. The water should also be autoclaved to prevent nuclease degradation of the pool.
4. It is helpful to add ethidium bromide to agarose gel and running buffers to visualize nucleic acids, as opposed to staining the gel after electrophoresis. Typically, 100 μg ethidium bromide per 200 mL liquid agarose gel or 10 mL 6X Orange dye is used.
5. Glass plates (17 × 15 cm) should be stringently cleaned prior to pool purification to prevent cross-contamination with other nucleic acids. Mix 20 g sodium hydroxide in 100 ml methanol for 10 min (not all the NaOH will dissolve) and pour over plates in an autoclave tub. Allow to soak for 30 min. Scrub the plates with a detergent like Alconox, rinsed with tap water and spray with 70% ethanol to dry. Apply a silanizing agent to one side of the small plate and mark the other side with labeling tape. The silane coating will make it easier to pry the plates apart after running the gel.
6. Allow acrylamide gels to fully polymerize. Unpolymerized acrylamide can be seen when shadowing for nucleic acids, and is typically present near the edges of the plate where oxygen is present. Unpolymerized acrylamide can inhibit enzymatic reactions.

7. Because of failed syntheses, there is rarely a discrete band for the pool, but rather a dark smudge atop a smear of smaller molecular weight sequences. Cut out the top centimeter of the smeared band, since this contains the full-length pool. In addition, many short deletion variants in this band can also be amplified. This will increase the sequence complexity of the library, but will also lead to some size heterogeneity.
8. Centri-Sep columns can be used to separate labeled primer from the unincorporated radiolabel. To hydrate these columns, add 800 μL diH_2O directly to the white powder, cap, vortex, and incubate at room temperature for 30 min (remove any air bubbles by tapping the column on a surface). At the end of the incubation period, remove the column caps, drain by gravity into a 2 mL wash tube, and discard the interstitial fluid. Spin columns at 450 *g* for 2 min, blot the tip dry. Immediately purify nucleic acid samples to prevent the column from drying out.
9. Assuming little primer was lost during the labeling reaction and subsequent purification, 80 pmol/20 μL (or 4 pmol/μL) of primer should still be present. While the assumption that no primer has been lost is likely unwarranted, for the extension assay, ssDNA pool will be used in excess relative to the radiolabeled primer.
10. The annealing temperature will differ depending on the pool and primer sequence, and is typically set at 8–10°C below the melting temperature (T_m). For the N62 pool the T_m calculated using OligoAnalyzer 3.0 (IDT) was 57.4°C.
11. The average molecular weight of different oligonucleotide bases (*see* **Table 1.1**). The molecular weight of the pool equals the number of each individual base times its molecular weight plus the random region length times the average molecular weight. For instance, the molecular weight of the N62 pool ssDNA = $(9_A \times 313.2) + (8_T \times 304.2) + (13_C \times 289.2) + (12_G \times 329.2) + (62_N \times 308.9)$. For double stranded DNA, multiply this number by two. Other less accurate references suggest using an average of 330 and 345 g/mol per base for ssDNA and RNA, respectively.
12. Magnesium is added separately in order to allow further optimization, as necessary. Typically, a final concentration of 1.5 mM is used and adjusted by 0.5–1 mM increments (up or down) for optimization.
13. The dilution allows pipetting on the small scale, and will have to be mimicked on the large scale; that is, 33 μL will be diluted to 125 μL.

Table 1.1
Oligonucleotide molecular weights per (deoxy)-nucleoside monophosphate

Molecular weight

DNA		DNA	
dA	313.2	A	329.2
dC	289.2	C	305.2
dG	329.2	G	345.2
dT	304.2	U	306.2

14. During the cycle course, the PCR machine should be paused within the final 10 s of the extension step. Following removal of 5 μL sample the PCR should be resumed. At the time the sample is taken, 1 uL 6X orange dye is added (*see* **Note 4**).
15. Taking only a single copy of the original pool will likely result in skewing of some sequences (no representatives of some, multiple representations of others), and thus it is frequently recommended starting with at least three pool equivalents in order to be assured of complete coverage.
16. When using the Epicentre kit, 8 μL of total sample plus water can be added per 20 μl reaction and should be added last. Guanosine should be added as the last nucleoside triphosphate since it is prone to precipitate. The reaction should be assembled at room temperature.

References

1. Nimjee, S.M., Rusconi, C.P. and Sullenger, B.A. (2005) APTAMERS: an emerging class of therapeutics. *Annu. Rev. Med.* **56**, 555–583.
2. Mairal, T., Cengiz Ozalp, V., Lozano Sanchez, P., Mir, M., Katakis, I. and O'Sullivan, C.K. (2008) Aptamers: molecular tools for analytical applications. *Anal. Bioanal. Chem.* **390**, 989–1007.
3. Gopinath, S.C. (2007) Methods developed for SELEX. *Anal. Bioanal. Chem.* **387**, 171–182.
4. Blank, M. and Blind, M. (2005) Aptamers as tools for target validation. *Curr. Opin. Chem. Biol.* **9**, 336–342.
5. Bartel, D.P. and Szostak, J.W. (1993) Isolation of new ribozymes from a large pool of random sequences. *Science*, **261**, 1411–1418.
6. Sambrook, J. and Russel, D.W. (2001) *Molecular Cloning: A Laboratory Manual.* 3rd ed. Cold Springs Harbor Laboratory Press, Cold Spring Harbor, NY.
7. Pollard, J., Bell, S.D. and Ellington, A.D. (2000) In Beaucage, S. L., Bergstrom, D.E., Glick, G.D. and Jones, R.A. (ed.), *Current Protocols in Nucleic Acid Chemistry.* John Wiley and Sons, New York, pp. 9.2.1–9.2.23.
8. Meyers, L.A., Lee, J.F., Cowperthwaite, M. and Ellington, A.D. (2004) The robustness of naturally and artificially selected nucleic acid secondary structures. *J. Mol. Evol.* **58**, 681–691.

Chapter 2

In Vitro Selection of ssDNA Aptamers Using Biotinylated Target Proteins

Günter Mayer and Thomas Höver

Abstract

Aptamers are single-stranded nucleic acids that bind specifically to a target molecule and thus often inhibit target-associated biological functions. Aptamers have been described for a series of target molecules including peptides, proteins, and even living cells. Besides RNA and 2′-modified RNA molecules also ssDNA molecules can be subjected to in vitro selection protocols aiming at the enrichment of ssDNA aptamers. ssDNA aptamers can be selected using the SELEX procedure (systematic enrichment of ligands by exponential amplification) from libraries of randomized single-stranded DNA with a diversity of up to 10^{16} different molecules. In repetitive selection cycles, the library is incubated with the target of choice and separation of non-binding sequences from bound sequences is achieved by distinct separation methods. The bound molecules are specifically eluted and amplified, thus representing the starting library for the next cycle. Thereby, an enriched population of aptamers is evolved. Here we describe a generalized in vitro selection experiment aiming at the enrichment of ssDNA aptamers using biotinylated target molecules. This procedure allows the application of streptavidin–biotin chemistry to separate bound from unbound DNA species during the selection process.

Key words: SELEX, aptamers, ssDNA, in vitro selection, biotin, streptavidin.

1. Introduction

Nucleic acids are able to fold into well-defined three dimensional structures *(1)* that enable them to interact with a great number of different targets. This ability has been exploited in the SELEX process (systematic evolution of ligands by exponential enrichment), which allows the selection of single-stranded nucleic acids that bind specifically and with high affinity to their cognate target molecules *(2, 3)*. These nucleic acids are termed as aptamers. In several cases, the interaction of the aptamer with its cognate target

Günter Mayer (ed.), *Nucleic Acid and Peptide Aptamers: Methods and Protocols, vol. 535*

DOI 10.1007/978-1-59745-557-2_2 Springerprotocols.com

molecule is accompanied by inhibition of target-associated biological functions *(4–6)*. The SELEX method, originally described for RNA aptamers, has consequently been adapted for ssDNA yielding DNA-based aptamers. In this manner aptamers against numerous targets including proteins, small molecules, viruses, and whole cells *(7–9)* have been identified and ssDNA aptamers that target thrombin have been developed into clinical trials as *anti*-thrombotic agents *(10, 11)*. During the SELEX process a starting library *(12)* of ssDNA molecules with a random region flanked by defined primer binding sites is incubated with the target molecule, either in solution or coupled to a solid matrix. In the following, non-binding sequences are removed by washing steps and the bound species are eluted and amplified by PCR. After denaturation of the dsDNA, the counter-strands will be removed to generate an enriched library of ssDNA. This library represents the starting point for the next selection cycle. Radioactive labelling of the DNA allows the detection of binding sequences in a filter-retention analysis assay. The coupling of target protein to solid supports can be achieved either by non-covalent or covalent attachment of the proteins to sepharose-based matrices, such as CNBr-activated sepharose or thiopropyl sepharose. Here we describe an alternative approach that makes use of biotinylation of a target protein and its subsequent coupling to magnetic beads coated with streptavidin. These beads can be implemented in selection schemes for the successful enrichment of aptamers. This selection scheme can be generalized and applied to a variety of target proteins amenable to the NHS-chemistry-based biotin modification.

2. Materials

2.1. Biotinylation

1. 10x phosphate-buffered saline (PBS): 1.37 M NaCl, 27 mM KCl, 65 mM Na_2HPO_4 and 14.7 mM NaH_2PO_4. Adjust to pH 7.4 with HCl and NaOH. Store at room temperature (*see* **Note 1**).
2. Sulfo-NHS-LC-Biotin (Pierce). Prepare a fresh 1.8 mM solution in H_2O as required.
3. Bio-Spin Chromatography Columns P6 (Bio-Rad).

2.2. SDS-Polyacrylamide Gel Electrophoresis (SDS-PAGE)

1. Separating buffer: 1.5 M Tris–HCl, pH 8.8. Store at room temperature.
2. Stacking buffer: 1 M Tris–HCl, pH 6.8. Store at room temperature.

3. 10% SDS solution. Store at room temperature.
4. 30% bis-acrylamide (Roth). Store at 4°C. Bis-acrylamide is a neurotoxin. Be careful and avoid direct contact.
5. 10% ammoniumperoxodisulphate solution (APS). Store at 4°C.
6. N,N,N',N'-Tetramethylethylendiamin (TEMED). Store at 4°C.
7. Isopropanol.
8. Running buffer: Prepare 10x glycine electrophoresis buffer: 250 mM Tris–HCl, pH 8.9, 2 M glycine, 1% SDS (w/v). Dilute 1:10 with water prior use.
9. Prestained molecular weight markers.
10. SDS-PAGE loading buffer: prepare 8 ml 4x buffer by mixing: 4.3 ml water, 0.5 ml stacking buffer, 0.8 ml glycerol, 1.6 ml 10% SDS solution, 0.4 ml 2-mercaptoethanol, a spatula tip bromophenol blue. Store at –20°C.

2.3. Coomassie Staining

1. Staining solution: 375 mg Coomassie R-250 (Bio-Rad), 125 ml isopropanol, 50 ml acetic acid, 300 ml water.
2. Destaining solution: 30% (v/v) isopropanol, 10% (v/v) acetic acid.
3. Whatman-paper (Schleicher & Schuell).

2.4. Dot-Blotting

1. 10x PBS: 1.37 M NaCl, 27 mM KCl, 65 mM Na_2HPO_4 and 14.7 mM NaH_2PO_4. Adjust to pH 7.4 with HCl and NaOH. Dilute 1:10 with water prior use. Store at room temperature.
2. Blocking buffer: 1x PBS supplemented with 0.1 mg/ml BSA. Store at 4°C.
3. Fluorescently labeled antibody: Monoclonal *anti*-biotin (mouse IgG1 isotype) FITC-conjugate (Sigma).
4. Nitrocellulose transfer-membrane, pore size 0.45 μm (Protran, Whatman).

2.5. Preparation of the Matrix

2.5.1. Pre-selection Matrix

1. Dynabeads M-280 Streptavidin (Invitrogen, Dynal Biotech).
2. Washing buffer: 1x PBS supplemented with 1 mM $MgCl_2$.
3. 5x selection buffer: 685 mM NaCl, 13.5 mM KCl, 32.5 mM Na_2HPO_4, 9 mM NaH_2PO_4, 7.35 mM $MgCl_2$ and 0.5% (w/v) bovine serum albumin (BSA). Dilute to 1x buffer by mixing one part buffer with four parts water. Adjust to pH 7.4 with HCl and NaOH. Store at 4°C.
4. Magnetic particle concentrator rack (Invitrogen, Dynal Biotech).

2.5.2. Selection Matrix

1. Dynabeads M-280 Streptavidin (Invitrogen, Dynal Biotech).
2. Washing buffer: 1x PBS with 1 mM $MgCl_2$.

3. Selection buffer.
4. Magnetic particle concentrator rack (Invitrogen, Dynal Biotech).

2.6. Strand Displacement

1. 2x Bind and wash buffer (B & W buffer): 10 mM Tris–HCl (pH 7.5), 1 mM EDTA, 2 M NaCl. Dilute to 1x B & W buffer by mixing one part with same amount of water. Store at room temperature.
2. 0.15 M NaOH.
3. 0.3 M HCl.
4. Magnetic particle concentrator rack (Invitrogen, Dynal Biotech).
5. 5x selection buffer: 685 mM NaCl, 13.5 mM KCl, 32.5 mM Na_2HPO_4, 9 mM NaH_2PO_4, 7.35 mM $MgCl_2$ and 0.5% (w/v) bovine serum albumin (BSA).

2.7. SELEX

2.7.1. First Cycle

1. 5x selection buffer: 685 mM NaCl, 13.5 mM KCl, 32.5 mM Na_2HPO_4, 9 mM NaH_2PO_4, 7.35 mM $MgCl_2$ and 0.5% (w/v) bovine serum albumin (BSA).
2. DNA-library D1: 5′- GCC TGT TGT GAG CCT CCT AAC (N49) CAT GCT TAT TCT TGT CTC CC - 3′ (Metabion, *see* **Note 2**).

2.7.2. Strand Displacement

1. 2x binding and washing buffer (B & W buffer): 10 mM Tris–HCl (pH 7.5), 1 mM EDTA, 2 M NaCl. Store at room temperature.
2. 0.15 M NaOH.
3. 0.3 M HCl.
4. Magnetic particle concentrator rack (Invitrogen, Dynal Biotech).
5. 5x selection buffer: 685 mM NaCl, 13.5 mM KCl, 32.5 mM Na_2HPO_4, 9 mM NaH_2PO_4, 7.35 mM $MgCl_2$ and 0.5% (w/v) bovine serum albumin (BSA). Dilute to 1x buffer by mixing one part buffer with four parts water.

2.7.3. Polymerase Chain Reaction (PCR) and Agarose Gel Electrophoresis

1. 10x $MgCl_2$-free Taq PCR-buffer (Promega).
2. 25 mM $MgCl_2$.
3. dNTPs (Sigma, Roche).
4. 100 µM Forward primer: 5′- GCC TGT TGT GAG CCT CCT AAC – 3’ (Metabion).
5. 100 µM Reverse primer with 5′ biotin tag: 5′(bio)- GGG AGA CAA GAA TAA GCATG - 3′ (Metabion).
6. Taq DNA-polymerase.

7. Agarose, electrophoresis grade (Invitrogen).
8. 10x TBE-buffer: 89 mM Tris–HCl (pH 8.0), 89 mM boric acid, 2 mM EDTA solution. Dilute 1:10 with water.
9. Ethidiumbromide (Roth). Ethidium bromide intercalates in DNA and is highly toxic.
10. Agarose gel-loading buffer: 50% glycerol, 50 mM Tris–HCl (pH 8.0), 50 mM EDTA (pH 8.0), optionally add one spatula tip of bromophenol blue and xylene cyanol.

2.7.4. SELEX Cycles 2-x

1. 5x selection buffer: 685 mM NaCl, 13.5 mM KCl, 32.5 mM Na_2HPO_4, 9 mM NaH_2PO_4, 7.35 mM $MgCl_2$ and 0.5% (w/v) bovine serum albumin (BSA).

2.8. 5′-End Labelling of ssDNA Molecules

1. T4 Polynucleotide-Kinase (New England Biolabs).
2. 10x PNK-Buffer (New England Biolabs).
3. γ-[^{32}P]-ATP (Perkin Elmer). This compound is radioactive and should be handled with great care and only with appropriate protection measures.

2.9. Polyacrylamide Gel Electrophoresis (PAGE)

1. Concentrated gel solution: 25% bis-acrylamide in 8.3 M urea. Bis-acrylamide is a neurotoxin. Be careful and avoid direct contact.
2. Thinner: 8.3 M urea.
3. Gel-buffer: 8.3 M urea in 10x TBE-buffer.
4. 10x TBE-buffer: 89 mM Tris–HCl (pH 8.0), 89 mM boric acid, 2 mM EDTA solution.
5. 10% ammoniumperoxodisulphate solution (APS). Store at 4°C.
6. TEMED. Store at 4°C.
7. PAGE-Gel loading buffer: 9 M urea, 50 mM EDTA (pH 8.0), a spatula tip of bromophenol blue and xylene cyanol. Store in aliquots at –20°C.

2.10. Filter-Retention Analysis

1. 5x selection buffer: 685 mM NaCl, 13.5 mM KCl, 32.5 mM Na_2HPO_4, 9 mM NaH_2PO_4, 7.35 mM $MgCl_2$ and 0.5% (w/v) bovine serum albumin (BSA). Dilute to 1x buffer by mixing one part buffer with four parts water.
2. 10 μg/μl tRNA from baker's yeast (Fluka).
3. Interaction assay mix: Mix 500 μl 5x selection buffer with 66 μl tRNA; add water to 1,900 μl. The mix is sufficient for 100 binding tests. Store at –20°C.
4. Washing buffer: 1x PBS with 1 mM $MgCl_2$.
5. 96 well Dot-blot unit (Minifold, Schleicher & Schuell).

6. Blotting paper (Whatman, Schleicher & Schuell).
7. Nitrocellulose transfer-membrane, pore size 0.45 μm (Protran, Whatman).

3. Methods

3.1. Biotinylation of Target Proteins

1. Mix 100 μg target protein with a threefold molar excess of sulfo-NHS-LC-biotin in a total volume of 100 μl 1x PBS (*see* **Note 3**).
2. Incubate on ice for 30 min and for further 15 min at room temperature.
3. Remove non-reacted sulfo-NHS-LC-biotin by gel filtration using P6 Spin-Columns. Safe a 5 μl aliquot of the reaction for SDS-PAGE and dot-blot analysis (*see* **Note 4**).

3.2. SDS-Polyacrylamide Gel Electrophoresis (SDS-PAGE)

1. Clean the glass plates with water and 70% ethanol before use.
2. Prepare a 12% separating gel by mixing 1,700 μl water, 1,250 μl separating buffer, 50 μl 10% SDS solution, 2,000 μl 30% bis-acrylamide, 25 μl 10% APS and 2.5 μl TEMED. Pour a 0.75 mm gel. Leave space for the stacking gel. Cover the gel with isopropanol and let polymerize for 30 min.
3. Pour off isopropanol and prepare a 4% stacking gel by mixing 1,220 μl water, 500 μl stacking buffer, 10 μl 10% SDS solution, 270 μl 30% bis-acrylamide, 10 μl 10% APS and 2.5 μl TEMED. Pour on top of the stacking gel and insert comb. Let polymerize for 30 min (*see* **Note 5**).
4. Carefully remove comb and assemble gel-running unit. Fill lower chamber with running buffer and remove any air-bubbles. Then fill upper chamber and rinse wells with running buffer with the help of a syringe.
5. Mix 4 μl sample with 1 μl 4x loading-buffer and heat 3 min at 95°C. Load gel immediately. Include one well with molecular weight markers.
6. Run the gel at 180 V until the dye-fronts reach the bottom of the gel.

3.3. Coomassie Staining

1. Remove the gel from the SDS-PAGE gel-running unit and separate the glass plates. Cut off and discard the stacking gel.
2. Remove the separating gel from the glass plate, place it into a small bowl and cover it with Coomassie staining solution. Put the gel onto a shaker and stain the gel for about 30 min.
3. Remove staining solution and add destaining solution to the gel. Destain the gel on the shaker for 2 h. Replace solution 2–3 times.

4. Put the gel on a Whatman-paper and dry it in a gel-dryer for 1 h (*see* **Fig. 2.1**).

3.4. Dot-Blot Analysis

1. Apply spots of 0.25, 0.5 and 1.0 μl of the biotinylated and purified target protein (*see* **Section 3.1**) onto a nitrocellulose membrane (4 × 5 cm in size). At the 1.0 μl spot first apply 0.5 μl and let dry for a minute. Then add the other 0.5 μl. This will prevent the spot from becoming too large. As a negative control apply spots of non-biotinylated protein. Cut off one edge of the membrane to ensure orientation.
2. Dry the membrane for 30 min in an incubator at 65°C.
3. Put the membrane into a small box and cover it with blocking buffer. Close the box and incubate the membrane for 3 h on a shaker (*see* **Note 6**).
4. Remove blocking buffer and wash the membrane in 1x PBS for 5 min.
5. Dilute the antibody 1:1,000 in blocking buffer and add this solution to the membrane. Wrap the box with the membrane in aluminium-foil. Incubate the membrane for 1 h on a shaker.
7. Remove the antibody solution and wash the membrane twice for 2 min in blocking buffer and then twice for 2 min in 1x PBS. Keep exposure to light as short as possible.
8. Remove the buffer and take the membrane out of the box with tweezers. Place the membrane onto a paper-towel. Place another towel on the membrane and dry the membrane

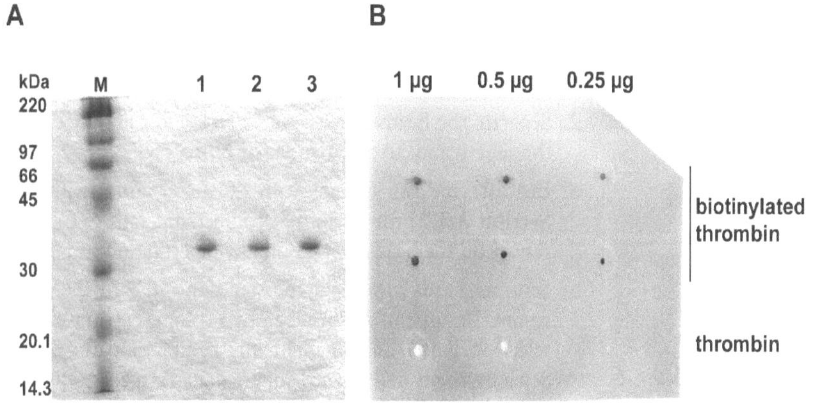

Fig. 2.1. Biotinylation of proteins. (**A**) 12% SDS-PAGE gel of human a-Thrombin after Coomassie staining. *M*: low range protein marker, *1*: 1 μg human α-Thrombin from stock, *2*: 1 μl biotinylation reaction before purification with P6 mirco-spin column, *3*: 1 μl biotinylated protein after purification with P6 micro-spin column. (**B**) Dot blot of human α-Thrombin. The indicated amounts of protein were spotted on a nitrocellulose membrane and were dried 30 min at 65°C and blocked 3 h in blocking buffer before incubation with monoclonal *anti*-biotin (mouse IgG1 isotype) FITC-conjugate.

between them by pressing the towels together with your hands. Do not wipe the membrane.

9. Place the membrane on a fluorescence-imager and read fluorescence (*see* **Fig. 2.1**).

3.5. Preparation of the Matrix

3.5.1. Pre-selection Matrix

1. Take 2.5 mg beads from stock. Separate the beads from storage buffer in a magnetic rack and discard supernatant.
2. Wash the beads twice with 250 μl 1x PBS and three times with 250 μl selection buffer. Resuspend the beads in 500 μl selection buffer.

3.5.2. Selection Matrix

1. Take 2.5 mg beads from the stock (10 mg/ml). Separate the beads from storage buffer in a magnetic rack and discard the supernatant.
2. Wash the beads twice with 250 μl 1x PBS and three times with 250 μl selection buffer. Resuspend the beads in 250 μl selection buffer.
3. Add 50 μl of the biotinylation reaction (*see* **Section 3.1**) to the prepared Dynabeads and incubate the suspension for 30 min at room temperature in a head-to-tail shaker.
4. Take off the supernatant and wash the beads twice with 250 μl selection buffer.
5. Resuspend the beads in 500 μl selection buffer. The beads can be stored at 4°C up to 1 week.

3.6. SELEX

3.6.1. First Cycle

1. Incubate 500 pmol of the ssDNA library in 80 μl selection with 80 μl of the pre-selection matrix (*see* **Section 3.5.1**) for 30 min at room temperature. Carefully resuspend the beads every 3 min by pipetting up and down.
2. Separate the beads in a magnetic rack and carefully transfer the supernatant to a new tube (*see* **Note 7**).
3. Incubate the supernatant with 80 μl selection matrix (*see* **Section 3.5.2**) for 30 min at room temperature. Carefully resuspend the beads every 3 min by pipetting up and down.
4. Separate beads in a magnetic stand and discard the supernatant. Resuspend the beads in two volumes (160 μl) of the selection buffer and incubate the resupension for 5 min. Separate beads and discard supernatant. Make sure that the buffer is completely removed.
5. Resuspend the beads in 100 μl water and heat for 3 min at 95°C.
6. Take off the supernatant and transfer it into a new tube. Discard the beads.

3.6.2. Polymerase Chain Reaction (PCR) and Agarose Gel Electrophoresis

1. Prepare ten 100 μl PCR reactions: 1x PCR-buffer, 3.5 mM $MgCl_2$, 0.2 mM dNTP-mix, 1.1 μM forward primer, 0.9 μM reverse primer with 5' biotin-tag, 0.5 μg/μl BSA, 2.5 U Taq. Add water to reach 90 μl total volume (*see* **Note 8**).
2. Add 10 μl of the eluted ssDNA species (*see* **Section 3.6.1 Step 6**) to each PCR reaction and amplify the DNA in a thermocycler with the following settings: 1 min 95°C, 1 min 54°C, 1.5 min 72°C (*see* **Notc 9**).
3. Prepare a 2.5% (w/v) agarose gel in 1x TBE-buffer and heat until the agarose is completely solved.
4. Add 0.1 μl/ml ethidiumbromide (10 mg/ml) and mix.
5. Pour the gel into a gel-casting chamber, insert the comb and let cool for 30 min.
6. Mix 2 μl PCR-product with 2 μl agarose-loading buffer and load the gel. Load one lane with a suitable DNA-ladder (e.g. 100 bp ladder).
7. Run the gel at 160 V with 1x TBE as running buffer for 20 min and visualize the dsDNA bands with a UV-lamp (λ = 254 nm).

3.6.3. Strand Displacement

1. Take 2.5 mg Dynabeads from the stock and remove storage buffer with the help of a magnetic rack. Wash the beads twice with 250 μl 1x B & W buffer.
2. Resuspend the beads in 500 μl 1x B & W buffer and add 500 μl 2x B & W buffer.
3. Add 500 μl PCR-product and incubate for 30 min at room temperature on a head-to-tail shaker (*see* **Note 10**).
4. Take off the supernatant and wash the beads three times with 250 μl 1x B & W buffer and once with 250 μl 2x B & W buffer.
5. Remove the buffer and resuspend the beads in 30 μl 0.15 M NaOH. Incubate for 3 min and separate beads using the magnetic rack.
6. Transfer the supernatant into a new tube and neutralize by adding 15 μl 0.3 M HCl. Control the pH by spotting a drop on pH-indicator paper with a pipette-tip and adjust pH with NaOH and HCl if necessary.
7. Add 16 μl 5x selection buffer and water to reach a total volume of 80 μl.

3.6.4. SELEX Cycles 2-x

1. Perform pre-selection and selection-step with the ssDNA following the instructions of the first SELEX cycle (*see* **Section 3.6.1**).
2. Raise the selection pressure by increasing the washing steps in each selection cycle: Wash two times in the second, four times in the third and eight times in the fourth cycle. Every four

washing steps the suspended beads can be incubated for 5 min at room temperature prior removal of the buffer. Reduce the amount of selection matrix used from the fifth cycle on (*see* **Note 11**).

3. After successful completion of the selection (monitoring of the selection can be accomplished by filter-retention analysis, *see* **Sections 3.7 and 3.9**) the library can be cloned and sequenced (Kits are available from various suppliers). A collection of representative sequences can be found in **Fig. 2.2**.

3.7. 5′-End Labelling of ssDNA Molecules

1. Take 200 µl of the PCR-product and perform the strand displacement using 100 µl streptavidin-coated beads. Elute ssDNA by adding 10 µl 0.15 M NaOH and neutralize the solution with 5 µl of 0.3 M HCl. Add water to reach a final volume of 20 µl.
2. Mix 10 µl ssDNA with 2 µl 10x PNK-Buffer. For control experiments prepare a second reaction using 10 pmol of the initial DNA library.
3. Add 2 µl of γ-[^{32}P]-ATP.
4. Add 20 U T4 polynucleotide-kinase. Add water to reach a total volume of 20 µl and incubate for 45 min at 37°C.
5. Add water to a final volume of 100 µl and purify the ssDNA by gel filtration using a G25 micro-spin column.

3.8. Polyacrylamide Gel Electrophoresis (PAGE)

1. Clean the glass plates with water and 70% ethanol before use.
2. Assemble the plates with the spacers and fasten them with several clips. Make sure the spacers are put neatly together without any gaps.
3. Prepare a 10% gel by mixing 20 ml concentrated gel solution, 25 ml thinner, 5 ml gel-buffer, 400 µl APS and 20 µl TEMED.
4. Pour the gel immediately and insert the comb. Lay the gel down horizontally and let polymerize for 45 min.
5. Remove the comb and wash away any gel fragments with water. Assemble the gel-running unit and fill the lower tank with 1x TBE. Remove any air-bubbles and fill the upper tank with 1x TBE.

```
2.5     ACTACGGGGCTGGCGTGGTTGGGTAGTTGCTGGAGGAGAGCATCTTGAT

2.2     GGCTGGTGGGTTGGAGGGGTGCTCAAATATAATTTGGGTCGCTTTGGGT

1.3     GGTTGGTGTGGGTTGGATGGAGGCTGCGCCGTCACAGTATGCTGAGGAT

05.4    GATTACTCGCTCCCCGTGGTTGGCTTCGGGTTGGTCTGGGTGCGCTATT
```

Fig. 2.2. Nucleic acid sequences of the DNA aptamers selected against human α-Thrombin 2.5, 2.2, 1.3 and 05.4, respectively. Shown are the initial random regions without the flanking primer regions.

6. Pre-run the gel 15 min at 380 V.
7. Mix 2 μl radioactive DNA with 18 μl loading buffer. Heat the samples to 95°C for 3 min.
8. Switch off the power and rinse the wells with 1x TBE with the help of a syringe.
9. Load the gel with each 15 μl of the samples.
10. Run the gel at 380 V with power limited to 25 W for about 1.5 h.
11. Take the gel out of the gel-running unit and remove one of the glass plates. Cover the gel with plastic foil.
12. Put the gel on a paper towel and put it in an X-ray film cassette.
13. Apply a phosphor-screen and close the cassette and expose the screen for 10 min. Analyse the screen using a phosphor-imager (e.g. Fuji FLA 3,000).

3.9. Filter-Retention Analysis

1. Dilute the radioactive labelled DNA 1:10 with water.
2. The diluted DNA is mixed 1:20 with binding-assay mix.
3. Prepare a concentration series of the target protein: Make a 2.5 μM solution in 1x PBS and dilute two times to 1.25 and 0.625 μM by mixing each one part protein solution with one part 1x PBS.
4. Mix in a well of a 96-well plate 20 μl of the prepared DNA in binding-assay mix with 5 μl protein solution. Prepare the samples for each protein-concentration at least in duplicate.
5. Cover the nitrocellulose membrane with 0.4 M KOH. Incubate samples and membrane for 20 min.
6. Assemble dot-blot unit. Equilibrate blotting-paper in binding-buffer and place it on the blotting unit. Take the membrane out of the KOH bath and shortly rinse it with water. Place it on top of the blotting-paper and remove any air-bubbles.
7. Complete the assembly of the unit and connect to vacuum pump.
8. Apply vacuum and wash the membrane twice with 200 μl binding-buffer using a multichannel-pipette. Drainage of the wells might be prevented by air-bubbles. To remove them, carefully tilt the blotting unit and tap it on the table.
9. With the help of a multichannel-pipette apply 20 μl of each sample on the membrane.
10. After filtration wash each well four times with 200 μl binding-buffer.
11. Disassemble the blot-unit and turn off the pump. Remove the membrane and dry it between paper towels.

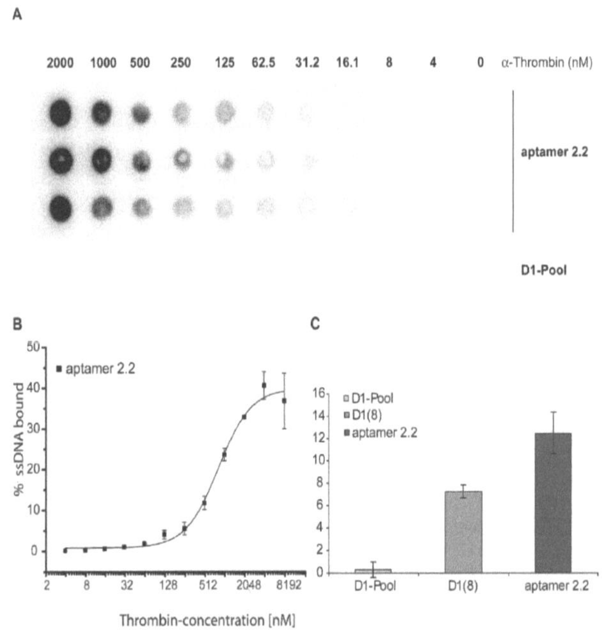

Fig. 2.3. Filter-retention analysis of the selected aptamer 2.2 after eight SELEX cycles. (**A**) Aptamer 2.2 was incubated with the indicated thrombin-concentrations for 20 min and filtrated through a nitrocellulose membrane (0.45 μm). The original D1-pool from which the aptamer has been selected shows no binding to thrombin. (**B**) Results from the filter-binding assays are evaluated with a non-linear logistic fit. The K_D-value of aptamer 2.2 was determined to be 816 nM. (**C**) Binding to human α-thrombin at 500 nM concentration. After eight SELEX cycles binding of the enriched pool (D1(8)) could be detected. Several sequences from the enriched pool were analysed of which aptamer 2.2 showed the highest affinity.

12. Place the membrane on a paper towel and cover it with foil. Cover the membrane with an X-ray film screen and close cassette.
13. Expose the screen for at least 1 h. Then take out the screen and read it in a phosphor-imager (*see* **Note 12**, *see* **Fig. 2.3**).

4. Notes

1. All solutions are prepared with de-ionized water purified with a resistivity of 18.2 MΩ and sterilized by filtration with a filter pore-size of 0.22 μm.
2. The D1-pool contains a random region consisting of 49 bases. We used this library in our lab with good results. DNA libraries with random regions are commercially available.
3. For the biotinylation the protein must not be solved in solutions containing Tris, as it reacts with the sulfo-NHS-LC-biotin.

Change buffer before the reaction. After the biotinylation Tris has no effect on the biotinylated protein.

4. Sometimes the biotinylation reaction can be inhibited by any impurities in the protein stock. If this is the case, purify the protein with a P6 spin-column before you add biotin to the reaction. P6 columns usually come in Tris storage-buffer, so be sure to change the buffer of the spin-column before use.
5. The blocking of the membrane in the dot blot can be reduced to 1 h to speed up the process, although this will lead to an increased background. For better quality block the membrane overnight.
6. SDS-PAGE-gels can be stored for 1–2 weeks at 4°C if you pack them in wet paper towels and wrap them in plastic foil.
7. The incubation with the pre-selection matrix removes sequences that bind to the matrix and not to the target.
8. A PCR-mastermix without DNA-polymerase can be prepared in advance and stored in aliquots at −20°C. The PCR-mix described here is optimized for the D1 library. Optimal concentrations of $MgCl_2$ and possible additives like BSA, glycerol or DMSO depend on the library and primers used.
9. Especially in the first SELEX cycle only a small fraction of the DNA library will bind to the target. Therefore, you may need much more PCR-cycles than usual. Control your product on an agarose gel and add more PCR-cycles if the product-yield is low, then check again. If you feel that your product is not amplified any more you may add fresh DNA-polymerase. The described PCR-conditions are optimized for the D1 library and depend on the used DNA library and primers.
10. The biotin-tag which has been inserted into the DNA by the PCR reaction binds to streptavidin. Thus, ssDNA can be eluted by denaturation while the counter-strand is restrained.
11. The number of SELEX cycles required for the selection of an aptamer varies from about 8 to 20 cycles and depends on the target and the DNA library.
12. Exposure-time of the binding-assay membrane to the X-ray film depends on the amount of radioactivity. In general, longer exposure leads to better contrast. Exposure overnight usually leads to good results.

References

1. Famulok, M., Hartig, J.S. and Mayer, G. (2007) Functional aptamers and aptazymes in biotechnology, diagnostics, and therapy. *Chem. Rev.* **107**, 3715–3743.
2. Ellington, A.D. and Szostak, J.W. (1990) *In vitro* selection of RNA molecules that bind specific ligands. *Nature (London)* **346**, 818–821.

3. Müller, J., Wulffen, B., Pötzsch, B. and Mayer, G. (2007) Multi-domain targeting generates a high affinity thrombin-inhibiting bivalent aptamer. *ChemBioChem* **8**, 2223–2226.
4. Mayer, G. and Jenne, A. (2004) Aptamers in research and drug development. *BioDrugs* **6**, 351–359.
5. Bock, L.C., Griffin, L.C. Latham, J.A., Vermaas, E.H. and Toole, J.J. (1992) Selection of single stranded DNA molecules that bind and inhibit human thrombin. *Nature* **355**, 564–566.
6. Mayer, G., Wulffen, B., Huber, C., Brockmann, J., Flicke, B., Neumann, L., Hafenbradl, D., Klebl, B.M., Lohse, M.J., Krasel, C. and Blind, M. (2008) An RNA molecule that specifically inhibits G-protein coupled receptor kinase 2 *in vitro*. *RNA* **14**, 524–534.
7. Schurer, H., Stembera, K., Knoll, D., Mayer, G., Blind, M., Forster, H., Famulok, M., Welzel, P. and Hahn, U. (2001) Aptamers that bind to the antibiotic moenomycin A. *Bioorg. Med. Chem.* **9**, 2557–2563.
8. Raddatz, M.-S.L., Dolf, A., Knolle, P., Endl, E., Famulok, M. and Mayer, G. (2008) Enrichment of cell-targeting and population-specific aptamers by fluorescent-activated cell-sorting. *Angew. Chem. Int. Ed.* **47**, 5190–5193.
9. Famulok, M., Mayer, G. and Blind, M. (2000) Nucleic acid aptamers – From selection *in vitro* to application *in vivo*. *Acc. Chem. Res.* **33**, 591–99.
10. Wang, K.Y., Krawczyk, S. H., Bischofberger, N., Swaminatham, S. and Bolton, P.H. (1993) The tertiary structure of a DNA aptamer which binds to and inhibits thrombin determines activity. *Biochemistry* **32**, 11285–11292.
11. Li, W.X., Kaplan, A.V., Grant, G.W., Toole, J.J. and Leung, L.L.K. (1994) A novel nucleotide-based thrombin inhibitor inhibits clot-bound thrombin and reduces arterial platelet thrombus formation. *Blood* **83**, 677–682.
12. Müller, J., El-Maarri, O., Oldenburg, J., Pötzsch, B. and Mayer, G. (2008) Monitoring the progression of the *in vitro* selection of nucleic acid aptamers by denaturing high-performance liquid chromatography. *Anal. Bioanal. Chem.* **390**, 1033–1037.

Chapter 3

Isolating Aptamers Using Capillary Electrophoresis–SELEX (CE–SELEX)

Renee K. Mosing and Michael T. Bowser

Abstract

SELEX (systematic evolution of ligands by exponential enrichment) is a process for isolating DNA or RNA sequences with high affinity and selectivity for molecular targets from random sequence libraries. These sequences are commonly referred to as aptamers. The process typically requires 10–15 cycles of enrichment, PCR amplification and nucleic acid purification to obtain high-affinity aptamers. We have demonstrated that using capillary electrophoresis (CE) as an enrichment step greatly improves the efficiency of the process. CE–SELEX is capable of isolating high-affinity aptamers in as little as 2–4 rounds of selection, shortening the process time from several weeks to as little as a few days.

Key words: CE–SELEX, SELEX, in vitro selection, in vitro evolution, capillary electrophoresis, aptamers.

1. Introduction

Aptamer development has grown exponentially since the introduction of SELEX in 1990 *(1–3)*. The high affinity and specificity aptamers possess for a variety of target molecules has fueled widespread interest in the unique applications of aptamers *(4–7)*. Although new aptamer applications continue to emerge, greater adoption of aptamers is limited by the length and difficulty of the aptamer selection process. Isolating aptamers using capillary electrophoresis (CE) has combined extreme resolving power with highly stringent selection conditions. The much simpler process (CE–SELEX) allows high-affinity aptamers to be obtained in significantly fewer rounds of selection *(8–10)*. This dramatically

Günter Mayer (ed.), *Nucleic Acid and Peptide Aptamers: Methods and Protocols, vol. 535*

DOI 10.1007/978-1-59745-557-2_3 Springerprotocols.com

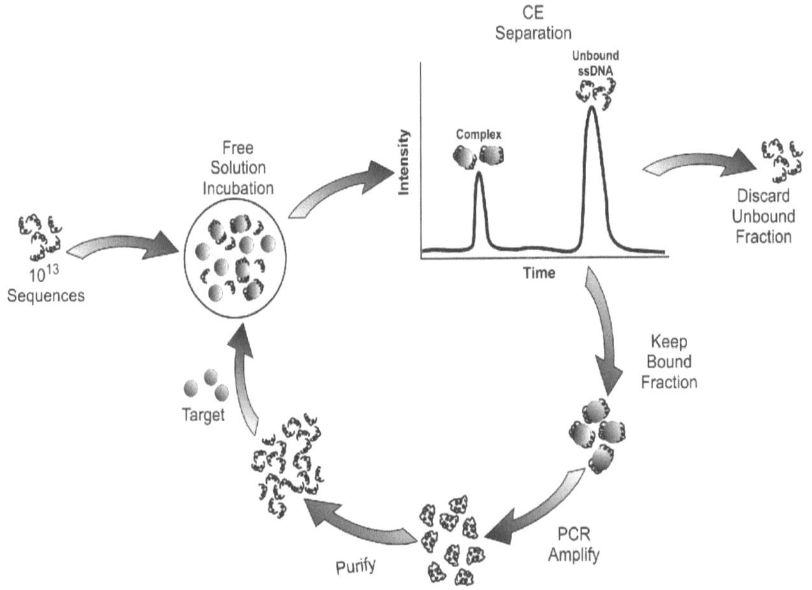

Fig. 3.1. Schematic of the CE–SELEX process. A random sequence DNA library is incubated with the target. Sequences bound to the target are separated using capillary electrophoresis, PCR amplified and made single stranded, generating a new pool suitable for further rounds of enrichment.

decreases the time requirement of the process compared to preexisting methods.

The CE–SELEX process is illustrated in **Fig. 3.1**. Briefly, a library of random nucleic acid sequences is incubated with the target molecule. Bound sequences are separated from non-binding sequences via CE. The bound sequences are collected, PCR amplified, made single stranded, and purified. The refined pool is used in the next selection round. Selections are stopped when no further improvement in affinity is observed. The procedural details of the selection process are discussed in the following sections.

2. Materials

2.1. Capillary Electrophoresis

The procedure described makes use of an automated P/ACE™ MDQ Capillary Electrophoresis instrument (Beckman Coulter, Fullerton, CA) equipped with all standard features plus a laser module to facilitate LIF detection. The procedure can easily be adapted to other commercially available automated CE systems.

1. Single stranded DNA library (*see* **Note 1**) (Genelink, Hawthorne, NY) diluted to 200 μM concentration in standard TE buffer and stored at –20°C.
2. Bare fused silica capillary 50 i.d., 360 o.d. (Polymicro Technologies, Phoenix, AZ).
3. Polypropylene-coated centrifuge tubes , 0.6 and 1.5 mL sizes (*see* **Note 2**) (Fisher).
4. Polypropylene-coated filter pipette tips (*see* **Note 3**) (Fisher).
5. TGK buffer consisting of 25 mM Tris–HCl, 192 mM glycine, and 5 mM KH_2PO_4, pH 8.3 (chemicals of highest purity available from Fisher Scientific).

2.2. Polymerase Chain Reaction (PCR)

1. FAM 5' labeled forward primer (Genelink, Hawthorne, NY) diluted to 60 μM concentration in standard TE buffer. Store at –20°C.
2. Unlabeled forward primer (Genelink, Hawthorne, NY) diluted to 60 μM concentration in standard TE buffer. Store at –20°C.
3. Biotin 5' labeled reverse primer (Genelink, Hawthorne, NY) diluted to 60 μM concentration in standard TE buffer. Store at –20°C.
4. Unlabeled reverse primer (Genelink, Hawthorne, NY) diluted to 60 μM concentration in standard TE buffer. Store at –20°C.
5. Solution of 10 mM deoxynucleotide triphosphates (dNTP's) (Invitrogen, Carlsbad, CA). Store at –20°C.
6. Thermo pol buffer (10X) and 5000 u/mL taq polymerase (New England Biolabs, Ipswich, MA). Store at –20°C.
7. Nuclease-free water (Invitrogen, Carlsbad, CA). Store at room temperature.
8. Solution of 25 mM $MgCl_2$ (Sigma). Store at 2–8°C.
9. Thin walled 0.5 μL PCR tubes (Eppendorf, Westbury, NY).

2.3. Agarose Gels

1. Low EEO agarose (Sigma).
2. 10X Tris–borate–EDTA (TBE) buffer (Invitrogen, Carlsbad, CA) diluted to 0.5X which consists of 50 mM Tris–HCl, 45 mM boric acid, and 0.5 mM EDTA. Store at room temperature.
3. Solution of 10 mg/ mL ethidium bromide (Sigma). Ethidium bromide is a known mutagen. Double gloves, lab coat, and goggles should be worn when handling. Store at room temperature.
4. 6X Blue-orange dye (New England Biolabs, Ipswich, MA) diluted to 1X in nuclease-free water. Store at –20°C.
5. 25 base pair DNA molecular weight ladder (Invitrogen, Carlsbad, CA). Store at –20°C.

2.4. Streptavidin Purification

1. Streptavidin agarose resin (Thermo Scientific). Store at 2–8°C.
2. Poly prep chromatography columns (Bio-Rad, Hercules, CA).
3. Solution of streptavidin-binding buffer containing 10 mM Tris–HCl, 50 mM NaCl, 1 mM ethylenediamine tetraacetic acid (EDTA). Store at room temperature.
4. Solution of 0.15 M NaOH. Store at 2–8°C.
5. Solution of 0.15 M acetic acid. Store at 2–8°C.

2.5. Ethanol Precipitation

1. Solution of 100% ice cold ethanol. Store at –20°C.
2. Solution of ice cold 70:30 ethanol/water. Store at –20°C.
3. Solution of 3 M sodium acetate. Store at room temperature.

3. Methods

The methods outlined in this section assume a bare fused silica capillary is used to perform the CE separations and that a DNA library is used. Modifications to the procedure resulting from using a capillary with surface coatings to eliminate or reverse electroosmotic flow (EOF) are discussed in the notes section.

3.1. Identification of the Collection Window

1. The migration of the library and target should be determined separately on CE (*see* **Note 4**). Common separation conditions are as follows: TGK incubation and separation buffer, 50.2 cm, 50 μm i.d., 360 μm o.d. bare fused silica capillary, 30 KV normal polarity separation, 1 psi, 4 s injection, LIF detection.
2. Modify the separation conditions to achieve an adequate separation between the unbound library and the bound sequences (*see* **Note 5**).
3. Determine the migration window (*see* **Fig.** 3.2) which should be approximately 1–2 min before the library begins to migrate off the capillary.

3.2. Fraction Collection using Capillary Electrophoresis

1. Heat the 200 μM ssDNA library to 72°C for 2 min and allow it to cool to room temperature to ensure DNA folding into stable room temperature conformations (*see* **Note 6**).
2. Incubate the 200 μM library with the target (50–500 pM) for 20 min at room temperature to allow binding to occur (*see* **Note 7**).
3. Inject an aliquot of this mixture on CE. A separation voltage is applied and the separation is performed directly into the outlet vial which should contain ~50 μL separation buffer (*see* **Note 8**). For most targets, when using an uncoated capillary

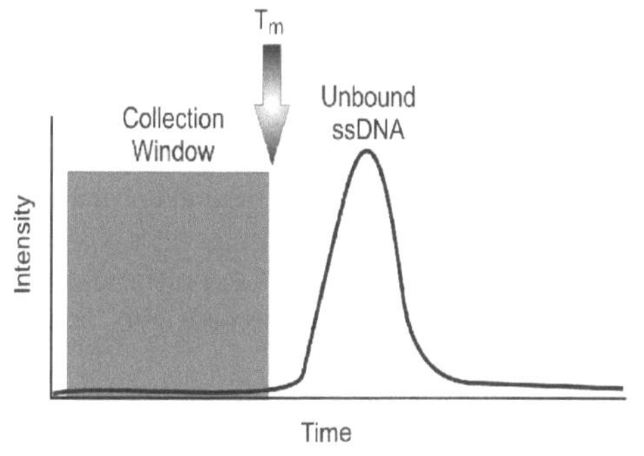

Fig. 3.2. Collection strategy when using an uncoated capillary. Aptamer–target complexes will generally be less negative than unbound sequences and will therefore migrate off the capillary first. Sequences migrating in a window ending 1–2 min before the leading edge of the unbound sequences should be collected. Equation (3.1) should be used to correct for the time required for sequences to migrate through the length of capillary after the detector.

the complex will migrate toward the negative electrode and normal polarity should be used.

4. Approximately 1–2 min before the library begins to migrate off the capillary, stop the separation. Since CE has on column detection, the following equation should be used to determine when the library is actually beginning to exit the capillary.

$$t_{\text{out}} = \frac{L_{\text{T}}}{L_{\text{D}}}(t_{\text{det}}) \tag{3.1}$$

where t_{out} is the time the library will begin to migrate off the capillary, L_{T} is total length of the capillary, L_{D} is the length of the capillary to the detector, and t_{det} is the time required for the analyte to reach the detector. The outlet vial should collect the bound fraction.

5. Rinse the unbound sequences to waste with a pressure rinse.

3.3. Polymerase Chain Reaction (PCR)

1. Using filter pipette tips, prepare a PCR master mix containing the following (*see* **Note 9**):

   408 μL nuclease-free water
   300 μL 25 mM $MgCl_2$
   100 μL 10X thermo pol buffer
   100 μL deoxynucleotide triphosphates (10 mM each dNTP mixture)
   8.5 μL 60 μM forward primer with 5' FAM label
   8.5 μL 60 μM reverse primer with 5' biotin label

2. Distribute 92.5 μL of the PCR master mix into nine thin walled PCR tubes using a filter pipette tip.

3. Add 6 μL nuclease-free water to one tube with a filter pipette tip. Label this tube as the control and set aside.

4. To each remaining tube, add 6 μL of the bound collection from CE using a filter pipette tip.
5. Place the PCR tubes in a thermocycler. Heat the samples to 94°C for 2 min. Keep the temperature at 94°C, add 1.5 μL of 5,000 u/mL taq polymerase to each sample, starting with the control using a filter pipette tip. Heat for an additional minute after the taq polymerase is added and run the desired PCR method according to standard protocol (*see* **Note 10**).

3.4. Agarose Gels

1. Prepare a 2% agarose gel (8 × 12 × 1 cm) by adding 2 g agarose to 100 mL 0.5X TBE buffer. Heat the mixture by placing in microwave for 2 min or until the agarose is in solution. Remove from microwave and allow cooling to approximately 60°C. Once cooled, add 1 μL of 10 mg/mL ethidium bromide. Do not add the ethidium bromide until the solution has cooled to 60° or less to prevent aerosol which is harmful if inhaled. Swirl the solution to mix and pour in gel setter. Immediately place comb in gel. Allow the gel to set for 20 min or until firm at room temperature or 5–7 min in the refrigerator.
2. Prepare the samples to be run on the agarose gel. First, add 5 μL 1X blue-orange dye to ten microcentrifuge tubes. Add 2 μL water and 2 μL 25 base pair DNA ladder to two of these tubes and set them aside. To the remaining eight tubes, add 4 μL of the respective PCR sample.
3. Carefully remove the comb from the gel. Submerge the gel which should still be on the UV transparent plastic tray in 0.5X TBE buffer and use a pipette tip to remove bubbles from the wells.
4. Load 9 μL of sample into each well. The ladders should be loaded in different places. Generally one is placed in the center of the samples and one on the end of the samples. This will help determine the positions of the lanes later.
5. Complete the gel apparatus assembly and plug into the power supply. Run at 200 V for 60 min or until a good separation between the yellow, purple, and blue dyes is achieved. Do not allow any of the dyes to migrate off the gel. The voltage connections should be confirmed by observing bubble formations at the electrodes. DNA will migrate toward the positive electrode which is colored red.
6. Once the gel run is complete, disconnect from the power supply. Carefully place the gel on a UV box for imaging. Do not look directly into the UV box without UV eye protection. The gel should have a solid band at the expected aptamer molecular weight. No additional bands should be observed such as primer dimer bands which would run lower on the

gel. Controls should be scrupulously carried out to confirm the DNA was a result of the collection, not contamination.

3.5. Streptavidin Column Purification

1. Shake the streptavidin agarose resin stock to evenly distribute the settled beads into solution. Place 300 μL streptavidin agarose resin into a poly prep chromatography column. Discard stabilizing solution by pushing the top cap onto the column until the liquid drains. Add 500 μL streptavidin binding buffer and all PCR samples except the control. Incubate for 30 min, vortexing periodically.
2. The column is rinsed with 500 μL streptavidin buffer approximately ten times to remove excess PCR reagents. Rinse once with 500 μL distilled water to rinse any residual salt from the streptavidin buffer.
3. Add 200 μL 0.15 M NaOH and heat to 37°C for 15–20 min to break the hydrogen bond network between the double stranded DNA without disrupting the biotin–streptavidin complex.
4. Elute the ssDNA sequences into a collection 1.5 mL centrifuge tube. Immediately add 200 μL 0.15 M acetic acid to neutralize the solution.
5. Repeat **Steps 3–4** one more time.
6. Initiate ethanol precipitation by adding 40 μL 3 M sodium acetate and 1 mL 100% ice cold ethanol to each centrifuge tube. Incubate the collection at −80°C for 1 h or until frozen to ensure complete precipitation.

3.6. Ethanol Precipitation

1. Without thawing the collections, centrifuge 14,000 rpm for 25 min at 4°C. The DNA will move to the bottom 50 μL of solution.
2. Carefully remove the supernatant from each vial leaving approximately 50–100 μL in the bottom of the vial. This should be done immediately to minimize diffusion of the DNA back into the rest of the solution. The vial should not be disturbed in any way by shaking or tilting which may also drive diffusion of the DNA back into the solution.
3. Add 1 mL ice cold 70:30 ethanol/water.
4. Repeat **Steps 1–3** two times to wash the DNA pellet with the exception that **Step 3** is not carried out after the final wash.
5. Dry the samples in a speed vac (60°C for 25 min) or until dry.
6. The resulting DNA pellet is re-suspended in 30 μL of TGK buffer. The vial is distributed in the following manner: 10 μL for the next selection, 10 μL for bulk K_d determination, and 10 μL for archiving in case cloning and sequencing is desired at a later date.

3.7. Bulk Dissociation Constant Measurements

1. Bulk dissociation constants can be determined using affinity capillary electrophoresis (ACE). Heat the ssDNA collection pool to 72°C for 2 min and allow it to cool to room temperature to ensure stable room temperature conformations.
2. Dilute 10 μL ssDNA collection into 100 μL buffer (*see* **Note 11**).
3. Prepare samples to titrate a constant amount of ssDNA (~1–5 nM) with increasing concentrations of target. Allow the samples to incubate for 20 min to allow binding to occur.
4. Analyze samples on CE. The free aptamer peak should drop as the target concentration increases as a result of an equilibrium shift toward the aptamer–target complex (*see* **Note 12**).
5. Assuming the aptamer concentration is much lower than the target concentration, dissociation constants (K_d) can be estimated by fitting the heights of the unbound peak to the following equation:

$$\frac{I_o - I}{I_o} = \frac{constant \times [\text{target}]}{K_d + [\text{target}]} \tag{3.2}$$

where I_o is the height of the unbound aptamer peak in the absence of target, I is the height of the unbound aptamer peak in the presence of target, and [target] is the concentration of the target.

3.8. Cloning and Sequencing

1. Amplify the final pool using eight PCR cycles. The same procedure described above should be used with the exception of the primers, which should be unlabeled.
2. dsDNA is transfected into DH5 *Escherichia coli* after ligation into a pGEM vector.
3. Colonies with individual clones are raised.
4. Plasmids from 30 (or more) colonies are randomly chosen.
5. Individual sequences are isolated using the T7 promoter sequence and sequenced using standard protocols.

3.9. Aptamer Characterization

1. Programs such as ClustalW identify conserved motifs in a pool of sequences. When using programs that identify motifs, it is necessary to remove the primer regions because the conservation of these regions will dominate the analysis.
2. Programs such as *m*-fold predict the secondary structure of ssDNA and RNA molecules.
3. Dissociation constants for individual aptamers can be determined using ACE as described for the bulk dissociation constants.

4. Notes

1. The library is generally made up of 70–120 base ssDNA or RNA molecules with a 30–80 base random region flanked by two PCR primer regions. As a general rule, the random region should contain at least 30 bases so that common structural motifs such as hairpins, bulges, pseudoknots, and g-quartets can form. Although longer sequences add more randomness to the pool, the process is limited by the mass volume of material necessary to have every possible sequence present in the selection. Additionally, the purity of sequences drastically declines at lengths greater than 100 bases.
2. Polypropylene-coated materials should be consciously used to prevent the target and/or DNA from adhering to the surface of the materials. This is especially important in CE–SELEX because minimal amounts of the target are used in each sample, and minimal amounts of DNA are collected after each selection.
3. Filter pipette tips are especially important in PCR to avoid contamination via the pipette.
4. Many proteins have non-specific affinity for the largely negative surface of bare fused silica capillaries. The shape of the target peak observed when identifying the migration window can be used to determine if significant wall interactions are taking place. Electropherograms in which the target peak exhibits significant tailing or a peak for the target is not observed due to irreversible wall adsorption suggest that a coated capillary should be used to reduce these surface interactions. Coated capillaries with pre burned windows are available through Beckman Coulter, Inc., Fullerton, CA. These capillaries generally eliminate EOF, requiring a negative polarity to be used during CE separation and reversing the order of peak migration. As a consequence, aptamer–target complexes should be collected after the DNA migrates off the capillary (*see* **Fig.** 3.3). Note that during selections the aptamer–target complex is generally not observed during the CE separation since the low concentration of target used limits the maximum amount of complex that can be formed. This is acceptable since the large library concentration makes it easy to observe unbound sequences and the position of the bound sequences in relation to this peak is easy to predict.
5. Modifications that may increase the resolution between the unbound and bound fractions include changing the separation voltage, capillary length and/or inner diameter, separation voltage, buffer components, and injection volume.

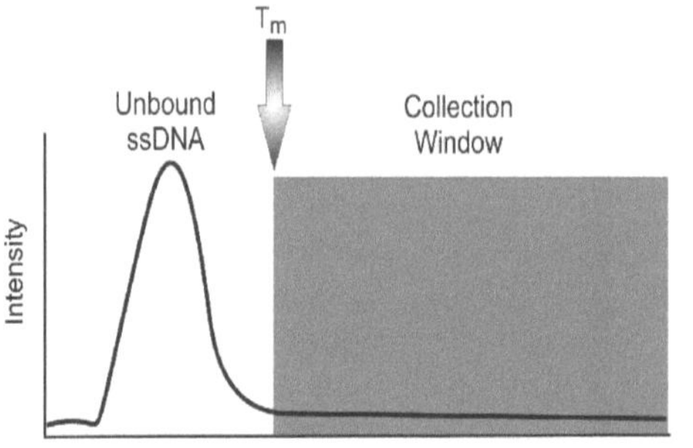

Fig. 3.3. Collection strategy when using a coated capillary that eliminates the EOF. Aptamer–target complexes will generally migrate slower than the unbound sequences and will come off the capillary last. Sequences migrating in a window beginning 1–2 min after the trailing edge of the unbound sequences should be collected. Equation (3.1) should be used to correct for the time required for sequences to migrate through the length of capillary after the detector.

6. The concentration of the DNA pool used in the second and subsequent rounds of selection is generally lower than that used in the first selection cycle due to the limited amount of material that can be produced in the PCR reaction. Therefore, a relatively high concentration stock solution for the target should be used to minimize dilution of the DNA pool during incubation in subsequent rounds of selection.
7. It is important to have the ssDNA concentration much higher than the concentration of the target to ensure competition for binding sites. It is advised to start with the lower concentration of target (50 pM) and only increase the concentration if DNA is consistently not present in the collection fraction.
8. If using a coated capillary, the library will generally migrate off the capillary first. In these cases, stop the separation after the library has completely migrated off the capillary. Rinse the bound fraction in a clean collection vial with buffer at 50 psi for 10 min. This will yield approximately 50 μL of the collection fraction. This can be distributed evenly into eight PCR tubes just like when a bare fused silica capillary is being used.
9. Great care should be taken to avoid contamination of PCR stock solutions with DNA. To prevent contamination, PCR stock solutions should be stored in a separate area from DNA. A new pipette tip should be used for every PCR component, even if the same component is going in several vials. Gloves that have never been worn in the presence of DNA should be worn at all times. It is recommended to prepare PCR master mixes before DNA is touched for the day. Pipette boxes should be used at all times when dealing with PCR components.

10. Formation of a complex with the target does not interfere with PCR amplification of aptamers for several reasons. First, most proteins irreversibly denature at 94°C which will prevent further binding to the DNA. Secondly, heating to 94°C eliminates all secondary and tertiary structure of the DNA which will release even high affinity DNA aptamers from their targets.
11. In most cases, a 1:10 dilution will yield enough DNA to determine a dissociation constant, but little enough DNA that it can still be assumed in most cases that the DNA concentration is lower than the target concentration. In rare cases, there is not enough DNA in the collection to perform K_d measurements. In these cases, more DNA needs to be PCR amplified and purified.

References

1. Joyce, G.F. (1989) Amplification, mutation, and selection of catalytic RNA. *Gene* **82,** 83–87.
2. Ellington, A. D. and Szostak, J. W. (1990) *In vitro* selection of RNA molecules that bind specific ligands. *Nature* **346,** 818–822.
3. Tuerk, C. and Gold, L. (1990) Systematic evolution of ligands by exponential enrichment: RNA ligands to bacteriophage T4 DNA polymerase. *Science* **249,** 505–510.
4. Famulok, M., Mayer G., and Blind, M. (2000) Nucleic acid aptamers-from selection *in vitro* to applications *in vivo*. *Acc. Chem. Res.* **33,** 591–599.
5. Hamula, C.L.A., Guthrie, J.W., Zhang, H., Li, X.-F. and Le, X. C. (2006) Selection and analytical applications of aptamers. *Trends. Analyt. Chem..* **25,** 681–691.
6. Bunka, D.H.J. and Stockley, P.G. (2006) Aptamers come of age – at last. *Natl. Rev. Microbiol..* **4,** 588–96.
7. Tombelli, S., Minunni, M. and Mascini, M. (2005) Analytical applications of aptamers. *Biosens. Bioelectron.* **20,** 2424–2434.
8. Mendonsa, S.D. and Bowser, M.T. (2004) *In vitro* evolution of functional DNA using capillary electrophoresis. *J. Am. Chem. Soc.* **126,** 20–21.
9. Mendonsa, S.D. and Bowser, M.T. (2004) *In vitro* selection of high-affinity DNA ligands for human IgE using capillary electrophoresis. *Anal. Chem.* **76,** 5387–5392.
10. Mosing, R.K., Mendonsa, S.D. and Bowser, M.T. (2005) Capillary electrophoresis-SELEX selection of aptamers with affinity for HIV-1 reverse transcriptase. *Anal. Chem.,* 77, 6107–6112.

Chapter 4

In Vitro Selection of Allosteric Ribozymes

Nicolas Piganeau

Abstract

In vitro selection techniques offer powerful and versatile methods to isolate nucleic acid sequences with specific activities from huge libraries. The present protocol describes an in vitro selection strategy for the de novo selection of allosteric self-cleaving ribozymes responding to virtually any drug of choice. We applied this method to select hammerhead ribozymes inhibited specifically by doxycycline or pefloxacin in the sub-micromolar range. The selected ribozymes can be converted into classical aptamers via insertion of a point mutation in the catalytic center of the ribozyme.

Keywords: Ribozyme, in vitro selection, allostery, aptazyme.

1. Introduction

The activity of catalytic RNAs can be regulated by small molecules. These so-called allosteric ribozymes or aptazymes can find applications in the field of basic biological research or applied biotechnology. For example, they can be employed as molecular sensors detecting the presence of the effector molecule *(1)*. Alternatively they can be inserted in genes and serve as synthetic switches for the control of gene expression *(2)*.

The first allosteric ribozymes were generated via rational design by the fusion of a constitutive ribozyme to an RNA aptamer *(3)*. Later, in vitro selection methods were developed to optimize the "communication module" between the aptamer domain and the catalytic portion of the aptazyme *(4)*. These methods depend on the pre-selection of an aptamer prior the creation of the allosteric ribozyme. During the selection of aptamers, the small ligand must be immobilized allowing affinity

Günter Mayer (ed.), *Nucleic Acid and Peptide Aptamers: Methods and Protocols, vol. 535*

DOI 10.1007/978-1-59745-557-2_4 Springerprotocols.com

chromatography for the separation step. The immobilization can be difficult to achieve and may mask a potential binding site of the target molecule important for the interaction with certain RNA species of the library.

However, it is also possible to select de novo an allosteric ribozyme by introducing a random sequence into the ribozyme and selecting for inhibition or activation of the catalytic activity via a small effector molecule *(5, 6)*. Using this method no immobilization procedure of the small ligand is required.

The method presented here describes the selection of allosteric hammerhead ribozyme variants, which are inhibited by virtually any drug of choice. It can be easily adapted for the selection of ribozymes activated by the effector molecule.

2. Materials

2.1. Pool Synthesis

1. Oligodeoxynucleotides.

Pnp-rev:	5'-ACG TCT CGA GGT AGT TTC GT
Pnp-1:	5'-CGC GTT GTG TTT ACG CGT CTG ATG
Pnp-pool:	5'-CGC GTT GTG TTT ACG CGT CTG ATG AGT NNN NNN NNN NNN NNN NNN NNN NNN NNN NNN NNN NNN NNN NAC GAA ACT ACC TCG AGA CGT
Pnp-2:	5'-AGC TGG TAC CTA ATA CGA CTC ACT ATA GGA GCT CGG TAG TGA CGC GTT GTG TTT ACG CGT CTG ATG
Pnp-3:	5'-AGC TGG TAC CTA ATA CGA CTC ACT ATA GGA GCT CGG TAG TCA CGC GTT GTG TTT ACG CGT CTG ATG

2. Double distillated water (ddH_2O). The water should be free of RNases. In our hand no further treatment was necessary. If required add 0.1% diethylpyrocarbonate (DEPC) to water, mix overnight, and autoclave 20 min to hydrolyze DEPC.
3. DAp Gold star DNA polymerase and Gold star Buffer, Eurogentec.
4. 25 mM $MgCl_2$, filtered through a 0.2 μm nitrocellulose filter.
5. 4 mM dNTP (each). Store at –20°C.
6. Phenol/chloroform/isoamyl alcohol (25:24:1) saturated with TE (10 mM Tris–HCl (pH 8.0), 1 mM EDTA). Store at 4°C protected from light.
7. 3 M Sodium acetate (pH 5.2) (adjust pH with glacial acetic acid).

8. Sephadex G50 medium (GE healthcare). Prepare 50% slurry according to manufacturer instructions.

2.2. Transcription

1. 5X T_7 reaction buffer: 200 mM Tris–HCl (pH 8.0), 40 mM $MgCl_2$, 250 mM NaCl, 10 mM spermidine, 150 mM DTT. Store at −20°C.
2. 25 mM NTP (each). Prepare aliquots and store at −20°C.
3. α-^{32}P CTP: 10 μCi/μl, 3,000 Ci/mmol.
4. 100 mM GMPS (Guanosine- 5'-*O*-monophosphorothioate) from emp biotech.
5. T_7 RNA polymerase (Ambion).
6. 6 M Ammonium acetate (pH 6.0) (adjust pH with glacial acetic acid).
7. DNase I (RNase-free) (Roche).

2.3. Polyacrylamide Gel Electrophoresis (PAGE)

1. PAGE loading solution: 9 M urea, 50 mM EDTA. For UV-shadowing the buffer should be free of dye. To follow electrophoresis add xylene cyanol and bromophenol blue (0.4% w/v each) on a separate lane.
2. 40% Acrylamide/bis solution (19:1) (this is a neurotoxin when unpolymerized and so care should be taken not to receive exposure) and N,N,N,N'-Tetramethyl-ethylenediamine (TEMED, Bio-Rad, Hercules, CA).
3. Ammonium persulfate: prepare 10% solution in water, store at 4°C for no more than 1–2 weeks.
4. Dichlordimethylsilane (5% in chloroform). Store and manipulate under a fume-hood.
5. 10X TBE: 1.1 M Tris–HCl, pH 8.3, 900 mM borate, 25 mM EDTA.
6. Thin-layer chromatography plates F_{254} (20 × 20 cm, Merck).

2.4. Selection

1. 10X Biotinylation buffer: 500 mM Tris–HCl, pH 8.3, 50 mM EDTA.
2. Iodoacetyl-LC biotin 4 mM in dimethyl formamide (DMF), Pierce. Prepare aliquots and store at −20°C.
3. Streptavidin agarose (Pierce) equilibrated in coupling buffer (PBS, 150 mM NaCl −50% slurry) according to manufacturer instructions.
4. W: 25 mM HEPES (pH 7.4), 1 M NaCl, 5 mM EDTA.
5. WB: 3 M urea, 5 mM EDTA.
6. 5X Selection buffer: 200 mM Tris–HCl, pH 8.0 (25°C), 250 mM NaCl, 10 mM spermidine. Store at −20°C. During the selection the 1X buffer should be supplemented with $MgCl_2$ (8 mM final concentration).

7. Glycogen (20 μg/μl) (Roche).
8. 5X RT-PCR buffer: 250 mM bicine/KOH, pH 8.2 (25°C); 575 mM K-acetate; 40% glycerol (v/v).
9. Tth DNA polymerase (Roche).
10. Taq reaction buffer (10X): 100 mM Tris (pH 8.3), 500 mM KCl, 0.01% gelatin.
11. Taq polymerase.

2.5. Analysis of Selected Clones

1. Calf Intestine Alkaline Phosphatase (Fermentas).
2. RNasin (Promega).
3. γ-^{32}P-ATP: 10 μCi/μl, 3,000 Ci/mmol.
4. T4 polynucleotide kinase (Ambion).

3. Methods

3.1. Construction of Initial Pool

The general design of the initial pool is depicted in **Fig. 4.1**.

1. Prepare three water-bathes pre-heated at the following temperatures: 94, 55, and 72°C.
2. In a total volume of 80 ml mix the following components: Pnp-pool 5 nmol, primers (Pnp-1 and Pnp-rev) 80 nmol, $MgCl_2$ 1.5 mM, dNTP 0.2 mM, Gold star reaction buffer 1X, DAp Gold star DNA polymerase 250 U. Aliquot PCR reaction into ten 15 ml tubes (8 ml each).
3. Perform five PCR cycles by transferring the tubes successively into the three water-bathes like following: 94°C, 5 min; 55°C, 5 min; 72°C, 7 min. Mix every 2 min by inversion. Take 5 μl aliquots after each cycle to follow amplification on a 2% agarose gel.
4. To purify the PCR reaction, add 7 ml phenol/chloroform/isoamyl alcohol to each tube, vortex strongly, and centrifuge for 10 min at 4,500 *g*. Transfer the aqueous phase to a new tube. Add 7 ml chloroform, vortex, and centrifuge as previously. Transfer aqueous phase to a new tube.
5. Pool PCR into six 50 ml tubes (13 ml each), add 1.3 ml 3 M sodium acetate and add 30 ml 100% ethanol. Incubate for 30 min at –20°C and centrifuge for 30 min at 4,500 *g* at 4°C. Remove supernatant, add 10 ml 70% ethanol, and centrifuge for 10 min at 4,500 *g* at 4°C. Remove supernatant, let pellets dry, and resuspend the PCR products into 1 ml ddH_2O (total volume).
6. Apply the PCR on a G50 column (0.7 × 20 cm) pre-equilibrated with ddH_2O and elute with ddH_2O. Collect

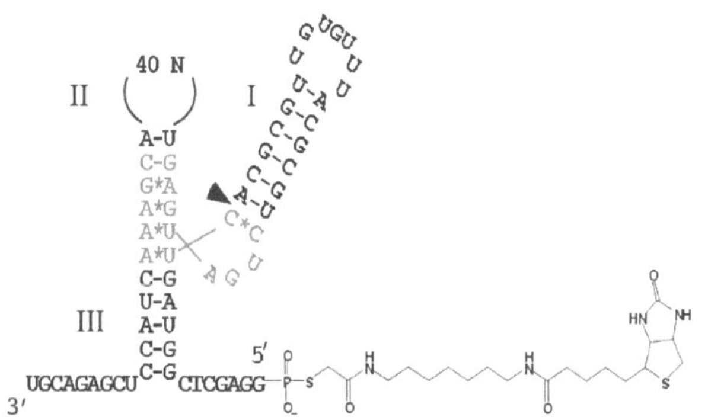

Fig. 4.1 Secondary structure of the transcripts from the initial pool. Helix II is shortened to two base pairs, and loop II is replaced with a 40 nt random region. *Gray*: Nucleotides of the catalytic center of the hammerhead ribozyme (HHR). *Black arrow*: Cleavage site. The chemical link between the biotin moiety and the RNA used during the selection is also shown.

1 ml fractions and measure absorption at 260 nm (*see* **Note 1**). Analyze DNA containing fractions on a 2% agarose gel and pool fractions containing the PCR product. Typical recovery rates should be around 20–40 nmol.

7. Repeat the whole large scale PCR procedure with 5 nmol of the amplified product using primers Pnp-3 and Pnp-rev.

3.2. *In Vitro* Selection

The following protocol describes a typical selection cycle (*see* **Note 2** and **Fig. 4.2**).

The reaction volumes and the concentration of the effector molecule should be adjusted during the selection to increase selection stringency. Conditions used during a successful selection are shown in **Table 4.1**.

3.2.1. Transcription

1. To produce GMPS-primed RNA mix following components on ice to a final volume of 100 μl: 5X T_7 reaction buffer, 20 μl; 25 mM NTP, 10 μl; alpha-P^{32} CTP, 3 μl; 100 mM GMPS, 20 μl; 200 pmol DNA template; 1 mM effector molecule; 250 U T_7 polymerase. Start the reaction by addition of the polymerase. Incubate 4 h at 37°C (*see* **Notes 3** and **4**).
2. The DNA template is then degraded by addition of 5 U DNAse I followed by 30 min incubation at 37°C. Stop enzymatic reaction by addition of 100 μl 0.25 M EDTA (pH 8.0).
3. Add 100 μl 6 M ammonium acetate and 900 μl 100% ethanol. Vortex. Incubate 5 min at room temperature before centrifugation at 15,000 *g* for 15 min at 4°C. Remove supernatant, add 1 ml 70% ethanol, and centrifuge at 15,000 *g* for 5 min at 4°C. Remove supernatant, open test tube and dry pellet for a few minutes at 37°C. Resuspend pellet in 100 μl PAGE loading buffer.

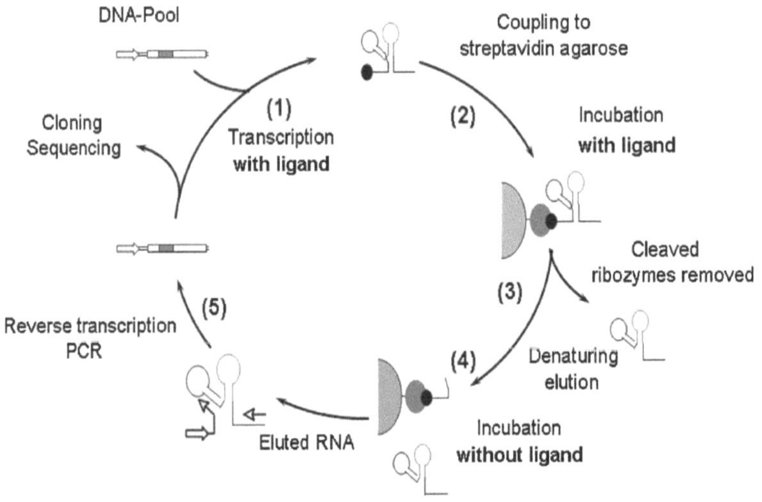

Fig. 4.2 Schematic representation of the selection procedure. The DNA-pool is transcribed using T_7 RNA polymerase in the presence of guanosine monophosphorothioate (GMPS) and 1.0 mM effector to avoid cleavage during transcription (1). To efficiently separate uncleaved from cleaved ribozymes, the whole RNA library is chemically biotinylated at the 5'-termini, immobilized on streptavidin agarose and incubated with the effector (2). Bound ligand and cleaved ribozyme products are removed by denaturing washing steps (3). Uncleaved, immobilized RNAs are incubated for cleavage without effector (4). Cleaved ribozymes are eluted, reverse transcribed, and amplified by PCR (5). The design of the PCR primers allows restoration of the 5' cleaved fraction of the HHR and the T_7 promoter. The resulting DNA is used for the next selection cycle.

3.2.2. Denaturing Polyacrylamide Gel Electrophoresis

1. Clean glass plates (16.5 × 22 cm), spacers (1.5 mm) and comb (ten wells, 1 cm each) thoroughly with water and 70% ethanol. If necessary (usually every 2–5 gels) treat one of the glass plates with dichlordimethylsilane by gently wiping a few milliliters on the plate under a fume hood and letting the surface dry. Assemble gel plates and spacers, seal with adhesive tape.
2. Prepare 8% polyacrylamide solution by mixing 5 ml 10X TBE, 10 ml 40% polyacrylamide solution and 25 g urea and ddH_2O to a final volume of 50 ml. After dissolution of urea, start polymerization with addition of 250 μl 10% APS and 25 μl TEMED. Cast gel immediately. If necessary remove air bubbles by knocking gently on glass plates. Place comb and wait until gel is polymerized (30 min to 1 h). Remove tape and comb; wash slots with water to remove polyacrylamide rests.
3. Assemble gel on electrophoresis apparatus with aluminum plate for heat dispersion and fill reservoirs with 1X TBE (dilute 100 ml 10X TBE to 1 l with ddH_2O in a cylinder, seal with parafilm and mix by inverting a few times). If needed remove air bubbles in the wells and at the bottom of the gel using a 50 ml syringe.
4. Connect gel to power-supply (minus electrode on top) and pre-run for 20 min at 300 V. Denature RNA probes by a short

Table 4.1

Selection conditions. The modifications of the basic selection procedure used during a successful selection are shown *(5)*. The actual conditions needed for a particular target may vary from the one presented here. If high ribozyme activity or high effector sensitivity is required a theoretical model can be employed to determine the optimal selection-parameters *(6)*

Cycle	Transcription volume	Biotinylation (RNA used) (nmol)	Streptavidin agarose (50% slurry) / RNA	Incubation volume	First incubation (h)	Effector concentration	Second incubation (min)
1[1]	4 ml (6 nmol template)	40	5 ml / 20 nmol	10 ml	2	1 mM	10
2	200 μl	4	500 μl / 2 nmol	1 ml	4	1 mM	5
3–5	100 μl	2	250 μl / 1 nmol	500 μl	4	1 mM	5
6 and 7	100 μl	2	250 μl / 1 nmol	500 μl	4 [2]	1 mM	5
8–10	100 μl	2	250 μl / 1 nmol	500 μl	4 [2]	100 μM	1
11[3]–13	100 μl	2	250 μl / 500 pmol	500 μl	4 [2]	10 μM	1
14[3]–16	50 μl	1	100 μl / 100 pmol	250 μl	4 [2]	1 μM	1

[1] Additional modification for this cycle: reverse transcription and PCR volumes are doubled.
[2] The selection cycle is modified according to **Section 3.3.2**.
[3] Before cycles 11 and 14, a mutagenic PCR is performed according to **Section 3.3.1**.

incubation at 95°C (1–2 min). Stop power-supply and rinse the wells thoroughly with 1X TBE using a 50 ml syringe with a 21-gauge needle to remove urea. Immediately load RNA on gel (two wells). Load loading dye in an adjacent well to follow electrophoresis. Run gel for 1 h to 90 min at 300 V.

5. Remove gel from apparatus and carefully separate glass plates using one of the spacers as lever. The gel should remain on one plate. Place wrapping foil on the gel, invert and remove second glass plate. Place wrapping foil on the other side.
6. Place gel on a thin-layer chromatography plate and illuminate with UV light (254 nm) the RNA should appear as a dark shadow. Two bands should be visible: intact ribozymes and cleaved products. Cut out band corresponding to the full-length ribozyme. Place gel piece containing RNA into a 2 ml reaction tube and crush against the tube wall with a blue tip. Add 600 μl 0.3 M sodium acetate (pH 5.4) and incubate for 90 min at 65°C with strong shaking.
7. Insert glass wool into a syringe and use the piston to press the solution containing the gel pieces through the glass wool into a new 2 ml tube. Fill the tube with 100% ethanol, incubate at –20°C for 20 min and centrifuge at 15,000 *g* for 15 min at 4°C. Remove supernatant and wash with 1 ml 70% ethanol.
8. Resuspend dried pellet into 50 μl ddH_2O. Use 5 μl of this solution in a total volume of 200 μl H_2O to determine optical density at 260 nm. Typical yield is 2–3 nmol RNA.

3.2.3. RNA Biotinylation

1. Mix the following component to a final volume of 1 ml: 100 μl 10X biotinylation buffer, 2 nmol GMPS-primed RNA and 100 μl iodoacetyl-LC-biotin. Incubate for 90 min at room temperature protected from light with occasional shaking.
2. Precipitate the reaction products by addition of 100 μl 3 M sodium acetate and 2.5 ml 100% ethanol, incubation at –20°C for 20 min, and centrifugation at 15,000 *g* for 15 min at 4°C. Remove supernatant and wash with 1 ml 70% ethanol.
3. The dried pellet are resuspended in 50 μl PAGE loading buffer and purified with PAGE (in one slot) as before. The final amount of recovered RNA is quantified on a photometer at a wavelength of $\lambda = 260$ nm.

3.2.4. Column Immobilization and Selection

1. 1 nmol Biotinylated RNA is incubated for 30 min at room temperature on 250 μl streptavidin agarose equilibrated in coupling buffer. The amount of RNA linked on the column typically ranges between 20 and 40%.
2. To eliminate unlinked species the column is washed thoroughly (six times with alternatively 1 ml WA and 1 ml WB)

and rinsed with water (five times 500 μl). Collect flow-through and wash fractions for analysis.

3. For the first selection incubation, the column material is incubated in 500 μl selection buffer with the appropriate amount of effector molecule at 37°C with gentle shaking for the appropriate time (*see* **Table 4.1**). The incubation is initiated upon addition of $MgCl_2$.
4. Repeat **Steps 2 and 3** without effector molecule. Adjust incubation time according to **Table 4.1**.
5. Finally, the cleaved RNA is eluted with WB (two times 500 μl). The different washing and eluting fractions are counted in a scintillation counter, and the amount of eluted RNA is determined.
6. The eluted RNA is purified with three phenol–chloroform–i-soamylalcohol extractions, one chloroform extraction and precipitated (sodium acetate) in the presence of glycogen (5 μg). The pellet is washed for two additional times with 70% ethanol before resuspension in 20 μl ddH_2O with 200 pmol of Pnp-rev primer.
7. The RNA–oligonucleotide mix is denatured (1 min, 95°C) and mixed with 80 μl reverse transcription mix (5 μl dNTP, 4 mM; 10 μl MnOAc, 25 mM; 20 μl 5X RT-PCR buffer; 43 μl ddH_2O; 2 μl Tth DNA polymerase (2–10 units)). The total mix is incubated for 30 min at 72°C (at the same time a control without RNA is performed).
8. The reverse transcription mix is then diluted into a 500 μl PCR reaction under standard conditions with primers Pnp-3 and Pnp-rev (including a negative control). The number of cycles is calculated using the following rule:

$$n \geq 1 + \ln\left(\frac{1000}{x}\right)/\ln(2)$$

where x is the amount of RNA eluted in pmol.

9. The PCR products are analyzed on a 2% agarose gel, before phenol/chloroform extraction and precipitation (sodium acetate). 25% of the resulting DNA is used for the next selection cycle.
10. The whole selection cycle should be reproduced up to 16 times.

3.3. Optional Selection Components

New mutations can be inserted into the selected pool via mutagenic PCR *(7)*. The following protocol should introduce on average one mutation per RNA molecule.

3.3.1. Mutagenic PCR

1. Prepare the following master-mix and make twenty 83 μl aliquots in PCR tubes 10X Taq reaction buffer, 210 μl;

10 mM dATP–dTTP, 42 µl; 10 mM dCTP–dTTP, 210 µl; 100 mM $MgCl_2$, 147 µl; 5 mM $MnCl_2$, 210 µl; 100 µM Pnp-3, 105 µl; 100 µM Pnp-rev, 105 µl; ddH_2O, 714 µl.

2. Add 15 µl of PCR product from the last selection cycle to the first aliquot. Pre-heat mix to 55°C before adding 2 µl Taq polymerase (10 U). Perform three PCR cycles (94°C 50 s, 55°C 1 min, 72°C 1 min). Let the block cool down to 55°C.
3. After the three cycles, 15 µl of the reaction mixture (15 µl) is transferred into a new aliquot pre-heated at 55°C. Add Taq polymerase and perform PCR cycles as before. Repeat procedure 20 times (60 PCR cycles). Verify regularly the DNA levels (every 15 PCR cycles) on a 2% agarose gel.
4. Fix the mutations using a standard PCR protocol (without manganese) for four cycles using the last 100 µl as template in a total volume of 800 µl. Purify PCR product by phenol/chloroform extraction and precipitation. Use 15% of the final product for the next selection cycle.

3.3.2. Counter-Selection of Misfolded Ribozymes

The selection procedure described above can also enrich ribozymes folding into several different states, some active and some inactive. To avoid these ribozymes to overcome the population a counter-selection can be performed. For this purpose replace the first incubation (with effector) during the selection (**Step 6**) with the following procedure.

1. The column material is incubated in 500 µl selection buffer with the appropriate amount of effector molecule (*see* **Table 4.1**) at 37°C with gentle shaking for 15 min. The incubation is initiated upon addition of magnesium.
2. The column is washed twice with 1 ml WB and three times with 1 ml ddH_2O. To allow denaturation and renaturation of the ribozymes.
3. **Step 1 and 2** are repeated ten times.
4. Incubate column material in 500 µl selection buffer with the appropriate amount of effector molecule (*see* **Table 4.1**) at 37°C with gentle shaking for final 90 min.
5. Continue selection with **Step** 7.

3.4. Analysis of Selected Allosteric Ribozymes

3.4.1. Cloning and Sequencing

1. Clone the selected DNA-pool after the last selection cycle into a vector of choice using standard molecular biology methods. For cloning of the pool with the T_7 promoter use the restriction sites KpnI and XhoI. For cloning of the sole ribozyme use the sites SacI and XhoI.
2. Sequence the inserts and identify individual sequences.
3. Perform PCR from a clone of interest with primers Pnp3 and Pnp-rev. Use PCR product for a transcription reaction as

described in **Section 3.2.1** in the absence of GMPS and radioactive nucleotide. Purify transcripts with PAGE (**Section 3.2.2**).

4. Dephosphorylate RNA with calf intestine alkaline phosphatase. 150 pmol RNA is incubated in 50 mM Tris–HCl (pH 8.5), 1 mM EDTA, RNasin (20 U) and calf intestinal alkaline phosphatase (10 U) in 50 μl for 30 min at 37°C and 10 min at 75°C after addition of 0.5 μl 0.5 M EDTA (pH 8.0) and vortexing. After the dephosphorylation, purify the RNA via a phenol/chloroform extraction and precipitate (with sodium acetate) in the presence of glycogen (5 μg). Finally, resuspend pellet in 20 μl ddH_2O.
5. 10 pmol of the dephosphorylated RNA (3 μl considering 50% loss during the purification), is then radioactively marked in a total volume of 20 μl in 70 mM Tris–HCl (pH 7.6), 10 mM $MgCl_2$, 5 mM dithiothreitol, with 10 U Rnasin, 30 μCi gamma-P^{32}-ATP (i.e., 10 pmol) and 20 U T4 polynucleotide kinase in the presence of 1 mM effector molecule. Incubate the reaction 30 min at 37°C before quenching with 1 μl 0.5 M EDTA (pH 8.0), and precipitating (ammonium acetate).
6. Perform PAGE purification as before. Instead of UV-shadowing the RNA band should be visualized using an X-ray film with 30 s exposure. Mark film position on the gel during exposure with waterproof marker. After film development replace the film under the gel at the position marked during exposure and cut-out band corresponding to uncleaved ribozymes. After elution resuspend RNA into 500 μl ddH_2O. Assuming a loss of 50% during purification the concentration of RNA should be 10 nM.
7. Incubate RNA (1 nM) in selection buffer with various amounts of effector molecule (typically between 5 nM and 5 μM). Start reaction upon addition of $MgCl_2$ and take 10 μl aliquots of the reaction every 20 s for the first 2 min of the reaction. Mix aliquots immediately with an equal volume of PAGE loading buffer on ice. Load 10 μl samples on an 8% polyacrylamide gel. For processing of large amounts of samples a sequencing-gel is recommended.
8. Separate plates and transfer gel on a Whatman paper. Cover with wrapping foil. Expose over night with a PhosphorImager screen. Develop screen and quantify bands corresponding to cleaved and uncleaved ribozymes. Determine the percentage of uncleaved ribozyme for each sample. Determine cleavage rate k_{obs} by fitting each reaction with a simple exponential decay $\left(e^{(-k_{obs}\cdot t)}\right)$. To determine the inhibition constant (K_i) fit the cleavage rates at different effector concentrations with the following formula (where k is the cleavage rate in the absence of effector):

$$k_{obs}([Drug]) = \frac{k}{1 + \frac{[Drug]}{K_i}}$$

3.4.2. Conversion of Allosteric Ribozymes into Aptamers

1. Perform PCR from a clone of interest with primers Pnp2 and Pnp-rev. Primer Pnp2 introduces a point mutation into the hammerhead ribozyme sequence abolishing self-cleaving activity.
2. Binding of the aptamer to the small molecule can be monitored via competition experiments with intact ribozymes.

4. Notes

1. Correspondence between absorbance at 260 nm (OD = 1) and concentration of nucleic acids: ssDNA: 33 μg/ml, dsDNA: 50 μg/ml, RNA: 40 μg/ml.
2. The protocol is tailored for the selection of aptazymes inhibited by the effector molecule. When selecting for activation instead of inhibition, remove the effector from all steps before the second incubation on the column and add it during this incubation.
3. When working with RNA special care should be taken to avoid contamination with RNAses. All solutions should be tested for the presence of RNase activity prior use, e.g., by incubation with radioactively labeled RNA followed by polyacrylamide gel electrophoresis and autoradiography.
4. Radioactive labeling of the RNA is optional but is recommended to monitor the progress of the selection procedure. It is advisable to perform every few cycles a control without effector to detect if the enrichment observed is due to the effector or to misfolded ribozymes. Counter-selection of misfolded ribozymes can be preformed as described in **Section 3.3.2**.

References

1. Breaker, R.R. (2002) Engineered allosteric ribozymes as biosensor components. *Curr. Opin. Biotechnol.* **13**, 31–39.
2. Yen, L., Svendsen, J., Lee, J. S., Gray, J.T., Magnier, M., Baba, T., D'Amato, R.J. and Mulligan, R.C. (2004) Exogenous control of mammalian gene expression through modulation of RNA self-cleavage. *Nature* **431**, 471–476.
3. Tang, J. and Breaker, R.R. (1997) Rational design of allosteric ribozymes. *Chem. Biol.* **4**, 453–459.
4. Koizumi, M., Soukup, G.A., Kerr, J.N. and Breaker, R.R. (1999) Allosteric selection of ribozymes that respond to the second messengers cGMP and cAMP. *Nat. Struct. Biol.* **6**, 1062–1071.

5. Piganeau, N., Jenne, A., Thuillier, V. and Famulok, M. (2001) An allosteric ribozyme regulated by doxycyline. *Angew. Chem. Int. Ed. Engl.* **40**, 3503.
6. Piganeau, N., Thuillier, V. and Famulok, M. (2001) *In vitro* selection of allosteric ribozymes: theory and experimental validation. *J. Mol. Biol.* **312**, 1177–1190.
7. Cadwell, R.C. and Joyce, G.F. (1994) Mutagenic PCR. *PCR Methods Appl.* **3**, S136–S140.

Chapter 5

Cell-Specific Aptamers for Targeted Therapies

Laura Cerchia, Paloma H. Giangrande, James O. McNamara, and Vittorio de Franciscis

Abstract

Many signalling proteins involved in diverse functions such as cell growth and differentiation can act as oncogenes and cause cellular transformation. These molecules represent attractive targets for cancer diagnosis or therapy and therefore are subject to intensive investigation.

Aptamers are small, highly structured nucleic acid molecules, isolated from combinatorial libraries by a procedure termed SELEX. Aptamers bind to a target molecule by providing a limited number of specific contact points imbedded in a larger, defined three-dimensional structure. Recently, aptamers have been selected against whole living cells, opening a new path which presents three major advantages: (1) direct selection without prior purification of membrane-bound targets, (2) access to membrane proteins in their native conformation similar to the in vivo conditions and (3) identification of (new) targets related to a specific phenotype. The ability to raise aptamers against living cells opens some attractive possibilities for new therapeutic and delivery approaches. In this chapter, the most recent advances in the field will be reviewed together with detailed descriptions of the relevant experimental approaches.

Key words: Aptamer, SELEX, ret, delivery, siRNA.

1. Introduction

1.1. Intact Cells as Targets

With the first description of SELEX in 1990 *(1, 2)* the therapeutic potential of aptamers was apparent. In particular, the potential application of RNA ligands as antagonists of clinically relevant protein targets and their advantages over other macromolecular technologies as antibodies or peptides was clear *(3, 4)*. The fact that the entire aptamer identification process is performed in vitro permits the researcher to raise aptamers against virtually any soluble protein. Indeed, several aptamers have been raised that target extracellular soluble proteins with potential therapeutic value.

Günter Mayer (ed.), *Nucleic Acid and Peptide Aptamers: Methods and Protocols, vol. 535*

DOI 10.1007/978-1-59745-557-2_5 Springerprotocols.com

These targets include growth factors, cytokines and coagulation factors, proteins that can be easily produced in huge amounts, and more importantly have the common advantage of being readily accessible to drug ligands that do not cross cell membranes. In fact, some of these aptamers have entered clinical trials, as for example, pegaptinib (Macugen) that targets one isoform of the vascular endothelial growth factor ($VEGF_{165}$) now approved for treatment of age-related macular degeneration *(5)*.

In addition, aptamers targeting intracellular proteins that act as signalling mediators have also been generated. However, a major obstacle to their use as therapeutics is the development of intracellular delivery approaches, as described for intramers *(6)*.

Aptamer selection approaches that target the cell surface have been developed more recently. In the first paper describing such an approach, Morris et al. demonstrated that SELEX could be used to simultaneously isolate RNA ligands to multiple targets bound to a biological membrane *(7)*. Purified red blood cell ghosts were used as target, with essentially the same SELEX protocol that is used with individual purified proteins. After 25 rounds of SELEX, they have isolated a set of different ligands each targeting distinct red blood cell proteins, thus demonstrating that the procedure could be considered as the resultant of multiple simultaneous and independent experiments. Furthermore, it was evident that increasing the stringency by reducing the target concentrations enriched the pool for the highest affinity ligands.

Homann and Goringer *(8)* confirmed that SELEX protocol can be performed with live cells and even without the knowledge of all elements of a target's surface. They were the first to apply the SELEX technology to a parasite system by addressing the question whether aptamers can be selected to recognize the surface of live parasite cells. By using African trypanosomes as a model system, an extracellular blood parasite with a very specific surface architecture, they identified an aptamer family that recognized an invariant surface component of bloodstream stage trypanosomes.

More recently we took advantage of these results to develop a strategy that allowed us to inhibit a transmembrane receptor tyrosine kinase (RTK) by targeting its extracellular region with a high-affinity ligand aptamer *(9)*. RTKs are privileged targets for cancer therapy, which is underscored by the promising outcome of clinical trials with small molecules or antibody inhibitors *(10)*. Indeed, these receptors are large molecules heavily modified by post-translational changes, as glycosylation and phosphorylation and thus purification of even a portion of these receptors may require a long and wasteful procedure. Furthermore these proteins are functional in their membrane-bound conformation therefore, using the extracellular portion as target may frequently lead to isolate ligands that are unable to recognize the functional native receptor *(11)*.

We validated a general strategy to isolate aptamers for an activated mutated transmembrane receptor tyrosine kinase, Ret. Germline mutations in the RET gene are responsible for constitutive activation of the receptor and for inheritance of multiple endocrine neoplasia (MEN) type 2A and 2B syndromes, and of familial medullary thyroid carcinoma *(12–16)*. The RET receptor constitutes a model system of choice *(16)* in that the transforming mutations, of MEN2A type, located in the extracellular domain simplify the issue of intracellular accessibility for a targeting molecule *(14, 15)*.

We based our experimental approach on the notion that targeting a complex target constituted of multiple proteins, as is the membrane surface of a cell, permits to isolate ligands for individual proteins provided that they are highly represented *(7, 9)*.

Therefore, we used as target of the selection procedure living mammalian cells growing in culture dishes engineered in order to express high levels of the RET mutant protein. These conditions are expected to expose a native protein to the selection procedure, thus best mimicking in vivo conditions. In order to deplete the pool expressing two different forms of the receptor tyrosine kinase Ret: one with a transforming mutation located in the extracellular domain leading to constitutively dimeric Ret, i.e. *the target (14)*, and one in which the receptor remains monomeric, with a transforming mutation located in the intracellular domain, i.e. *the sham (15, 17)*.

Molecular-level differentiation of neoplastic cells is essential for accurate and early diagnosis, but effective molecular probes for molecular analysis and profiling of neoplastic cells are not yet available. The intact cell-based SELEX strategy is generally applicable to different cell types and holds a great promise in developing specific molecular probes for cancer biomarker discovery and for cancer diagnostic and therapeutic applications (*see* **Fig. 5.1**).

1.2. Cell Internalizing Aptamers as Therapeutic Reagents

The development of aptamers as therapeutics has primarily involved aptamers that bind and inhibit the activity of their protein targets. Another promising application of aptamers is to use them to deliver a variety of secondary reagents specifically to a targeted cell population. Once delivered, the secondary reagents would then impart their therapeutic effect to this subset of cells within the treated individual. Because non-targeted cells would not be exposed to the secondary reagent, the potential for unwanted side-effects such as death of normal cells as occurs with the use of many cancer therapeutics is substantially reduced.

This approach utilizes the cell-type specific expression of cell surface proteins on cell populations of therapeutic value. The idea here is to develop an aptamer to the extracellular portion of such a

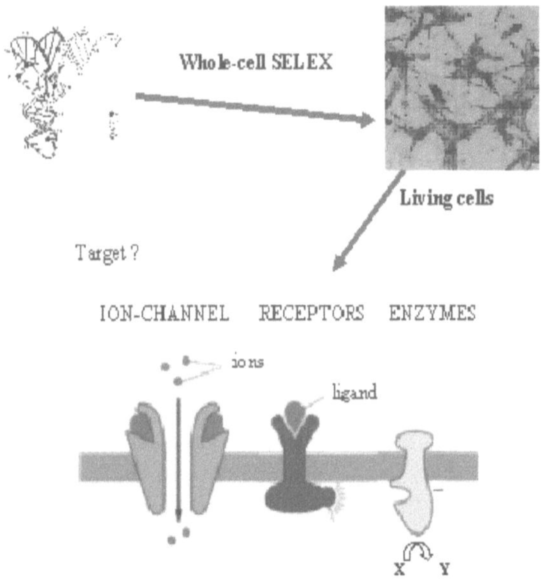

Fig. 5.1. The whole-cell SELEX technology allows identifying unique molecular features of cancer cells by selecting aptamers in a physiological context, and, most importantly, it can be done without prior knowledge of the target molecules.

protein and to then use the aptamer to deliver the secondary reagent to the targeted cell population via binding the targeted protein on the surface of the targeted cell type. Because this binding in some cases also results in the endocytosis of the aptamer/secondary reagent complex, this approach can be used to deliver reagents such as siRNAs that depend on delivery to intracellular compartments for their proper function.

Antibodies and other protein-based reagents have previously been developed to serve comparable roles in targeting therapeutics to specific cell types *(18, 19)*. However, aptamers have a number of important advantages over proteins as therapeutic reagents. A number of these advantages stem from the fact that proteins must be produced in cell culture while aptamers can be chemically synthesized. The production of proteins is thus expensive and complicated by batch-to-batch variability in activity, resulting in a more complicated regulatory approval process. In addition, aptamers can be readily chemically modified to enhance their bioavailability and pharmacokinetics. Another important advantage of RNA aptamers over proteins is the fact that RNA is much less immunogenic than proteins. Therapeutics made from RNA are thus likely to be safer when repeated administrations are necessary. RNA made with pyrimidines modified at the 2'-position, which renders them resistant to extracellular nucleases are even less immunogenic than natural RNA *(20)*.

Aptamers targeting the prostate-specific membrane antigen (PSMA) have been used to deliver both nanoparticles and siRNAs to prostate cancer cells in therapeutic proof of concept studies *(21–23)*. Nanoparticles containing docetaxel or siRNAs targeting cancer cell survival genes, when targeted with PSMA-binding aptamers were internalized by PSMA-expressing prostate cancer cells and resulted in cancer cell death in vitro and retarded tumour growth in vivo. The ability of aptamers to specifically deliver secondary therapeutic reagents has thus been demonstrated for both nanoparticles and siRNAs. It seems likely that aptamers targeting membrane proteins of other therapeutic target cell populations will also prove to be useful reagents in other clinically relevant contexts.

Here, we provide protocols for the approach we used *(23)* to deliver secondary therapeutic siRNAs specifically to PSMA-expressing cells. This approach entails the annealing of two distinct strands of RNA, one strand that consists of the aptamer with an extended tail that makes up the upper strand of the siRNA and a second strand that consists of the lower strand of the siRNA (*see* **Fig. 5.2**).

Using this approach, we showed that when added to cells expressing the aptamer target receptor on the surface, the aptamer–siRNA chimeras are rapidly internalized. Importantly, we showed that internalization of the chimera results in silencing of the siRNA target, by an RNAi-mediated mechanism, resulting in the death of the targeted cancer cell (*see* **Fig. 5.3**). Advantages of this approach include the facts that this reagent (an aptamer–siRNA chimera) consists only of RNA, which has important advantages (see above) as a therapeutic material, and that it can easily be carried out in labs that have the reagents and equipment to carry out basic molecular biology procedures.

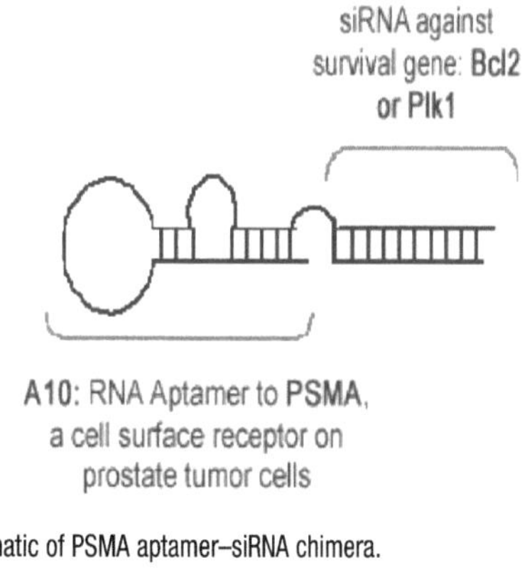

Fig. 5.2. Schematic of PSMA aptamer–siRNA chimera.

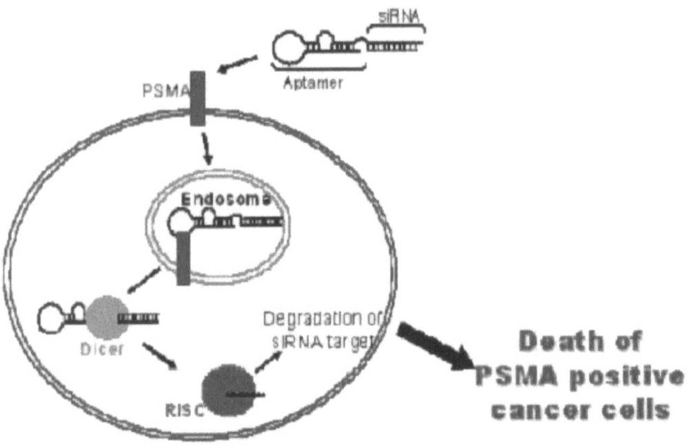

Fig. 5.3. Mechanism of aptamer–siRNA chimera-mediated targeting and silencing. Target specificity by the RNA chimera can be achieved both at level of the aptamer (PMSA-specific) as well as at the level of the siRNA (by silencing cancer cell-specific survival factors). This approach leads to selective killing of cancer cells that express both the cell surface receptor PSMA and prostate cancer-specific survival factors (e.g. Plk1 and Bcl2).

2. Materials

2.1. RNA Transcription and Purification

1. Transcription buffer (5X): 0.2 M Tris–HCl (pH 7.5), 30 mM $MgCl_2$, 50 mM NaCl and 10 mM spermidine.
2. Transcription mix: Transcription buffer (1X) with 1 mM 2'F-Py (2'F-2'-dCTP and 2'F-2'-dUTP, TriLink Biotech, San Diego, CA), 1 mM ATP, 1 mM GTP (Amersham Pharmacia Biotech), Uppsala Sweden, 10 mM dithiothreitol (DTT) (Sigma, St. Louis, MO), 0.5 u/μl RNAse inhibitors (Amersham Pharmacia Biotech), 5 μg/ml inorganic pyrophosphatase (Roche, Germany).
3. Loading solution: prepare a solution of 480 μl of formamide, 10 μl water, 10 μl EDTA and Bromophenol Blue (BBF, Bio-Rad, Hercules, CA).
4. Denaturing polyacrylamide gel (8% final concentration): 40% acrylamide/bis solution (37.5:1) (Bio-Rad) dissolved in Tris–borate–ethylenediamine tetraacetic acid (EDTA) buffer (TBE) containing 7 M urea and N,N,N',N' tetramethyl-ethylendiamine (TEMED, Bio-Rad) (*see* **Note 1**) and ammoniumpersulfate (Sigma).
5. Ammoniumpersulfate: Prepare 10% solution in water and immediately freeze in aliquots at –20°C.
6. Elution buffer: 300 mM NaOAc with 200 mM EDTA.
7. ^{32}P-αUTP (3,000 Ci/mmol, Amersham Pharmacia Biotech); T7 RNA/DNA polymerase (the mutant $T7^{Y639F}$ RNA polymerase) (Epicentre), DNase I (Amersham Pharmacia Biotech).

2.2. Cell Culture

1. PC12/MEN2A and PC12/MEN2B are PC12 cells stably expressing Ret9C634Y and Ret9M918T proteins, respectively.
2. Growth medium for PC12 cells: RPMI 1640 (Gibco/BRL, Bethesda, MD) with 10% heat-inactivated horse serum (HS, Gibco/BRL), 5% heat-inactivated fetal bovine serum (FBS, Gibco/BRL), 2 mM glutamine.
3. Growth medium for PC12/MEN2A and PC12/MEN2B: the same medium for PC12 cells but supplemented with HAT medium supplement 50X (Sigma), 250 μg/ml xantine (Sigma) and 25 μg/ml micophenolic acid (Sigma).
4. Xantine is dissolved in 0.1 N NaOH at 5 mg/ml and adjust pH of 10.8 with 3 N HCl and pH 10.5 with 1 N HCl, filtered and stored at dark and at room temperature (*see* **Note 2**).
5. Micophenolic acid (Sigma) is dissolved at 25 mg/ml in EtOH and stored at room temperature.
6. Solution of trypsin and EDTA from Gibco/BRL.

2.3. Counter-Selection and Selection Steps

1. Buffer of incubation for the RNAs: RPMI 1640 without serum
2. Washing buffer: RPMI 1640 without serum.
3. Total yeast RNA from Sigma.
4. Total RNA extraction kit from Ambion Inc. (Texas, USA).

2.4. Restriction Fragment Length Polymorphism (RFLP) Analysis

1. Polymerase chain reaction (PCR) buffer (10X): 100 mM Tris–HCl (pH 8.3), 15 mM $MgCl_2$, 500 mM KCl.
2. PCR mix: PCR buffer (1X) with 200 μM dATP, 200 μM dGTP, 200 μM dCTP, 200 μM dTTP (Amersham Pharmacia Biotech), 2 μM primers, DNA of each cycle of selection and Taq polymerase (0.02 U/μl) (Roche, New Jersey, USA).
3. [γ-^{32}P]ATP (3,000 Ci/mmol, Amersham Pharmacia Biotech).
4. REact 1 (10X) for digestion: 500 mM Tris–HCl (pH 8.0), 100 mM $MgCl_2$, 500 mM NaCl.
5. RsaI, AluI, HaeIII, HhaI enzymes from Invitrogen.
6. Denaturing polyacrylamide gel (6% final concentration): 40% acrylamide/bis solution (37.5:1) dissolved in TBE buffer containing 7 M urea.

2.5. Binding Analysis

1. Dephosphorylation buffer (10X): 500 mM Tris–HCl (pH 8.5), 1 mM EDTA.
2. Buffer for phosphatase alkaline (PA) inactivation: 200 mM EGTA.
3. Phosphorylation buffer (10X): 500 mM Tris–HCl (pH 8.2), 100 mM $MgCl_2$, 1 mM EDTA, 50 mM DTT, 1 mM spermidine.

4. Buffer of incubation of RNAs: RPMI 1640 without serum.
5. Washing buffer: RPMI 1640 without serum.
6. Recovering buffer: 0.6% sodium dodecyl sulphate (SDS).
7. PA for dephosphorylation from Boehringer Mannheim; T4 Polynucleotide Kinase for phosphorylation from Roche; [γ-^{32}P]ATP (6,000 Ci/mmol, Amersham Pharmacia Biotech).

2.6. Cell Lysis and Western Blotting for Functional Analysis of Selected Aptamers

1. Lysis solution: 50 mM Tris–HCl (pH 8.0) with 150 mM NaCl, 1% Nonidet P-40, 2 μg/ml aprotin, 1 μg/ml pepstatin, 2 μg/ml leupeptin (Roche) and 1 mM Na_2VO_4(Sigma).
2. Separating buffer (4X): 1.5 M Tris–HCl (pH 8.7), 0.4% SDS.
3. Stacking buffer (4X): 0.5% Tris–HCl (pH 6.8), 0.4% SDS.
4. Denaturing polyacrylamide gel (10% – final concentration): 40% acrylamide/bis solution (37.5:1).
5. Running buffer (5X): 125 mM Tris, 960 mM glycine, 0.5% SDS.
6. Laemmli buffer: 2% SDS, 5% β-mercaptoethanol, 0.001% bromophenol blue, 10% glycerol.
7. Pre stained molecular weight marker: Kaleidoscope markers from Bio-Rad.
8. Supported polyvinylidenedifluoride (PVDF) membrane from Millipore, Bedford, MA, and 3 MM chromatography paper from Whatman, Maidstone, UK.
9. Transfer buffer: 25 mM Tris, 190 mM glycine, 20% methanol, 0.05% SDS (*see* **Note 3**).
10. Tris-buffered saline with Tween (T-TBS): prepare 10X stock with 1.37 M NaCl, 27 mM KCl, 250 mM Tris–HCl (pH 7.4); dilute 100 ml of TBS 10X with 900 ml water and add Tween at the concentration required for use.
11. Blocking buffer: 5% nonfat dry milk in the T-TBS required.
12. Primary antibody dilution buffer: 5% nonfat dry milk in the T-TBS required.
13. Enhanced chemiluminescent (ECL) reagent from Amersham Pharmacia Biotech and Bio-Max ML film (Kodak).
14. Stripping buffer: 62.5 mM Tris–HCl (pH 6.8), 2% SDS, 100 mM β-mercaptoethanol.

2.7. Generation of Transcription Template

High quality DNA oligonucleotides can be obtained desalted, from many sources such as Integrated DNA Technologies, Oligos, etc., and Promega. Longer oligos (>50 nucleotides) should be ordered PAGE purified.

2.8. In Vitro Transcription

1. 10X NTP mix: 30 mM 2'F-CTP, 30 mM 2'F-UTP, 10 mM 2'OH-ATP, 10 mM 2'OH-GTP, dissolved in water. 2'-fluoro modified NTPs can be obtained from Trilink Inc. Unmodified NTPs can be obtained from a number of sources including New England Biolabs and Roche.
2. Mutant (Y639F) T7 RNA Polymerase (Epicentre), Inorganic pyrophosphatase (Roche),
3. 5X T7 RNA polymerase buffer: 20% w/v PEG 8000, 200 mM Tris–HCl (pH 8.0), 60 mM $MgCl_2$, 5 mM spermidine HCl, 0.01% w/v triton X-100, 25 mM DTT.
4. 5'-FAM-G can be custom-synthesized by Trilink Inc.

2.9. Gel Purification

1. Urea (Sigma), acrylamide (Bio-Rad), 10x TBE (Sigma), formamide (Sigma), xylene cyanol (Sigma), bromophenol blue (Sigma), Centrex spin filters, Centricon YM-30 filtration units (Millipore), Dulbecco's phosphate-buffered saline RNAse-free DNAse (NEB).
2. For 2x formamide gel loading buffer, combine: 0.01 g xylene cyanol, 0.02 g bromophenol blue, 500 μl 10x TBE, 9.5 ml formamide.
3. For 10% acrylamide/urea gel solution, combine: 115 g urea, 62.5 ml 40% acrylamide/bis (29:1), 12.5 ml 10x TBE and water for a final volume of 250 ml. Heat to dissolve urea, filter-sterilize and store at 4°C, protected from light (*see* **Note 6**).

2.10. RNA Oligonucleotides

High quality RNA oligonucleotides can be obtained from a number of commercial vendors including Dharmacon and Promega.

The following RNA oligos will anneal to the upper sense strands of the chimeric RNAs that can be produced with the protocol included below: A10-Plk1 Antisense siRNA: 5'GCA-CUUGGCAAAGCCGCCCdTdT3'. A10-CON Antisense siRNA: 5'ACGUGACACGUUCGGAGAAdTdT3'.

2.11. Buffers for Cell Surface Binding Assays

1. Binding buffer: 20 mM HEPES (pH 7.4), 150 mM NaCl, 2 mM $CaCl_2$, 0.02% BSA. Fix solution: DPBS plus 1% formaldehyde.

2.12. Reagents for In Vitro siRNA Activity Assays

1. RIPA buffer
2. An antibody for Plk1 is available from Zymed. An antibody for Bcl-1 is available from DykoCytomation.
3. PERM/FIX and PERM/WASH buffers are available from Pharmacia.

3. Methods

3.1. The Rationale of the SELEX Methodology

The SELEX method includes steps of: (i) incubating the library with the target molecule; (ii) partitioning unbound nucleic acids from those bound specifically to the selector cell type; (iii) dissociating the nucleic acid–target complexes; and (iv) amplifying of the nucleic acids pool enriched for specific ligands. Usually the positive selection is preceded by a negative or counter-selection step against the capture system in the absence of the desired target. This ensures the selection of aptamers directed against the target but not the other elements of the aptamer capturing moiety. To select aptamers against a transmembrane protein it is useful to perform a counter-selection step with the non-expressing cell line or otherwise with a cell line expressing the target protein in a different conformation state (as for example, monomeric vs. dimeric for a tyrosine kinase transmembrane receptor).

After reiterating these steps (the number of rounds of selection necessary is determined by both the type of library used as well as by the specific enrichment achieved per selection cycle), the resulting oligonucleotides are subjected to DNA sequencing. The sequences corresponding to the variable region of the library are screened for conserved sequences and structural elements indicative of potential binding sites and subsequently tested for their ability to bind specifically to the target cell.

3.2. Preparation of RNA Libraries

The starting RNA library pool is a library of 2'F-Py RNA molecules containing a 50 nt random sequence flanked by two fixed regions for the amplification reaction. The PCR amplifications of this library are performed using the following set of primers:

P20: 5'TCCTGTTGTGAGCCTCCTGTCGAA3'

P10: 5'TAATACGACTCACTATAGGGAGACAAGAATAA ACGCTCAA3'

The complexity of the starting pool was roughly 10^{14} 2'F-Py RNAs (1–5 nmol).

1. The transcription reactions are performed at 37°C for 12 h in the transcription mix with 10 μCi/μl ^{32}P-αUTP, 1 pmol/μl DNA and 2.5 u/μl of the mutant form of T7, T7^{Y639F} RNA polymerase (*see* **Notes 4 and 5**).

2. Following transcription, the RNA is treated with DNase I to remove contamination of ssDNA. Ten units of DNase I are added at end of transcription and incubated at 37°C for 20 min. Large volume RNA transcriptions are concentrated and desalted with phenol/chloroform/isoamyl alcohol (25:24:1), ethanol precipitated in the presence of 0.1 mg/ml of linear acrylamide (Ambion) and centrifuged at 14,000 rpm for 30 min at 4°C. The pellet is suspended in 20 μl of loading solution and purified by 8% denaturing polyacrylamide gel.

Gel purification of full-length RNA is important to prevent artefacts. RNA is passively eluted from the gel at 42°C in elution buffer and the concentration is determined spectrophotometrically assuming that one A_{260} unit is equal to 40 µg/ml of RNA.

3.3. Selection Strategies

In each cycle two counter-selection steps were performed.

1. Denaturation/renaturation step: 2'F-Py RNAs (1–5 nmol) are heated at 85°C for 5 min in 3 ml of RPMI 1640 serum free, snap-cooled on ice for 2 min, and allowed to warm up to 37°C, before incubation with the cells.
2. Following the denaturation/renaturation step, the pool of 2'F-Py RNAs (resuspended in 3 ml of RPMI 1640) is first incubated for 30 min at 37 °C with 10^7 PC12 cells in order to eliminate non-specific binders of the PC12 cell surface.
3. The unbound sequences are recovered by centrifugation and incubated for 30 min at 37°C with 10^7 adherent PC12/MEN2B cells that express an allele of RET (RET/M918T) mutated in the intracellular domain.
4. For the selection step, the unbound sequences from the second counter-selection are recovered and incubated with 10^7 adherent PC12/MEN2A cells expressing the RETC634Y mutated in the extracellular domain, for 30 min at 37°C in the presence of total yeast RNA as non-specific competitor RNA. Finally, the bound sequences are recovered after several washings with 5 ml of RPMI by total RNA extraction.
5. During the selection process, the selective pressure is progressively increased by increasing the number of washings (from one for the first cycle up to five for the last three cycles) and the amount of non-specific RNA competitor (100 µg/ml in the last three cycles), and by decreasing the incubation time (from 30 to 15 min from round 5) and the number of cells exposed to the aptamers (5×10^6 in the last three cycles).

The approach used for cell-SELEX is reported in **Fig. 5.4** (adapted from *(9)*).

To monitor the evolution of the pool the appearance of four-base restriction sites in the population (by RFLP analysis) is analysed, which reveals the emergence of distinct families in the library. The PCR product of each cycle (about 500 ng) are end-labelled with [γ-^{32}P] ATP and digested with a mix of four restriction enzyme. The endonuclease used are: RsaI, AluI, HaeIII, HhaI (10 units/enzyme) in the buffer REactI (Invitrogen Life Technologies) for 1 h at 37°C.

Following ethanol precipitation, the digested samples are loaded onto 6% denaturing polyacrylamide gel. The gel is wrapped and an autoradiography film is exposed.

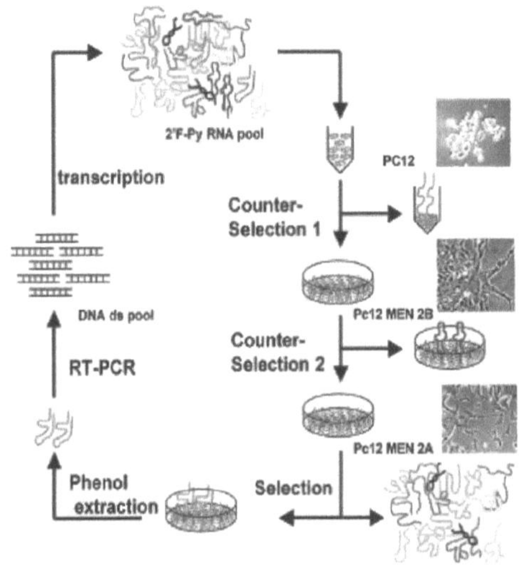

Fig. 5.4. Schematic protocol for the selection of PC12/MEN2A cell-specific aptamers.

3.4. Cloning and Sequencing

After 15 rounds of selection, sequences are cloned with TOPO-TA cloning kit (Invitrogen, Carlsbad, CA, United States) and analysed. About 100 clones are usually analysed in a SELEX protocol by using bioinformatics alignment tools. In general, SELEX-derived sequences contain regions of strong sequence conservation separated by regions of high variability. This hampers the usage of global alignment tools and resulted in the application of modified alignment protocols specifically tailored to identify and score sequence patterns in aptamers. This usually hallows the identification of conserved and variable nucleotide positions and permits the grouping of the various sequences into quasi-phylogenetic families. Conserved motifs within a sequence family are frequently candidates for specific target recognition elements.

3.5. Binding Analysis and K_d-Determination

1. To determine the binding of individual aptamers (or the starting pool as a control) to PC12 cells and derivatives, the aptamers are dephosphorylated with PA at 37°C for 1 h and end-labelled with T4 polynucleotide kinase in presence of [γ-^{32}P] ATP at 37°C for 30 min.
2. Binding of individual 5'-^{32}P-labelled RNAs is performed in 24-well plates in triplicate. 10^5 cells per well are incubated with various concentrations of individual aptamers for 10 min at 37°C in the presence of 100 μg/ml polyinosine as a non-specific competitor. After extensive washings (5 × 500 μl of RPMI 1640), bound sequences are recovered in 350 μl of 0.6% SDS, and the amount of radioactivity recovered is normalized

to the number of cells by measuring the protein content of each well.

3. Apparent K_d values for each aptamers are determined by Scatchard analysis according to the equation:

$$[\text{bound aptamer}]/[\text{aptamer}] = -(1/K_d) \times [\text{bound aptamer}] + ([T]_{tot}/K_d)$$

where $[T]_{tot}$ represents the total target concentration. The aptamers exhibit an affinity for the target in the low nanomolar range.

3.6. In Vitro and In Vivo Functional Analysis

Once isolated the sequences with the best binding properties, they are tested for the ability to block RET dependent intracellular signaling pathways.

To assess the effects of aptamers on RET activity, PC12/MEN2A cells (160,000 cells per 3.5-cm plate) are serum starved for 2 h and then treated for 16 h with 150 nM RNA aptamer, or the starting RNA pool after a short denaturation–renaturation step. Cell lysates are analysed by immunoblotting. The primary antibodies used were: anti-Ret (H-300), anti-Erkl (C-16) (Santa Cruz Biotechnology Inc., Santa Cruz, CA); anti-(Tyr-phosphorylated) Ret, anti-phospho-p44/42 MAP Kinase, also indicated as pERK (E10; Cell Signaling, Beverly, MA). An example of the results produced is shown in **Fig. 5.5** (adapted from *(9)*).

A clear example of the importance to choose the best selection procedure for generating aptamers specifically binding a transmembrane protein is illustrated by the comparison of the different strategies carried out to select RNA aptamers against the RETC634Y receptor *(9, 11)*. While several aptamers selected against the recombinant extracellular domain of transmembrane proteins recognize their targets on the cell surface *(24–27)*, in the case of RET, however, a SELEX protocol performed on the

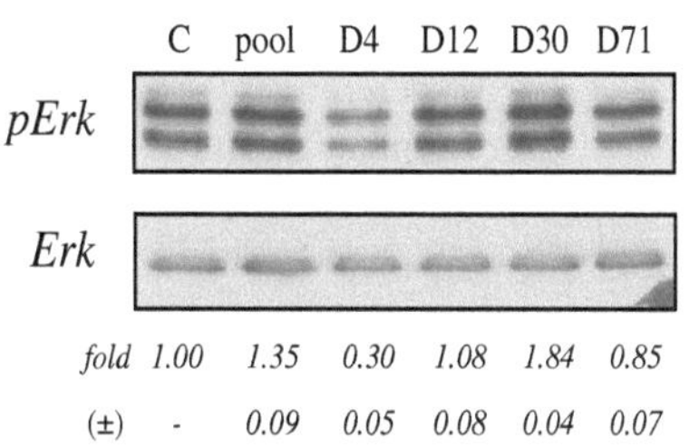

Fig. 5.5. PC12/MEN2A cells were either left untreated or treated with the indicated RNA aptamer, or the starting RNA pool (pool). Cell lysates were immunoblotted with anti-pErk antibody, then stripped and reprobed with anti-ERK antibody to confirm equal loading. Values below the blots indicate signal levels relative to untreated controls.

recombinant extracellular domain of RETC634Y did not provide aptamers able to recognize the protein in a cell surface environment (Cerchia et al. *(9)*), which most likely implies a different mode of recognition for the native and recombinant proteins. Furthermore, a different SELEX against whole-living cells without counter-selection against PC12/MEN2B, and a crossover SELEX alternating RETC634Y expressing cells and the recombinant purified extracellular domain of the RETC634Y protein as targets, were performed. The crossover SELEX leads to a higher enrichment of aptamers against RET. However, the selected aptamers using pure whole-living cells SELEX display a better apparent K_d. This study brings new insights into the respective advantages of each of these different methods for the selection of aptamers targeting membrane proteins.

3.7. Generation of Transcription Template

1. Transcription templates for use with the T7 RNA polymerase are produced by generating double-stranded DNA that encodes a T7 promoter sequence in the 5'-end. The aptamer sequence followed by the sequence of the upper strand of the siRNA are then encoded in the 3'-end of the template. In the case of the PSMA aptamer siRNA chimera targeting Plk1, the following DNA oligos can be used to generate such a template with PCR.

   A10 template primer:
   5'GGGAGGACGATGCGGATCAGCCATGTTTACGTCA CTCCTTGTCAATCCTCATCGGCAGACGACTCGCCC GA-3'
   Plk1 siRNA 3'-primer:
   5'AA**GCACTTGGCAAAGCCGCCCTT**TCGGGCGAG TCGTCTG3'
   A105'-primer: 5'TAATACGACTCACTATAGGGAGGAC GATGCGG3'

   In this case, the template oligo is used as the template (in trace amounts) in a PCR with the A10 5'-primer and Plk1 siRNA 3'-primer as amplification oligos. The choice of the 3'-primer determines the sense siRNA sequence that is appended to the aptamer sequence. Substitution of the Plk1 siRNA 3'-primer with the following control primer will produce the A10 aptamer with a control siRNA sequence in place of that targeting Plk1: Control siRNA 3'-primer: 5'AA**ACGTGACACGTTCGGAGAATT**TC GGGCGAGTCGTCTG3'
2. The generation of the correct-sized PCR product should be confirmed by running a sample on a 2% agarose gel with a DNA size ladder. This PCR product can then be purified (for instance, with the Qiagen PCR cleanup kit) in preparation for transcription (*see* **Note 7**).

3.8. In Vitro Transcription

For a 250 μl reaction, combine: 50 μl 5x T7 RNA polymerase buffer, 25 μl 10x NTP mix, 2 μl IPPI, 125 pmol DNA transcription template, 3 μl T7 (Y639F) polymerase and water for a final volume of 250 μl. Incubate at 37°C for 4–6 h. 5'-FAM-G can be added for a final concentration of 4 mM to produce RNA labelled with FAM only at the 5' position (*see* **Section 2.8.**).

3.9. Gel Purification

1. Add 1 μl RNAse-free DNAse (Roche) to the transcription reaction and incubate at 37°C for 10 min.
2. Extract reaction twice with an equal volume of chloroform. Then concentrate with a YM-30 Centricon spin filtration unit until volume is less than 100 μl. Add an equal volume of 2x formamide gel loading buffer and heat to 65°C for 5 min.
3. Transfer to ice and then load on a pre-run 10% acrylamide/urea gel. Gel is prepared by adding 75 μl 10% ammonium persulfate and 25 μl TEMED to 25 ml of 10% acrylamide/urea gel solution (*see* **Note 8**).
4. Separate plates and transfer gel to a piece of plastic wrap over a UV-shadowing screen. Use short wavelength handheld UV-light to visualize RNA (*see* **Note 9**).
5. Excise the piece of gel with the RNA and transfer to a 15 ml centrifuge tube with 2.5 ml of DPBS. Incubate on a rotator at 37°C for 4 h. Spin liquid with eluted RNA through a Centrex spin filter to remove any small pieces of gel that may be have been carried over in the liquid and then concentrate with a YM-30 Centricon spin filter unit. Wash twice by adding 2.5 ml DPBS and repeating spin.
6. Quantify RNA by measuring OD260 of a 1:50 dilution of the recovered material.

3.10. Annealing Reaction

The in vitro transcribed aptamer/siRNA upper strand is diluted to 10 μM and the complementary siRNA lower strand is diluted to 20 μM in DPBS (with calcium and magnesium) and this mixture is heated to 65°C for 5 min to denature the oligos. The reaction is then cooled to 37°C for 10 min to allow the two RNAs to anneal (*see* **Note 10**).

3.11. Cell Surface Binding Assay

Binding of the aptamer–siRNA chimeras to prostate cancer cells expressing PSMA can be assessed with flow cytometry. First, trypsinize PC-3 or LNCaP cells, wash twice with 500 μL PBS, and fix in 400 μL of fix solution for 20 min at room temperature. Wash cells with PBS to remove formaldehyde, resuspend in binding buffer and incubate at 37°C for 20 min. Then pellet cells and resuspend in 100 μL 37°C binding buffer containing 400 nM FAM-labelled A10 aptamer or FAM-labelled aptamer–siRNA chimera (*see* **Note 11** regarding the volume and RNA concentration of the labelling reaction). Incubate cells at 37°C for 30–40 min, wash three times with 500 µl 37°C binding buffer and resuspend in 400 µL 37°C fix solution. Incubate at 37°C for 5–10 min and then measure cellular fluorescence with flow cytometry.

3.12. Cell Culture (for siRNA Activity Assays)

Plate LNCaP or PC-3 cells in 6-well dishes at a density of 60% confluency. Incubate with aptamer siRNA chimeras or transfect with corresponding siRNAs (without aptamers). For siRNA transfections, transfect with 400 nM siRNA with Superfect (Qiagen) following manufacturer's instructions 1 and 3 days after plating cells. Add aptamer siRNA chimeras at the same concentration directly to the culture media of appropriate wells 1 day after plating cells. Remove media from these wells and replace with fresh media also supplemented with 400 nM of the appropriate chimera 3 days after plaling cells. Grow cells for an additional 2 days and then process as detailed below for either immunoblotting or flow cytometry.

3.13. siRNA Activity Assay (Immunoblotting)

Trypsinize cells transfected with siRNAs or treated with aptamer–siRNA chimeras as described above and then wash with PBS. Pellet cells and resuspend in RIPA buffer. Incubate on ice for 20 min. Pellet cells again and transfer supernatants to fresh tubes. Quantify protein concentration in supernatants with a Bradford assay. Run 50 µg of each on an SDS-PAGE gel (8.5% acrylamide for Plk1, 15% acrylamide for Bcl-2). Transfer protein to a PVDF membrane via electrophoresis. Block membranes with 5% milk in PBS. Then incubate membranes in block plus 1:1,000 of either anti-Plk1 or anti-Bcl2 antibodies diluted in 5% milk in PBS.

3.14. siRNA Activity Assay (Flow Cytometry)

Trypsinize cells transfected with siRNAs or treated with aptamer–siRNA chimeras, wash three times with 500 µl PBS and then count with a hemacytometer. Resuspend 400,000 cells in 400 µl FIX/PERM buffer for a final concentration of 5 × 10^5 cells/ml and incubate at room temperature for 20 min. Pellet cells, resuspend in 500 µl PERM/WASH buffer, wash three times with 500 µl PERM/WASH buffer and then resuspend in 50 µl PERM/WASH buffer plus 20 µg/ml anti-human Plk1, anti-human Bcl2 or the appropriate isotype-matched control antibody. Incubate cells at room temperature for 40 min, wash three times with 500 µl PERM/WASH buffer and then incubate for 30 min at room temperature in 50 µl PERM/WASH buffer with 1:500 diluted anti-mouse IgG-APC. Wash cells three times with PERM/WASH buffer and then resuspend in PBS. Measure cellular fluorescence with flow cytometry.

3.15. In Vivo Testing of PSMA-Targeting Aptamer–siRNA Chimeras

Because the A10 aptamer binds specifically to the human orthologue of PSMA, chimeras made with the A10 aptamer will only target cells of human origin or cells engineered to express human PSMA. One approach for in vivo testing is to grow human prostate cancer tumours in nude mice as described below. The use of a prostate cancer cell line that does not express PSMA (PC-3) can serve as a negative control, while the PSMA-expressing prostate cancer cell line LNCaP can produce the aptamer-targeted tumour. Culture PC-3 cells in Ham's F12-K medium supplemented with 2 mM L-glutamine, 1.5 g/L sodium bicarbonate, and 10% FBS. LNCaP cells were propagated in RPMI 1640 medium containing L-glutamine supplemented with 1.5 g/L sodium bicarbonate, 4.5 g/l glucose, 10 mM HEPES, 1.0 mM sodium pyruvate, and 10% FBS. Trypsinize cells, wash with DPBS and resuspend in DPBS. Count with a hemacytometer. Pellet cells again and resuspend in DPBS + 50% Matrigel with a cell concentration of 5×10^7 cells per milliliter. Inject 100 μl subcutaneously into the flanks of nude mice. Monitor tumour growth by examining animals and measuring any visible tumours every other day. Allow tumours to reach 0.5–1.0 cm in diameter and then inject with 200 pmol of chimeras diluted in 75 μl DPBS every other day. Continue to monitor tumour growth by measuring tumours every 2–3 days using a caliper and sacrifice animals if tumours grow excessively large (>2.0 cm in diameter).

4. Notes

1. TEMED is best stored at room temperature in a desiccator. Buy small bottles as it may decline in quality (gels will take longer to polymerize) after opening.
2. The solution precipitates when exposed to light, if this occurs you need to prepare a new solution.
3. Transfer buffer can be used for up to five transfers within 1 week so long as the voltage is maintained constant for each successive run (the current will increase each time). Adequate cooling to keep the buffer no warmer than room temperature is essential in order to prevent heat-induced damage to the apparatus and the experiment.
4. $T7^{Y639F}$ RNA polymerase is used to improve yields.
5. 2'F-Py RNAs are used because of their increased resistance to degradation by seric nucleases.
6. If urea precipitates, warm to 37°C to re-dissolve prior to pouring gel.

7. It is recommended that an aliquot of this DNA duplex be sequenced from both 5' and 3' ends to confirm the correct sequence.
8. Run gel until the first dye front is close to the bottom.
9. Only observe through UV-blocking eyeglasses because UV light is harmful to eyes.
10. Because the lower strand is in excess, there should be residual, unpaired lower strand RNA in the mixture following annealing. For many applications, this is not a concern. For instance, if the chimera is to be applied to cells in culture in the presence of serum, this RNA, which does not include 2'-fluoro modified pyrimidines, will be rapidly degraded by serum nucleases. However, if necessary, this RNA can be removed by purifying the aptamer–siRNA chimera on a non-denaturing acrylamide gel.
11. Because generation of FAM-labelled RNAs is expensive, it is usually desirable to minimize the volumes of these labelling reactions in order to conserve RNA. However, the concentration of RNA used must be rather high because the incorporation efficiency of FAM in the in vitro transcriptions is probably less than 50%. For this reason, it may be necessary to increase the concentration of RNA into the micromolar range to achieve binding saturation.

Acknowledgements

This work was supported by the European Molecular Imaging Laboratory (EMIL) Network (LSHC-2004-503569) and by the MIUR-FIRB Grant (#RBIN04J4J7).We wish to thank C.L. Esposito, B. Tavitian, F. Duconge and D. Libri for fruitful discussions.

References

1. Tuerk, C. and Gold, L. (1990) Systematic evolution of ligands by exponential enrichment: RNA ligands to bacteriophage T4 DNA polymerase. *Science* **249**, 505–510.
2. Ellington, A.D. and Szostak, J.W. (1990) *In vitro* selection of RNA molecules that bind specific ligands. *Science* **346**, 818–822.
3. Bock, L.C., Griffin, L.C., Latham, J.A., Vermaas, E.H. and Toole, J.J. (1992) Selection of single-stranded DNA molecules that bind and inhibit human thrombin. *Nature* **355**, 564–566.
4. Osborne, S.E. and Ellington, A.D. (1997) Nucleic acid selection and the challenge of combinatorial chemistry. *Chem. Rev.* **97**, 349–370.
5. Ruckman, J., Green, L.S., Beeson, J., Waugh, S., Gillette, W.L., Henninger, D.D., Claesson-Welsh, L. and Janjic, N. (1998) 2'-Fluoropyrimidine RNA-based aptamers to the 165-amino acid form of

vascular endothelial growth factor (VEGF165). Inhibition of receptor binding and VEGF-induced vascular permeability through interactions requiring the exon 7-encoded domain. *J. Biol. Chem.* **273**, 20556–20567.
6. Famulok, M. and Mayer, G. (2005). Intramers and aptamers: applications in protein-function analyses and potential for drug screening. *ChemBioChem.* **6**, 19–26.
7. Morris, K.N., Jensen, K.B, Julin, C.M., Weil, M., Gold, L. (1998) High affinity ligands from *in vitro* selection: complex targets. *Proc. Natl. Acad. Sci. U.S.A.* **95**, 2902–2907.
8. Homann, M. and Goringer, H.U. (1999) Combinatorial selection of high affinity RNA ligands to live African trypanosomes. *Nucleic Acids Res.*27, 2006–2014.
9. Cerchia, L., Ducongé, F., Pestourie, C., Boulay, J., Aissouni, Y., Gombert, K., Tavitian, B, de Franciscis, V. and Libri, D. (2005) Neutralizing aptamers from whole-cell SELEX inhibit the RET receptor tyrosine kinase. *PLoS Biol.* **3**, e123.
10. Gschwind, A., Fischer, O.M. and Ullrich, A. (2004). The discovery of receptor tyrosine kinases: targets for cancer therapy. *Nat. Rev. Cancer.* **4**, 361–370.
11. Pestourie, C., Cerchia, L., Gombert, K., Aissouni, Y., Boulay, J., de Franciscis, V., Libri, D., Tavitian, B. and Duconge, F. (2006) Comparison of different strategies to select aptamers against a transmembrane protein target. *Oligonucleotides* **16**, 323–335.
12. Jhiang, S.M. (2000) The RET proto-oncogene in human cancers. *Oncogene* **19**, 5590–5597.
13. Ichihara, M., Murakumo, Y. and Takahashi, M. (2004). RET and neuroendocrine tumors. *Cancer Lett.* **204**,197–211.
14. Hansford, J.R. and Mulligan, L.M. (2000) Multiple endocrine neoplasia type 2 and RET: from neoplasia to neurogenesis. *J. Med. Genet.* **37**, 817–827.
15. Takahashi, M. (2001) The GDNF/Ret signaling pathway and human diseases. *Cytokine Growth Factor Rev.* **12**, 361–373.
16. Putzer, B.M. and Drosten, M. (2004) The RET proto-oncogene: a potential target for molecular cancer therapy. *Trends Mol. Med.* **10**, 351–357.
17. Maniè, S., Santoro, M., Fusco, A. and Billaud, M. (2001) The RET receptor: function in development and dysfunction in congenital malformation. *Trends Genet.* **17**, 580–589.
18. Pirollo, K.F., Zon, G., Rait, A., Zhou, Q., Yu, W., Hogrefe, R. and Chang, E.H. (2006) Tumor-targeting nanoimmunoliposome complex for short interfering RNA delivery. *Human Gene Ther.*17, 117–124.
19. Song, E., Zhu, P., Lee, S., Chowdhury, D., Kussman, S., Dykxhoorn, D.M., Feng, Y., Palliser, D., Weiner, D.B., Shankar, P., Marasco, W.A. and Lieberman, J. (2005) Antibody mediated *in vivo* delivery of small interfering RNAs via cell-surface receptors. *Nat. Biotechnol.* **23**, 709–717.
20. Sioud, M. (2006) Single-stranded small interfering RNA are more immunostimulatory than their double-stranded counterparts: A central role for 2'-hydroxyl uridines in immune responses. *Eur. J. Immunol.* **36**, 1222–1230.
21. Farokhzad, O.C., Cheng, J., Teply, B.A., Sherifi, I., Jon, S., Kantoff, P.W., Richie, J.P. and Langer, R. (2006) Targeted nanoparticle-aptamer bioconjugates for cancer chemotherapy *in vivo*. *Proc. Natl. Acad. Sci. U.S.A.* **103**, 6315–6320.
22. Chu, T.C., Twu, K.T., Ellington, A.D. and Levy, M. (2006) Aptamer mediated siRNA delivery. *Nucleic Acids Res.* **34**, e73.
23. McNamara, J.O. 2nd, Andrechek, E.R., Wang, Y., Viles, K.D., Rempel, R.E., Gilboa, E., Sullenger, B.A. and Giangrande, P.H. (2006) Cell type-specific delivery of siRNAs with aptamer-siRNA chimeras. *Nat. Biotechnol.* **24**, 1005–1115.
24. Chen, C.H., Chernis, G.A., Hoang, V.Q. and Landgraf, R. (2003). Inhibition of heregulin signaling by an aptamer that preferentially binds to the oligomeric form of human epidermal growth factor receptor-3. *Proc. Natl. Acad. Sci. U.S.A.* **100**, 9226–9231.
25. Lupold, S.E., Hicke, B.J., Lin, Y. and Coffey, D.S. (2002). Identification and characterization of nuclease-stabilized RNA molecules that bind human prostate cancer cells via the prostate-specific membrane antigen. *Cancer Res.* **62**, 4029–4033.
26. Mi, J., Zhang, X., Giangrande, P.H., McNamara, J.O., 2nd, Nimjee, S.M., Sarraf-Yazdi, S., Sullenger, B.A. and Clary, B.M. (2005). Targeted inhibition of alphavbeta3 integrin

with an RNA aptamer impairs endothelial cell growth and survival. *Biochem. Biophys. Res. Commun.* **338,** 956–963.

27. Mori, T., Oguro, A., Ohtsu, T. and Nakamura, Y (2004). RNA aptamers selected against the receptor activator of NF-kappaB acquire general affinity to proteins of the tumor necrosis factor receptor family. *Nucleic Acids Res.* **32,** 6120–6128.

Chapter 6

Aptamers Targeting RNA Molecules

Marguerite Watrin, Eric Dausse, Isabelle Lebars, Bernard Rayner, Anthony Bugaut and Jean-Jacques Toulmé

Abstract

Oligonucleotides complementary to RNA sequences interact poorly with folded target regions. In vitro selection of oligonucleotides carried out against RNA structures have led to aptamers that frequently differ from antisense sequences, but rather take advantage of non-double-stranded peculiarities of the target. Studies along this line provide information about tertiary RNA architectures as well as their interaction with ligand of interest. We describe here a genomic SELEX approach and its application to the recognition of stem–loop structures prone to the formation of kissing complexes. We also provide technical details for running a procedure termed 2D-SELEX that takes advantage of both in vitro selection and dynamic combinatorial chemistry. This allows selecting aptamer derivatives containing modified nucleotides that cannot be incorporated by polymerases. Last we present in vitro transcription conditions under which large amounts of RNA, suitable for NMR structural studies, can be obtained. These different aspects of the SELEX technology have been applied to the *trans*-activating responsive element of the human immunodeficiency virus type 1, which is crucial for the transcription of the retroviral genome.

Key words: RNA structures, hairpins, TAR RNA, genomic SELEX, dynamic combinatorial chemistry, NMR structure.

1. Introduction

In vitro selection of aptamers (also named SELEX, for systematic evolution of ligands by exponential enrichment) can be carried out against a wide range of targets *(1, 2, 3)*. Aptamers have been identified that recognize nucleic acids *(4)*. This is of interest when non-single-stranded polynucleotide chains are targeted. Indeed, SELEX has been used for the identification of sequences or motifs that bind to double-stranded DNA *(5)* and to RNA structures *(6–8)*. This last application is attractive as such targets are not appropriate for binding antisense sequences or siRNAs. Due to

Günter Mayer (ed.), *Nucleic Acid and Peptide Aptamers: Methods and Protocols, vol. 535*

DOI 10.1007/978-1-59745-557-2_6 Springerprotocols.com

the competition with intramolecular RNA–RNA interactions, such antisense–sense duplexes are characterized by low affinity and consequently give rise to weak regulatory effects *(9, 10)*. In addition to providing informations on tertiary RNA architecture aptamers also offer an alternative to complementary (antisense, siRNA, ribozymes) oligonucleotides for targeting RNA structures that play a role in gene expression. Moreover, aptamers have been shown to display increased specificity compared to complementary oligonucleotides for recognizing folded targets *(11)*.

Various stem–loop (hairpin) structures have been used as target for in vitro selection. This includes tRNA *(12)* and RNA motifs of the Human Immunodeficiency Virus (HIV) *(6)* as well as the Hepatitis C Virus (HCV) RNA genomes *(8, 13, 14)*. Two types of aptamers have been characterized. Firstly, hairpins that bind to the target hairpin through apical loop–apical loop interactions, thus yielding so-called kissing complexes *(6, 12)*. Secondly, in some cases the loop of the target RNA structure interacts with an internal loop of the aptamer *(8, 13)*. This internal loop is invariably flanked by GC rich stems on either or both sides. The contribution of these stems to the stability of such apical loop–internal loop (ALIL) complex is presently unknown. Interestingly, both kissing and ALIL interactions have been described in natural RNA–RNA complexes *(15, 16)*. Aptamers targeted to the *trans*-activating responsive (TAR) element of HIV-1 *(17)* or to a domain of the internal ribosome entry site of the HCV RNA *(8)* have demonstrated inhibitory effects in vitro. Furthermore, post-SELEX modifications of the RNA aptamers selected against the TAR element have been introduced, giving rise to derivatives of longer lifetime, increased affinity and/or improved biological properties *(18–21)*.

We previously described procedures for the selection of aptamers against RNA structures *(22)*. In this chapter we describe the procedure for identifying human transcripts that generate loop–loop complexes with the TAR RNA element of HIV-1 (genomic SELEX). This is a first step toward the analysis of the repertoire of RNA hairpins prone to kissing interactions, some of them might be biologically relevant. In addition the production of chemically modified aptamers is of prime interest for use in biological media. Yet, only a tiny number of modified nucleotides are incorporated by polymerases, and post-SELEX modification is a tedious and risky process as this frequently results in conformation changes that in turn alter the properties of the aptamer. We describe here an alternative termed 2D-SELEX according to which dynamic combinatorial chemistry (for a recent review *see (23)*) allows for selecting chemically modified aptamers that could not be obtained by standard SELEX. Last, the determination of the structure of the RNA–RNA aptamer–target complexes reveals the crucial parameters driving the recognition between the two partners and beyond for rationally modifying the aptamer. To this end Nuclear Magnetic Resonance (NMR) is a powerful method, which, however, requires milligram amounts of the RNA partners. We describe a method routinely used in our laboratory for the preparation of large quantities of RNA molecules by in vitro transcription. A brief description of the NMR methods used to analyze RNA secondary and tertiary structures is also presented.

2. Materials

2.1. Genomic SELEX

1. Genomic RNA library.
2. Oligonucleotide primers.
3. Ampli Taq gold™ (PE Applied Biosystems).
4. DYNAL Magnetic Particle concentrator.
5. Streptavidin MagneSphere Promega or Dynabeads M-280.
6. Sephadex™ G-25 Fine from GE Healthcare.
7. M-MLV Reverse transcriptase, RNase H Minus, point mutant (Promega).
8. Ampliscribe T7 high yield transcription kit (Epicentre Technologie).
9. Nucleospin® Extract II from Macherey-Nagel.
10. TOPO TA cloning kit (Invitrogen).
11. BigDye® Terminator v3.1 Cycle Sequencing Kit (PE Applied Biosystems).
12. PAGE equipment.
13. R buffer : 20 mM HEPES, 20 mM sodium acetate, 140 mM potassium acetate and 3 mM magnesium acetate, pH 7.4.
14. Speed-Vac apparatus.
15. Biacore 3000 apparatus.
16. Biacore SA sensorship.
17. Expedite 8908 (Millipore) nucleic acid synthesizer.
18. Reagents for oligonucleotide synthesis, including 2'-*O-ter*-butyldimethylsilyl-ribonucleoside phosphoramidites (U, iBu-G, Bz-A and Bz-C) were from Glen Research.

2.2. Two-Dimension SELEX

In addition to items 2–5, 10–12, 14, 17 and 18 listed in the previous section, the following items are needed.

1. Chemically synthesized DNA library: the starting random DNA library is synthesized on a Expedite 8908 DNA synthesizer, using an equimolar mixture of the four deoxynucleoside

phosphoramidites at every position. Cycles and deprotection procedure were performed as recommended by the manufacturer.

2. Superscript^{TM} II RNase H minus reverse transcriptase (Life Technologies).
3. T7-MEGAshortscript^{TM} kit (Ambion).
4. 2'-Amino-UTP (Ambion).
5. Buffer 1 × SE : 20 mM NaCl, 140 mM KCl and 3 mM $MgCl_2$ in 20 mM sodium phosphate buffer at pH 6.0.
6. 2'-Trifluoroacetamido-2'-deoxyuridine phophoramidite (Glen Research).
7. Slide-A-Lyser Mini Dialysis units, 3500 MW cut-off (Pierce).
8. HPLC apparatus including a GP50 gradient pump, a PDA100 photodiode-array detector (Dionex) and an Uptisphere 5O DB 5 μm C18-column (250 × 4.6 mm) (Interchim, France).
9. Snake venom phosphodiesterase (Amersham Biosciences).
10. MALDI-ToF spectrometer (Reflex III, Brucker). A 1:1 mixture of 2,4,6-trihydroxy-acetophenone (30 mg/ml in ethanol) and 100 mM aqueous ammonium citrate (pH 9.4) is used as a matrix.

2.3. Preparation of RNA Samples for Structural Studies by NMR

1. DNA templates (Eurogentec or MWG Biotech) are resuspended in H_2O and stored at –20°C.
2. Nucleotide triphosphates (NTPs, Sigma or Spectra Stable Isotopes for $^{13}C/^{15}N$ labeled NTPs) are dissolved in water at 100 mM, the pH is adjusted at 5.0 with NaOH 0.1 M at room temperature. Aliquots are stored at –20°C.
3. T7 RNA polymerase ("home-made") is stored in aliquots at –20°C in 20 mM sodium phosphate pH 7.7, 1 mM DTT, 1 mM EDTA, 100 mM NaCl and 50% glycerol.
4. Buffer for transcription *(24–26)* freshly prepared: 40 mM Tris–HCl pH 8.1, 1 mM spermidine, 0.01% (v/v) Triton X-100, 5 mM DTT, 80 mg/ml polyethylene glycol (PEG 8000), 4 mM each NTP, 400 nM DNA strands and 0.02 mg/ml T7 RNA polymerase. Magnesium concentration is adjusted for each DNA template. Stocks solutions 2 M Tris–HCl pH 8.1, 1 M DTT (stored in single use aliquots), 0.1 M spermidine, 50% (w/v) PEG (8000), 1 M $MgCl_2$ are stored at –20°C. Triton X-100 is stored at room temperature.
5. EDTA 0.5 M, pH 8.0, stored at room temperature.
6. Phenol/chloroform saturated solution pH 4.7 stored at 4°C (SIGMA).

7. Ammonium acetate 3 M, pH 5.5, stored at 4°C.
8. Analytical and preparative 20% polyacrylamide (19:1, acrylamide/*bis*-acrylamide) gel electrophoresis under denaturing conditions (7 M urea) and N,N,N',N'-tetramethyl-ethylene-diamine (TEMED, Sigma).
9. Ammonium persulfate: prepare a 10% solution (w/v) in water and immediately freeze in single use aliquots at −20°C.
10. Stains-all stock solution: 0.1% in formamide, stored at 4°C in a dark bottle. Ready-to-use solution: 30 ml stock solution, 80 ml water and 90 ml formamide, stored at 4°C in a dark bottle.
11. Gel-running buffer: 45 mM Tris–borate, 1 mM EDTA, pH 8.3. Stored at 15–25°C.
12. PAGE equipment.
13. Oligonucleotides are eluted from the gel slice using an electroelution apparatus (Elutrap, Schleicher & Schuell) at 4°C. The elution buffer is 45 mM Tris–borate, 1 mM EDTA, pH 8.3, stored at 15–25°C.
14. Dialysis tubes (Spectra Por 3, 18 mm, MW cut-off 3,500).
15. Dialysis buffer: 10 mM sodium phosphate, pH 6.4.
16. Lyophilization apparatus.
17. Shigemi advanced 5 mm NMR microtube (Sigma).
18. Deuterium oxide 100% (Eurisotop).

3. Methods

3.1. Genomic SELEX Against RNA Target

Genomic SELEX consists in the in vitro selection, within a RNA pool generated by in vitro transcription of genomic sequences (exonic and intronic sequences as well as intergenic sequences), of the molecules with the best affinity for a pre-determined target, generally a protein *(27–29)*. It allows the identification of RNA fragments that might reflect natural interactions or provide artificial RNA ligands. The method can be easily adapted to RNA–RNA interactions. We describe below a procedure for genomic SELEX against the TAR RNA element of HIV-1, which is involved in the *trans*-activation of the transcription of the retroviral genome *(30)*. We wanted to address the question of whether the TAR RNA element of HIV-1 could interact with a transcript of the human genome.

3.1.1. Library Synthesis, Primers

The original library contains double-stranded fragments of genomic DNA extracted from human placenta. Each fragment is flanked by 5′ and 3′ primer regions. These primers enable the PCR amplification of the library, the T7-transcription reaction, the reverse transcription and the PCR amplification of selected sequences *(28, 29, 31)*.

The two different primers are:

Fixfor: 5′- CCAAGTAATACGACTCACTATAGGGGAATTCG GAGCGGG-3'

Fixrev: 5′- CGGGATCCTCGGGGCTG-3'

3.1.2. The TAR RNA Target Element

HIV-1 TAR element was restricted to nucleotides C18 to G44 (miniTAR). MiniTAR RNA stem loop was biotinylated at its 3'-end. Two GC pairs were added at the bottom of the stem. The miniTAR sequence $^{5'}$CGCCAGAUUUGAGCCUGGGAGCUC UCUGGCG$^{3'}$ was synthesized on an Expedite 8909 synthesizer (Applied Biosystems) and purified by electrophoresis on a denaturing gel (20% polyacrylamide, 7 M urea).

3.1.3. Transcription of the Genomic Library

DNA genomic library (6 μg) is transcribed at 37°C for 2 h in a final volume of 40 μl using the Ampliscribe T7 high yield transcription kit from Epicentre Technologie.

Add 2 μl of RNase-free DNase I (at 1 U/μl) at 37°C for 30 min.

Extract the transcription products with an equal volume of phenol (pH 4.3) and chloroform.

Collect the upper aqueous phase and precipitate with 1:10 volume of sodium acetate 3 M pH 5.3 and three volumes of ethanol for 1 h at –80°C.

Centrifuge the tubes for 30 min at 14,000 *g* at 4°C.

Wash the pellet with 75% ethanol and dry in a Speed-Vac.

Redissolve the pellets in 50 μl H_2O. Purify the transcription products on Sephadex^{TM} G-25 Fine from GE Healthcare 5 (*see* **Note 1**).

Quantify RNA amount by absorbance at 260 nm.

Genomic RNA candidates can then be used for the first round of selection in the R selection buffer (20 mM HEPES, 20 mM sodium acetate, 140 mM potassium acetate and 3 mM magnesium acetate, pH 7.4).

3.1.4. Genomic SELEX Procedure

The entire procedure is described in **Fig. 6.1**.

3.1.4.1. Counter-Selection

Prior to each round of positive selection, a counter-selection step enables to get rid of non-specific binders.

Fold the candidates by heating the pool of RNA candidates (in water) at 95°C for 1 min, and cooling it down at 4°C for 1 min. Add 5X R buffer (1X final) and hold the tube at room temperature for 5 min.

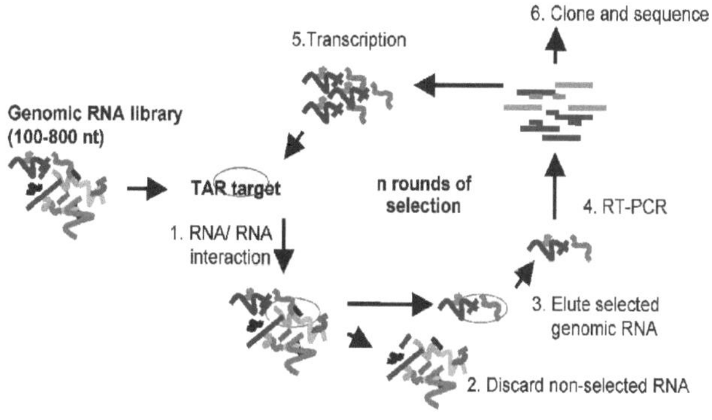

Fig. 6.1. Genomic SELEX procedure applied to RNA/RNA interaction.

Incubate 250 pmol of the RNA library (or different quantities of selected RNA in the successive rounds of selection) (*see* **Note 2**) in 100 μl of R buffer for 15 min at room temperature (~ 23°C) with streptavidin beads (100 μg of streptavidin MagneSphere Paramagnetic Particles from Promega or 1 mg of Dynabeads M-280) previously equilibrated in R buffer for 10 min.

Recover the supernatant in a tube.

Elute RNA candidates non-specifically retained on the beads in 80 μl water by heating at 80°C for 45 s.

Quantify non-specifically associated RNA by absorbance at 260 nm (*see* **Note 3**).

RNA candidates not retained by the beads are then submitted to the positive selection step.

3.1.4.2. Positive Selection

Heat the library and the target separately at 95°C for 1 min and 65°C for 3 min, respectively. Chill in ice for 1 min and then keep at room temperature for 5 min in R buffer.

Mix in a final volume of 100 μl R buffer, 100 μg of streptavidin magnetic beads from Promega (1 mg for beads from Dynabeads) with 10 pmol of the biotinylated TAR target.

Let the interaction occurs at room temperature for 10 min.

Magnetically separate the beads and the supernatant. Discard the supernatant.

Add 250 pmol of the counter-selected library to the beads.

Incubate at room temperature in a final volume of 100 μl of R buffer for 15 min.

Magnetically separate the beads and the supernatant. Discard the supernatant.

Wash the beads once with 100 μl of R buffer to eliminate candidates with poor affinity for TAR.

Elute the target-bound candidates in 80 μl of water by heating at 80°C for 45 s.

Reduce to dryness in a Speed-Vac, resuspend in 10 μl of water.

3.1.4.3. Reverse Transcription and Amplification

The RNA candidates need then be reverse-transcribed prior to amplification. Anneal the candidates to 2 μM of the downstream primer at 70°C for 10 min and copy the RNA into cDNA with 240 units of M-MLV reverse transcriptase, RNase H minus, point mutant (Promega) for 50 min at 50°C in a final volume of 20 μl of RT commercial buffer.

PCR-amplify the cDNA candidates using 40 units of Ampli Taq gold™ (Applied Biosystems) in 1 ml of the Taq buffer containing in addition 200 μM of dNTP mix, 0.5 mM of Mg^{2+}, 2% of DMSO, 2 μM of each primer. Then subject the reaction mixture to repeated cycles : (i) 94°C 10 min in order to activate the AmpliTaq gold and provide a Hot Start, (ii) 94°C 40 s / 55°C 40 s / 72°C 1 min for ten cycles, (iii) 72°C 10 min for one final cycle.

After phenol (pH 8)/chloroform extraction and precipitation, the PCR product is resuspended in 100 μl of water.

Purify the PCR product with nucleospin® Extract II from Macherey-Nagel in order to remove primers and free nucleotides. Elute twice with 25 μl Tris (–HCl or borate) 10 mM, pH 8.

Quantify by absorbance at 260 nm.

3.1.4.4. Transcription Reaction

Use 2 μg of purified PCR product to perform the transcription reaction as described in **Section 3.1.3** for the genomic library.

After DNase treatment, phenol (pH 4.3)/chloroform and precipitation, resuspend the pellet in 100 μl of water.

Purify the transcription product on Sephadex™ G-25 Fine to eliminate non-incorporated nucleotides (*see* **Note 1**).

Quantify RNA amount by absorbance at 260 nm.

RNA candidates can then be used for the next round of selection.

3.1.5. Evaluation of the Affinity of the RNA Pool

In order to evaluate the evolution of the selected population against TAR, Surface Plasmon Resonance measurements (SPR) have been performed with a Biacore 3000 apparatus.

Run all experiments on BIAcore with the R buffer.

Immobilize 2000 resonance unit (RU) of biotinylated mini-TAR (50 nM) on a SA sensorship coated with streptavidin and activated with three pulses of 50 mM NaOH, 1 M NaCl.

Prepare the library and each round of the selected populations in the R buffer at 500 nM.

Heat samples at 95°C for 1 min, cool to 4°C for 1 min, and then leave at room temperature for 5 min.

Inject 80 μl at a flow rate of 20 μl/min for 240 s at 23°C.

Let the dissociation occur for 600 s.

After each round, regenerate the target with one pulse of 20 mM EDTA (20 μl) and one pulse of R buffer (20 μl).

Wash the needle and the IFC with the R buffer between each injection.

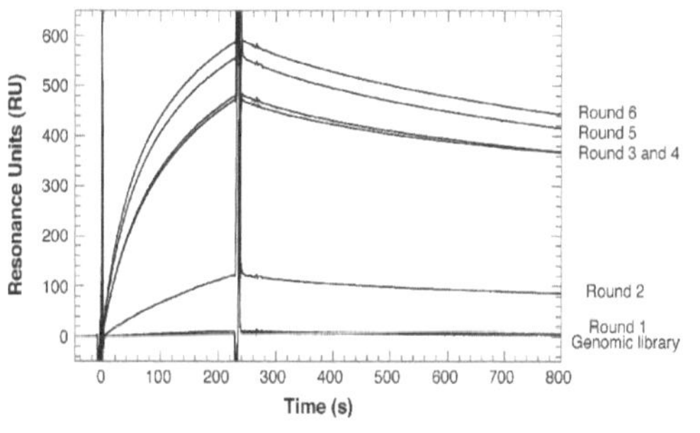

Fig. 6.2. Evolution of genomic selection against TAR. Sensorgram obtained by Surface Plasmon Resonance shows the affinity evolution of the selected genomic RNA during the selection against TAR.

Analyze the sensorgrams with the BIAeval software 2.2.4 (*see* **Fig. 6.2**) (*see* **Note 4**).

3.1.6. Cloning and Sequencing

After six cycles of selection against miniTAR target, selected sequences were cloned and sequenced. Following reverse transcription, amplify the cDNA pool as described above (*see* **Section 3.1.4.3.**). At the end of the PCR a 10 min step at 72°C enables the addition deoxyadenosine at the 3'end of the PCR product in order to clone it directly into the vector of the TOPO TA cloning kit from Invitrogen. Transform the *E. coli* TOP10 One Shot™ cells according to the manufacturer's instructions. Sequence the clones with the BigDye® Terminator v3.1 Cycle Sequencing Kit from PE Applied Biosystems according to the manufacturers' instructions. Sequences were analyzed on the four peaks 1.7.2 software (*see* **Note 5**).

3.1.7. Identification of a Genomic RNA Aptamer Against TAR of HIV-1 Element

This genomic selection was performed to identify genomic RNA fragments showing affinity for the TAR RNA element. Aptamers showing affinity for TAR were selected and cloned. About 41 sequences have been sequenced, all of them displaying a consensus region, 5'-CCCAG-3' complementary to a part of the TAR apical loop. Seven clones were similar to aptamers RII-17 and R06 previously identified by standard in vitro selection using libraries of 78 and 98 nt long candidates, respectively *(6, 32)*. Stem of the selected genomic hairpins was slightly different from the R06 and RII-17 aptamers. The genomic sequence RI-11 is presented in **Fig. 6.3**. The affinity of the genomic aptamers against TAR was determined by SPR (*see* **Note 4**). About 350 RU of miniTAR were immobilized and samples were prepared as described in **Section 3.1.4.2**. Immobilize on another channel a non-relevant RNA hairpin as a negative control. Prepare aptamer solutions at different concentrations ranging from 500 nM to 4 μM in the R buffer.

Treat each sample as described in **Section 3.1.5**.

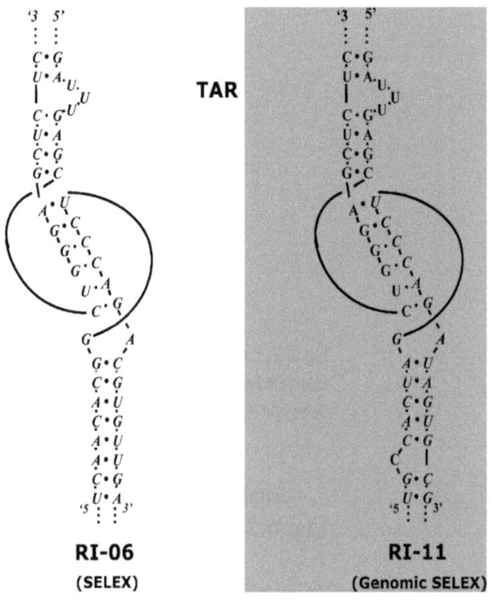

Fig. 6.3. RNA aptamers against the TAR HIV-1 element. Schematic representation of two RNA/RNA complexes selected either with a combinatorial library *(6)* (*left*) or with a genomic library (*right*).

3.2. Two-Dimension SELEX (2D-SELEX) Against the TAR RNA Element of HIV-1

2D-SELEX is designed to explore a much larger chemical space than that exhibited by nucleic acids and to select chemically modified aptamers with improved properties. This methodology rests on the use of a library of random RNA sequences containing 2'-amino-pyrimidine nucleotides instead of their "natural" counterparts. In a first round of selection, the random library of 2'-amino-RNAs is incubated with a set of chemically diverse aldehyde molecules (*see* **Fig. 6.4**). Reversible reaction between 2'-amines present on each RNA candidate and the aldehydes yields to the formation of a dynamic combinatorial library of 2'-imino-RNAs, where the motifs borne by the aldehydes are covalently linked to the RNAs (*see* **Fig. 6.4, Step 1**). Addition of the target to this dynamic library acts as a template and induces an equilibrium shift towards the preferential formation of the most fitted 2'-imino-RNAs, which are bound to the target molecules. Partitioning of ligand–target complexes from unbound candidates is performed (*see* **Fig. 6.4, Step 2**). Ligands are then eluted from the target, causing concomitant hydrolysis of the imine linkages (*see* **Fig. 6.4, Step 3**). After removal of the released aldehydes, selected 2'-amino-RNA scaffolds are isolated, reverse-transcribed and amplified by PCR (*see* **Fig. 6.4, Step 4**). Resulting double-stranded DNAs are then transcribed into 2'-amino-RNAs and another round of selection can be carried out. Repetition of this selection and amplification process progressively leads to a

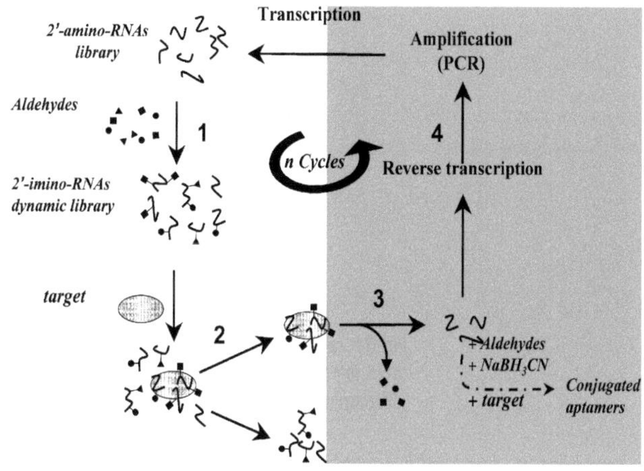

Fig. 6.4. Schematic representation of the in vitro 2D-SELEX.

population of 2'-amino-RNA scaffolds that have evolved in the presence of the set of aldehydes and the target to provide high affinity conjugated 2'-imino-RNA ligands. At the end of the selection process, remaining sequences are identified by cloning and sequencing. Selected 2'-amino-RNA scaffolds are then resynthesized and individually incubated with the set of aldehydes, the target molecule and sodium cyanoborohydride ($NaBH_3CN$) which selectively reduces the imine bonds. Thus, conjugated aptamers displaying the highest affinity for the target are preferentially in situ synthesized and converted into chemically stable analogues.

3.2.1. Library, Primers and Target

The RNA library is obtained by PCR amplification and transcription from the DNA template library: 5'-GGGAGGACGA AGCGG$(N)_{14}$CAGAAGACACGCCCGA-3'. The sequence of the reverse primer is 5'-TCGGGCGTGTCTTCTG-3' for the hybridization to the 3' end of the library. The sequence of the forward primer complementary to the 5' end of the complement DNA library is 5'-TAATACGACTCACTATAGGAGGACG AAGCGG-3' and includes the T7 polymerase transcription promoter. After PCR amplification, the transcription of the DNA library yields the 2'-amino-RNA library: 5'-GGGAGGACG AAGCGG$(N)_{14}$CAGAAGACACGCCCGA-3', where N stands for A, C, G or 2'-amino-uridine (*see* **Fig. 6.5**). The 3' biotinylated miniTAR RNA stem–loop was restricted to nucleotides C18 to G44 of TAR. The sequence 5'-CCAGAUUUGAGCCUGG GAGCUCUCUGG-3' was chemically synthesized in the laboratory on an Expedite 8909 synthesizer (Applied Biosystems) and purified by electrophoresis on a denaturing gel (20% polyacrylamide, 7 M urea).

a)

A, C, G, 2'-NH_2-U

5'-GGG AGG ACG AAG CGG N(14) CAG AAG ACA CGC CCG A-3'

b)

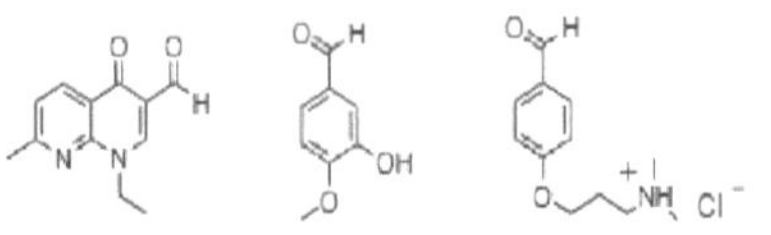

Fig. 6.5. (**a**) Library of randomized 2'-amino-RNAs, (**b**) aldehydes used during 2D-SELEX (from *left* to *right*): nalidixic aldehyde, 3-hydroxy-4-methoxybenzaldehyde and 4-[3-(dimethylamino)propoxyl]benzaldehyde hydrochloride.

3.2.2. Transcription Reaction

Perform the transcription reaction at 37°C for 4 h in a final volume of 40 μl using the T7-MEGAshortscript kit (Ambion) with 7.5 mM rATP, 7.5 mM rGTP, 7.5 mM rCTP and 10 mM 2'-amino-UTP (Ambion).

Add 2 μl of RNase-free DNase I (at 2 units/μl) for 15 min at 37°C.

Purify the transcription pool by electrophoresis on a 20% denaturing polyacrylamide gel and extract the band corresponding to the library molecular size. Quantify amount of purified RNA by absorbance at 260 nm.

3.2.3. In Vitro Selection Procedure

A set of three aldehydes: 4-[3-(dimethylamino)propoxyl]benzaldehyde hydrochloride (1 mM), 3-hydroxy-4-methoxybenzaldehyde (1.2 mM) and nalidixic aldehyde (200 μM) is used (*see* **Fig. 6.5** and **Note 6**) during the selection steps. Thereafter, set of aldehydes or mixture of aldehydes will refer to the above-mentioned mixture of aldehydes.

3.2.3.1. Counter-Selection

Prior each round of selection, a negative selection against the streptavidin-coated magnetic beads is performed.

Incubate the 2'-amino-RNA library (or selected sequences in the successive rounds of selection) for 5 min at room temperature with aldehydes in 90 μl of buffer 1 × SE.

Add 10 μl of a 5 mg/ml solution of streptavidin-coated magnetic beads (pre-washed several times with buffer 1 × SE) in buffer 1 × SE and incubate the mixture for 30 min at room temperature. Stir manually the mixture every 5 min to resuspend the beads.

Collect the supernatant for the positive selection step.

3.2.3.2. Positive Selection

Add 3' biotinylated miniTAR (*see* **Note 7**) to the supernatant (100 μl) recovered after the counter-selection step and incubate the mixture at room temperature for 25 min.

Add to the previous mixture 10 μl of a 5 mg/ml solution of streptavidin-coated magnetic beads in buffer 1 × SE for 5 min.

Remove the supernatant and wash the beads with a solution (100 μl) of aldehydes in buffer 1 × SE.

Elute the target-bound candidates in 60 μl of water at 80°C for 1 min. Repeat once this step.

Pool the eluates and ethanol-precipitate 2'-amino-RNAs.

3.2.4. Reverse Transcription

Denature recovered 2'-amino-RNA candidates in 10 μl of water at 70°C for 10 min and anneal at 42°C for 2 min to 2 μM of the reverse primer in enzyme buffer containing dNTPs.

Copy the 2'-amino-RNAs into cDNA with 240 units of Superscript II RNase H minus reverse transcriptase in a final volume of 20 μl at 50°C for 50 min and 70°C for 5 min.

3.2.5. PCR Reaction

Carry out PCR reactions with 1 unit of AmpliTaq Gold DNA polymerase in 50 μl of the Taq buffer containing in addition 200 μM of each dNTP, 7.5 mM $MgCl_2$, 1.5 μM of the forward primer and 1.5 μM of the reverse primer.

Subject the reaction mixture to repeated cycles: (i) 95°C for 10 min to activate the AmpliTaq Gold and provide a hot start; (ii) 95°C for 30 s, 55°C for 10 s, 72°C for 1 min for 12 cycles; (iii) 72°C for 5 min for one final cycle. Extract PCR products with phenol/chloroform and precipitate with 1:10 volume of 3 M sodium acetate pH 5.3 and three volumes of ethanol for 1 h at –80°C.

After amplification, transcription and purification, 2'-amino-RNA candidates can be used for the next round of selection.

3.2.6. Cloning and Sequencing

After seven cycles of selection against miniTAR, selected 2'-amino-RNAs are cloned and sequenced according to the same protocol as described in **Section 3.1.6**.

3.2.7. Identification of a 2'-Amino-RNA Scaffold

Analysis of the sequences led to sequence A30 as the most represented sequence (7 out of 18 sequences) (*see* **Fig. 6.6**). A30 exhibits a sequence complementary to the top part of miniTAR and can possibly form a hairpin structure displaying the interaction region into the loop. The dissimilarity between R06, an unmodified RNA aptamer previously identified for the miniTAR target *(6)* and A30 indicates that aldehydes and 2'-amino-uridines present in the RNA library have influenced the outcome of the present selection. The 2'-amino-RNA population could have evolved to provide a particular 2'-amino-RNA scaffold with which the aldehydes react for producing conjugated aptamers with high affinity for miniTAR.

Then, a 19-nucleotide truncated form of A30 (A30sl, *see* **Fig. 6.6**) was employed. A30sl consists of the 2'-amino-RNA hairpin (T_m = 64°C in buffer 1 × SE) that retains the affinity of

MiniTAR 5'-CCAGAUUUGAGCCUGGGAGCUCUCUGG-3'

A30 3'-agcccgcacagaagacGCCGGACCCUCUUAggcgaagcaggaggg-5'

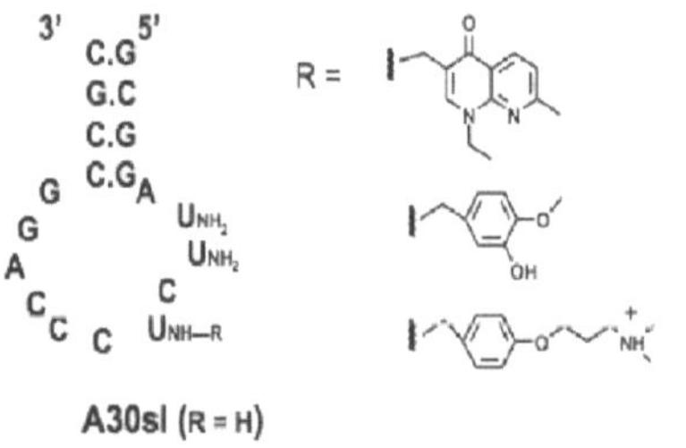

Fig. 6.6. Sequences of the selected 2'-amino-RNA scaffold A30, its truncated form A30sl and the conjugated aptamers selected against the miniTAR target. Watson–Crick complementarity between miniTAR and A30 is underlined. For the A30 sequence, *U* indicates 2'-amino-uridine and the fixed regions are denoted in *lower case*.

the full length A30 for miniTAR target (K_d (A30/miniTAR) = 38 nM; K_d (A30sl/miniTAR) = 23 nM; determined by electrophoresis mobility gel shift assays in buffer 1 × SE). A30sl contains three 2'-amino-uridine residues at position 6, 7 and 9 (*see* **Fig. 6.6**) and thus three reactive 2'-amino groups that can potentially lead, in the presence of three aldehydes, to the formation of 63 mono-, bi- or tri-conjugated aptamers.

3.2.8. Identification of the Conjugated Aptamers

Identification of the best-fitted conjugated aptamers is performed by comparing the distribution of the products obtained when truncated 2'-amino-RNA scaffold A30sl is reacted with the set of aldehydes in buffer 1 × SE and sodium cyanoborohydride either in the absence or in the presence of the miniTAR target. They correspond to products that are preferentially formed (or chemically amplified) in presence of the target.

Incubate A30sl (10 μM) with the set of aldehydes and $NaBH_3CN$ (5 mM) in 100 μl of buffer 1 × SE for 24 h at room temperature.

Repeat the reaction described above in presence of 10 μM of miniTAR target.

Dialyze individually both reaction mixtures (Slide-A-lyzer Mini Dialysis Units, 3,500 MW cut-off, Pierce) in 3 L of water for 16 h and analyze them by reverse-phase HPLC.

Collect the fractions corresponding to the few peaks which present a substantial increase upon addition of the target.

Submit collected products to MALDI-ToF spectrometry analysis and time-dependent snake venom phosphodiesterase digestion followed by MALDI-ToF *(33, 34)*. This allows for the identification of the appended groups and their precise location along the conjugated aptamer (*see* **Fig. 6.6**).

3.3. Structural Characterization of Aptamers and Their Complexes by NMR

RNAs molecules play crucial roles in a wide variety of biological functions from carrying genetic information to regulation. This diversity of functions relies on the formation of numerous three-dimensional structures. Structural studies aim to elucidate the mechanism in which the RNA structure is involved. Nuclear magnetic resonance is a powerful tool for studying RNA structures and their interaction with their partners such as proteins, DNA or RNA ligands. Information about structure, dynamics and interactions can be derived from NMR data for RNA molecules up to 100 nucleotides. Furthermore the development of multi-dimensional NMR spectroscopy and methods such as distance geometry allows the determination of numerous 3D structures *(35–40)*.

Structural studies of RNA oligonucleotides by NMR require the preparation of milligram amounts of RNA. Unlabeled and $^{13}C/^{15}N$ labeled RNAs are generally synthesized from DNA templates by in vitro transcription using T7 RNA polymerase *(24–26)*. RNA oligonucleotides can also be produced by chemical synthesis that is a convenient method for small and unlabeled RNAs *(41–43)*. Various methods have been developed for improving the yield and the homogeneity of the sample *(44–47)*. Here, we describe the preparation of RNA samples by in vitro transcription.

3.3.1. Preparation of RNA Samples by In Vitro Transcription Using T7 RNA Polymerase

The procedure is described in **Fig. 6.7**.

3.3.1.1. Optimization of In Vitro Transcription Conditions

Optimal concentration of $MgCl_2$ depends on the template sequence. For each oligonucleotide, conditions are optimized using 50 μl reaction mixtures. The ratio $[MgCl_2]/[NTPs]$ is varied from 0.4 to 2.8. All reactions are incubated for 4 h at 37°C, and then 5 μl of 0.5 M EDTA, pH 8.0 are added. Transcription mixtures are then loaded on a 20% polyacrylamide denaturing gel (7 M urea). After migration, the transcripts are revealed by Stains-all coloration. The optimum $[MgCl_2]/[NTPs]$ ratio is determined by comparing the transcription yields.

3.3.1.2. Preparative Transcription

Transcription is run at 37°C for 4 h in 10 ml reaction volume at the optimum $[MgCl_2]/[NTPs]$ ratio. About 1 ml of 0.5 M EDTA, pH 8.0 is added and the solution is mixed. Then, to remove proteins, 5 ml of phenol:chloroform (pH 4.7) are added and the solution is vortexed. After centrifugation over 10 min at room temperature, the aqueous phase is retained and 5 ml of

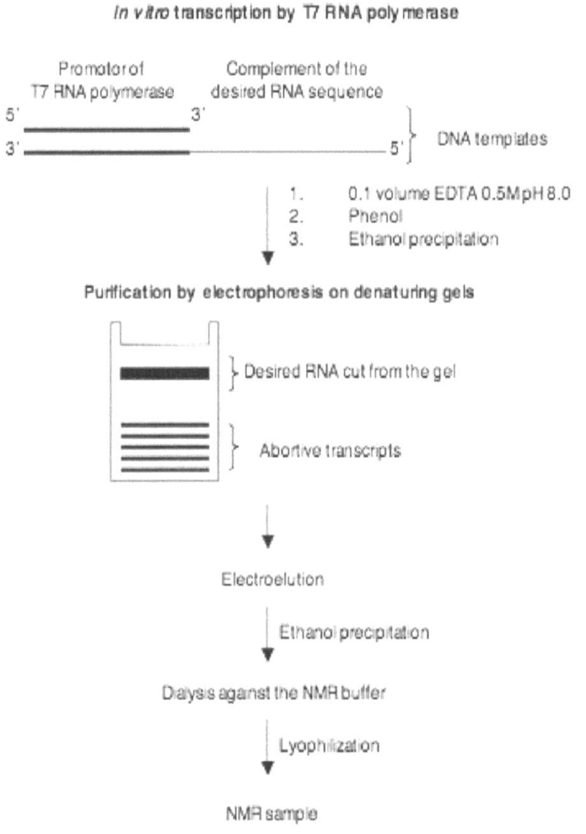

Fig. 6.7 Schematic representation of the preparation of RNA samples by in vitro transcription for NMR studies. For detailed explanations, *see* **Sections 2.3** and **3.3.1**.

water are added to the phenol fraction. This solution is mixed and after centrifugation (10 min, room temperature) the aqueous fractions are combined. Ammonium acetate (pH 5.5) is then added to a final concentration of 0.3 M and 2.5 volumes of ethanol are added to precipitate the RNA at −20°C overnight. After centrifugation for 30 min at 4°C, ethanol is removed and the pellet is suspended in about 1 ml of a 7 M urea solution.

3.3.1.3. Purification of RNA Transcript

RNA molecules are purified using electrophoresis on denaturing (7 M urea) 20% polyacrylamide gels. Gel is run at a constant power at about 50°C. Next, RNAs are visualized by shadowing the gel with UV light over a silica chromatography plate. The band corresponding to the transcript of expected length is cut from the gel and the desired RNA is eluted using an electroelution apparatus. The oligonucleotide is recovered in the sample chamber every hour, until there is no RNA left in the gel slices. RNA is precipitated by addition of 3 M ammonium acetate pH 5.5 (0.1 volume) and ethanol (2.5 volumes) at –20°C overnight. After 30 min centrifugation at 4°C, the pellet is suspended in water and dialyzed 48 h against 10 mM sodium phosphate buffer, pH 6.4.

3.3.1.4. NMR Sample Preparation

Sample is concentrated by lyophilization and resuspended in 90:10 H_2O/D_2O for experiments involving exchangeable protons or 100% D_2O for non-exchangeable proton experiments. The sample is refolded by heating at 95°C (2 min) and snap-cooling at 4°C. Complexes between the aptamer and the target are formed by titration monitoring the imino protons region of one-dimensional spectra.

3.3.2. Structural Studies of RNA Aptamers and Their Complexes with a Target by NMR

3.3.2.1. Secondary Structure: Identification of Base-Pairing Pattern

1. One-dimensional NMR spectra (spectral width of 20 ppm) recorded in 90:10 H_2O/D_2O at low temperature (above 10°C), in order to analyze the imino protons region found between 9.0 and 15.0 ppm (*see* **Note 8**)(*see* **Fig. 6.8A, B**).
2. 2D-NOESY experiments (spectral width of 20 ppm) recorded in 90:10 H_2O/D_2O at low temperature (above 10°C) (*see* **Note 9**).
3. Discriminate A:U Watson–Crick base-pair, G:C Watson–Crick base-pair and non-canonical base-pair by the analysis of 2D NOESY experiments, with the help of (^1H,^{15}N)-HSQC experiment *(48)* (*see* **Note 10**)(*see* **Fig. 6.8C, D**).
4. Assignment of resonances for specific base pairs accomplished by the analysis of 2D NOESY experiment that correlates imino protons of neighboring base pairs (*see* **Note 11**).
5. Identification of the secondary structure by following the primary sequence.

3.3.2.2. Experiments Recorded on the NMR Spectrometer for 3D Structure Determination

NMR experiments have to be recorded at the same temperature in 100% D_2O.

1. 2D NOESY experiments (spectral width of 8 ppm), at various mixing times (50, 150, 200, 300, and 400 ms) (*see* **Note 12**)(*see* **Fig. 6.9A**).
2. 2D TOCSY experiment *(49, 50)*: observation of H5–H6 cross-peaks (*see* **Note 13**).
3. 2D DQF–COSY experiment *(51)*: measurement of ^1H–^1H coupling constants (*see* **Note 14**).
4. 2D (^{13}C,^1H) HSQC experiments: correlate protons to their attached carbon (*see* **Note 15**).
5. 3D HCCH–TOCSY experiment *(52, 53)*: identify individual sugar spin systems (*see* **Note 16**).
6. 3D (^1H,^1H,^{13}C)-NOESY–HSQC experiment *(54)*: to complete the assignment (*see* **Note 17**).

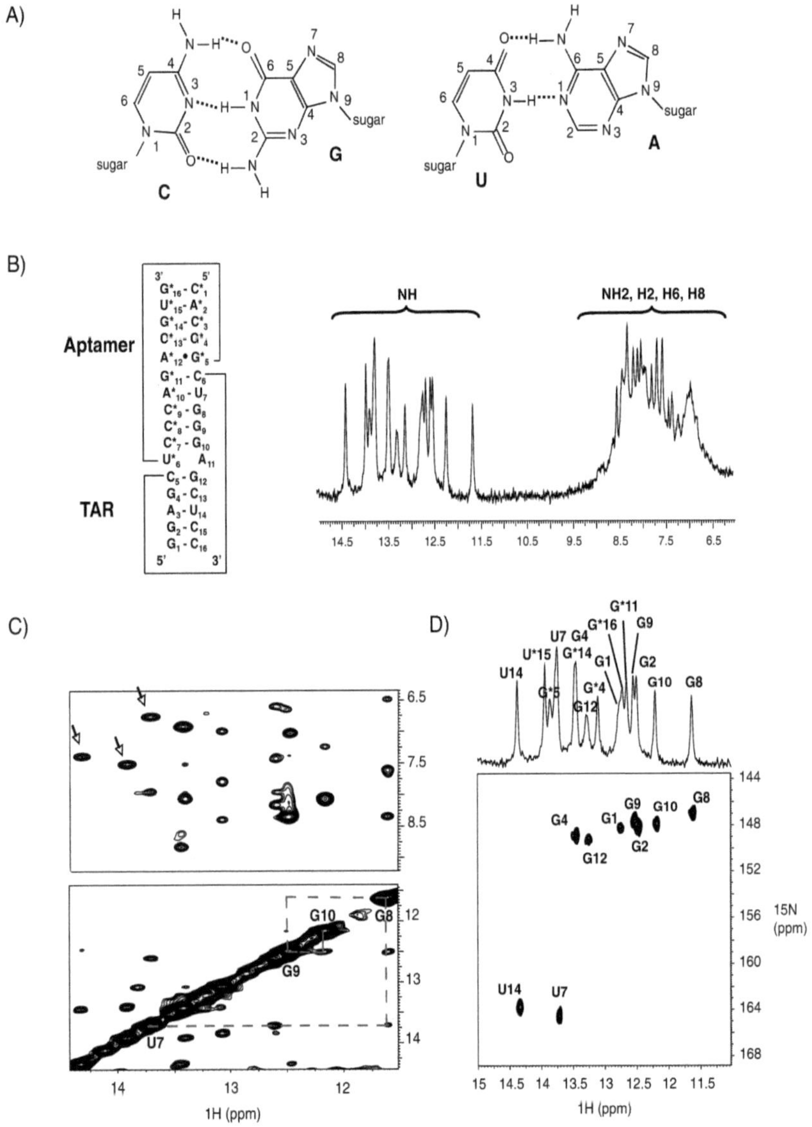

Fig. 6.8. Secondary structure, identification of base-pairing. (**A**) Representation of the Watson–Crick base pairs A:U and C:G. *Dashed lines* indicate hydrogen bonds. (**B**) *On the left*: secondary structure of the complex formed between the TAR element of HIV-1 and an aptamer, based on analysis of proton NMR data. *On the right*: one-dimensional NMR spectrum recorded in 90:10 H_2O/D_2O. Solvent suppression is achieved using the WATERGATE and the "Jump and Return" sequences. *See* **Note 8**. (**C**) NOESY experiment recorded in 90:10 H_2O/D_2O at 15°C. *On the top*: Imino protons connectivities with aminos, H2, H6, H8 region, where arrows indicate H3–H2 correlations (*see* **Note 10**). *On the bottom*: H3–H1 region where sequential and inter-strands connectivities between imino protons are observable. (**D**) $^1H/^{15}N$ HSQC spectrum recorded at 15°C in 90:10 H_2O/D_2O (*see* **Note 10**).

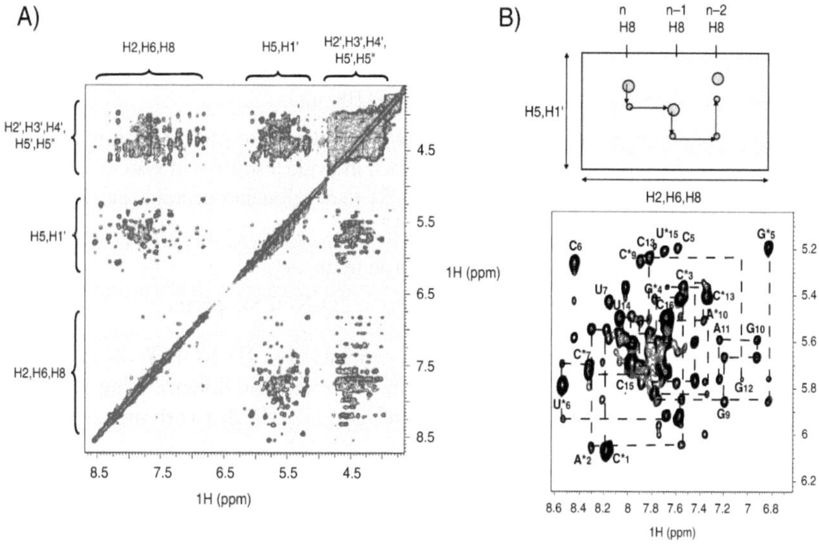

Fig. 6.9 Sequence specific assignment. (**A**) NOESY experiment recorded in 100% D_2O. The solvent suppression is achieved using low-power pre-saturation. Chemical shifts for each type of proton are indicated on the *left* and *bottom*. (**B**) *On the top*: Schematic representation of the sequential assignment achieved in the region, where H2, H6, H8 correlates with H5, H1' (*see* **Section 3.3.2.3**). *On the bottom*: example of sequential assignment for the complex between TAR and the aptamer represented in **Fig.** 6.8.

7. 2D HP–COSY experiment *(55)*: measurement of $^1H_{5',5''}$–^{31}P coupling constants (*see* **Note 18**).
8. 3D HCP experiment *(56)*: measurement of ^3J(Pi-C4'i), ^3J (Pi-C5'i) and ^3J(Pi-C2'i-1) coupling constants (*see* **Note 19**).

3.3.2.3. Sequence Specific Assignment: Assign Resonances to a Particular Type of Proton

1. Distinguish uracils and cytosines (H6) from adenines and guanines (H8, H2) by analysis of 2D NOESY and 2D TOCSY simultaneously (*see* **Note 20**).
2. Distinguish H2 protons from H8 protons with the help of 2D (^1H,^{13}C)-HSQC experiment (*see* **Note 15**).
3. Establish internucleotide connectivity pathways for double-stranded region by observation of aromatic to H1' protons ($H1'_{n-1}$-> $H6/8_n$ -> $H1'_n$). For single-stranded region, the assignment of resonances is not straightforward and requires analysis of heteronuclear experiments (*see* **Note 21**)(*see* **Fig. 6.9**).
4. In an A-form helix, identify the H2' protons by the very strong sequential NOE with H6/8 protons ($H2'_{n-1}$ –> $H6/8_n$), observable in 2D-NOESY experiment at low mixing time (50 ms).

5. In an A-form helix, identify the H3' protons by the two strong NOEs with their own H6/8 and with the one of the preceding base, observable in 2D-NOESY experiment at longer mixing time (100 ms).
6. Once assignment of resonances to each H1' proton is completed, identify each individual sugar spin system by analysis 3D HCCH–TOCSY to get all sugars protons (i.e., H2', H3', H4', H5' and H5").
9. Report every proton frequency.
10. Report assignment in 2D-NOESY spectra.

3.3.2.4. Constraints for Structure Calculation

1. Derived distance restraints from 2D-NOESY: the intensity of the cross-peaks are converted into distance using as internal standard the correlation H5–H6 that corresponds to a distance of 2.4 Å.
2. Derived sugar conformation from the inspection of both DQF–COSY and TOCSY experiments: nucleotides with no COSY and no TOCSY cross-peaks between H1' and H2' are restrained in *C3'-endo* conformation.
3. Torsion angles χ are derived from the observation of intra-residue H6/8-H1' cross-peak volumes.
4. Torsion angles β and ε are derived from ^{31}P-NMR experiments (*see* **Notes 19–20**).
5. Torsion angle γ are derived from DQF–COSY experiment (*see* **Note 22**).
6. Hydrogen bonding restraints are determined from the base-pairing pattern.

3.3.2.5. Structure Calculation

1. Generate the unfolded starting structure as random as possible with software packages such as X-PLOR *(57)* or CNS *(58)*.
2. Perform structure calculation using a protocol that generates the structure of the RNA molecule from the starting structure under experimental NMR constraints (distances, angles, hydrogen bonds) (*see* **Note 23**).
3. Analyze the "output structure" to judge its quality (*see* **Note 24**).

4. Notes

1. Use 0.5 g Sephadex™ G-25 Fine for each purification.

   Autoclave the required quantity of Sephadex in water. In a 2 ml syringe, introduce some glass wool, then add 2 ml of hydrated Sephadex.

Wait until water goes through. Complete up to 2 ml Sephadex.
Put a 1.5 ml tube under the column.
Centrifuge 1 min at 3,000 *g*. Discard flow through.
Put another 1.5 ml tube under the column. Load your preparation on the column.
Centrifuge 1 min at 3,000 *g*.

2. Although the stringency needs be low enough to retain in the selected pool the few sequences that display affinity for the target, the selection pressure is adjusted by increasing the stringency. This procedure enables to select the sequences that possess the highest affinity for TAR. The number of washing steps or the composition of the washing buffer can be modulated. The candidate/target ratio, the candidate and the target concentrations are other parameters that can be changed. During the selection, the concentration of the candidates is decreased at each round and the candidate/target ratio is kept high to promote the competition.
3. For each round of selection, quantify non-specific interaction of candidates on beads. Ideally the background should remain stable and low (less than 1%) all along the selection. If the background slightly increases, the kind of beads can be exchanged (from Promega to Dynabeads for instance).
4. Surface Plasmon Resonance measurements reveal an evolution of the affinity of the selected sequences for the target. The second round of selection was already enriched in sequences that display affinity for TAR. The affinity of the population increased from round 3 to 6. Kinetics parameters of the complex RI-11/TAR were determined by Surface Plasmon Resonance analysis. This genomic complex displays an equilibrium dissociation constant (K_d) of 10 nM, that is comparable to that of R06-24/TAR (K_d = 17 nM).
5. Four peaks by A. Griekspoor and Tom Groothuis, *mekentosj.com*
6. Aldehydes different of those reported can be used providing that they are soluble enough in the buffer used for the selection steps and they exhibit sufficient difference in molecular weights for allowing their identification by mass spectrometry when conjugated to the aptamer. In addition, it is recommended to adjust the concentration of the aldehydes present in the set to compensate for their difference in reactivity with 2'-amino-uridine and to provide comparable proportions of conjugated products in the absence of any target.
7. An increasing selection pressure was applied during the seven rounds of the SELEX process by applying decreasing concentrations of 3'-biotinylated miniTAR (from 0.30 to 0.01 μM) and decreasing library concentrations (5–0.1 μM).
8. Imino protons of guanine and uracil residues constitute a good probe of the overall secondary structure of the RNA

molecule. Their observation points out their protection from exchange with water and therefore their involvement in hydrogen bond formations. The resonances corresponding to those protons found between 12.0 and 15.0 ppm are characteristic of imino protons involved in Watson–Crick base pairs, whereas those found upfield (between 9.0 and 12.0 ppm) indicate the formation of non-canonical base-pairing. Assignment of these resonances provides thus precious information about the folding of the RNA.

9. 2D-NOESY (nuclear Overhauser effect spectroscopy) experiments correlate proton within a distance of 5 Å and allows the determination of interproton distances. The cross-peaks intensities in NOESY experiments vary as $1/r^6$, where r is the distance between two protons. Thus, distances are derived from the cross-peaks intensities, using the pyrimidine H5–H6 cross-peaks as an internal standard, which corresponds to a distance of 2.4 Å.

10. As a starting point, an A:U Watson–Crick base-pair can easily be discriminated from G:C base-pair by the strong correlation between the uracil H3 imino proton and the H2 proton of adenine. In a G:C Watson–Crick base-pair, two strong NOEs occur between the guanine H1 proton and the cytosine amino protons. The G:U wobble base-pair is identified from the very strong NOE between the H1 guanine imino proton and the H3 uracil imino protons. As the chemical shift of the resonances depends on the chemical environment (ring current shift, base stacking, RNA conformation and hydrogen bonding), it can also be helpful for NOEs assignment. Indeed, the chemical shift of imino protons involved in non-canonical base-pairs, is generally upfield and can be used as a starting point for assignment. Moreover, with the help of 15N labeling, imino protons of uracils and guanines can easily be discriminated. The chemical shifts of their attached nitrogens are separated by 10 ppm: N1 guanosine nitrogen generally resonates at 145–150 ppm, whereas N3 uracil nitrogen resonates at 160–165 ppm. Elucidation of base-pairing patterns has also been improved by the development of heteronuclear experiments such as HNN-COSY, which allow the direct identification of donor and acceptor nitrogen atoms involved in hydrogen bonds *(59, 60)*. These heteronuclear experiments are very helpful when resonances overlap and/or when imino protons are involved in non-canonical structures.

11. Sequential and inter-strands connectivities are observed between imino protons of neighboring base pairs. The sequential NOEs can be assigned following the primary sequence of the RNA.

12. 2D NOESY experiments recorded in 100% D_2O allow the observation of non- exchangeable protons. Exchangeable protons are not observable in these conditions.
13. 2D TOCSY (total correlated spectroscopy) experiments recorded in 100% D_2O allow the observation of H5–H6 correlations alone in the aromatic region and, thus the discrimination of H5–H6 cross-peaks among all correlated protons in the 2D-NOESY spectra. 2D-TOCSY experiment also gives information about the sugar conformation: sugars with no H1'–H2' cross-peaks can be restrained in a C3'-endo conformation.
14. A 2D COSY–DQF experiment (double quantum-filtered correlated spectroscopy) allows the measurement of ^1H–^1H coupling constants, giving informations about the sugar conformation (δ dihedral angle) and the phosphodiester backbone (γ torsion angle). $^3J_{H1'H3'}$ and $^3J_{H3'H4'}$ coupling constants depend strongly on the sugar conformation and γ torsion angle is related to $^3J_{H4'H5'}$ and $^3J_{H4'H5''}$ constants.
15. A 2D HSQC experiment (heteronuclear single quantum coherence) correlates an hydrogen to its attached carbon. This experiment is helpful to distinguish C8 and C6 carbons from C2 carbons that resonate about 15 ppm downfield. The sugar carbon resonances are also separated: C1', C4', C2', C3' and C5' downfield to upfield (from 90 to 60 ppm). 2D HMBC experiment is also helpful to correlate H2 and H8 protons *(61)*.
16. 3D HCCH–TOCSY experiment has been developed to overcome the overlapping in the region of sugar protons and allows the complete assignment of sugars atoms. The starting point is the assignment of H1' proton that is correlated to the other sugar resonances via magnetization transfer. This experiment allows the identification of individual sugar spin system.
17. A 3D (^1H,^1H,^{13}C)-NOESY–HSQC is used to complete assignments. This experiment combines the 2D-NOE and the HSQC. Slice through the 3D experiment are equivalent to filtered 2D NOESY spectra at a particular frequency in ^{13}C dimension.
18. A 2D HP–COSY experiment allows the measurement of $^1H_{5',5''}$ –^{31}P coupling constants. β angles restraints are derived from this NMR experiment for structure calculation. In an A-type helix, $^3J(^1H_{5',5''}-^{31}P)$ is weak, that corresponds to a *trans* conformation for β.
19. A 3D HCP experiment allows the measurement of ^3J(Pi-C4'i), ^3J(Pi-C5'i) and ^3J(Pi-C2'i-1) coupling constants. ε angles restraints are derived from this NMR experiment for structure calculation. In an A-type helix, ^3J(Pi-C4'i) is strong; ^3J(Pi-C5'i) and ^3J(Pi-C2'i-1) are weak.

20. H6 can be distinguished from H8 and H2 by analysis of the 2D TOCSY spectra that allows the observation of H5–H6 correlations alone in the aromatic region.
21. In an A-type RNA helix, each aromatic proton H6 or H8 is correlated to its own H1' proton and to the one of the preceding base. In a double-stranded region, the H2 of adenine can be identified in the 2D NOESY experiment recorded in 90:10 H_2O/D_2O by the strong NOE with the H3 of its paired uracil. In a A-form, H2 proton exhibits a sequential intrastrand NOE with the H1' proton of the following nucleotide and an interstrand NOE with the H1' proton of the base pair 3' to adenine. All these elements can be used as a starting point for structure elucidation. In a single-stranded region, assignment is not straightforward as the sequential pathway and the H2 of adenine are generally not observable. Heteronuclear experiments have been developed to overcome this problem (for reviews, *see (35–40)*).
22. A 2D DQF–COSY experiment allows the measurement of $^1H_{5',5''}$–$^1H_{4'}$ coupling constants. γ angle restraints are derived from this NMR experiment for structure calculation. In an A-type helix, $^3J(^1H_{5',5''}–^1H_{4'})$ is weak, that corresponds to a *trans* conformation for γ.
23. RNA stereochemistry, bond lengths, bond angles, base planarity, proper chirality, non- bonded interactions (Van der Waals and electrostatics contacts) are described in the software package. NMR constraints are introduced in additional term into the forcefield defined by the software. Several protocols have been developed to calculate RNA structures *(36, 57–58)*
24. Structures have to be examined individually and only acceptable solutions are kept. Structures with no agreement with the data are rejected. Structures with no violation on NOE distances and dihedral angles restraints and with the lowest energy are selected. The precision of the chosen ensemble of structures is evaluated by the root mean square deviation (r.m.s.d.) between different structures. The r.m.s.d. quantifies the degree by which the position of a specific atom differs on average between structures. In order to calculate r.m.s.d., structures have to be superimposed. The more defined the structures, the lower the r.m.s.d.

Acknowledgments

We are grateful to Frederike von Pelchrzim and Renée Schroeder for the gift of the genomic library. This work was supported in part by the Conseil Régional d'Aquitaine. We thank Ms N. Pierre for skillful technical assistance.

References

1. Osborne, S.E. and Ellington, A.E. (1997) Nucleic acid selection and the challenge of combinatorial chemistry. *Chem. Rev.* **97**, 349–370.
2. Toulmé, J.J., Di Primo, C. and Boucard, D. (2004) Regulating eukaryotic gene expression with aptamers. *FEBS Lett.* **567**, 55–62.
3. Nimjee, S.M., Rusconi, C.P. and Sullenger, B.A. (2005) Aptamers: an emerging class of therapeutics. *Annu. Rev. Med.* **56**, 555–583.
4. Toulmé, J.-J., Di Primo, C. and Moreau, S. (2001) Modulation of RNA function by oligonucleotides recognizing RNA structure. *Progr. Nucleic Acid Res. Mol. Biol.* **69**, 1–46.
5. Soukup, G.A., Ellington, A.D. and Maher, L.J. (1996) Selection of RNAs that bind to duplex DNA at neutral pH. *J. Mol. Biol.* **259**, 216–228.
6. Ducongé, F. and Toulmé, J.J. (1999) *In vitro* selection identifies key determinants for loop–loop interactions: RNA aptamers selective for the TAR RNA element of HIV-1. *RNA* **5**, 1605–1614.
7. Boiziau, C., Dausse, E., Yurchenko, L. and Toulmé, J.J. (1999) DNA aptamers selected against the HIV-1 TAR RNA element form RNA/DNA kissing complexes. *J. Biol. Chem.* **274**, 12730–12737.
8. Kikuchi, K., Umehara, T., Fukuda, K., Kuno, A., Hasegawa, T. and Nishikawa, S. (2005) A hepatitis C virus (HCV) internal ribosome entry site (IRES) domain III-IV-targeted aptamer inhibits translation by binding to an apical loop of domain IIId. *Nucleic Acids Res.* **33**, 683–692.
9. Shao, Y., Chan, C.Y., Maliyekkel, A., Lawrence, C.E., Roninson, I.B. and Ding, Y. (2007) Effect of target secondary structure on RNAi efficiency. *RNA* **13**, 1631–1640.
10. Yoshinari, K., Miyagishi, M. and Taira, K. (2004) Effects on RNAi of the tight structure, sequence and position of the targeted region. *Nucleic Acids Res.* **32**, 691–699.
11. Darfeuille, F., Reigadas, S., Hansen, J.B., Orum, H., Di Primo, C. and Toulme, J.J. (2006) Aptamers targeted to an RNA hairpin show improved specificity compared to that of complementary oligonucleotides. *Biochemistry* **45**, 12076–12082.
12. Scarabino, D., Crisari, A., Lorenzini, S., Williams, K. and Tocchini-Valentini, G.P. (1999) tRNA prefers to kiss. *EMBO J.* **18**, 4571–4578.
13. Aldaz-Carroll, L., Tallet, B., Dausse, E., Yurchenko, L. and Toulmé, J.J. (2002) Apical loop–internal loop interactions: a new RNA-RNA recognition motif identified through *in vitro* selection against RNA hairpins of the hepatitis C virus mRNA. *Biochemistry* **41**, 5883–5893.
14. Da Rocha Gomes, S., Dausse, E. and Toulme, J.J. (2004) Determinants of apical loop–internal loop RNA–RNA interactions involving the HCV IRES. *Biochem. Biophys. Res. Commun.* **322**, 820–826.
15. Tomizawa, J.I. (1990) Control of ColE1 plasmid replication: intermediates in the binding of RNA I and RNA II. *J. Mol. Biol.* **212**, 683–694.
16. Ferrandon, D., Koch, I., Westof, E. and Nusslein-Volhard, C. (1997) RNA-RNA interaction is required for the formation of specific bicoid mRNA 3'UTR-STAUFEN ribonucleoproteic particles. *EMBO J.* **16**, 1751–1758.
17. Kolb, G., Reigadas, S., Castanotto, D., Faure, A., Ventura, M., Rossi, J.J. and Toulme, J.J. (2006) Endogenous expression of an anti-TAR aptamer reduces HIV-1 replication. *RNA Biol.* **3**, 150–156.
18. Darfeuille, F., Arzumanov, A., Gait, M.J., Di Primo, C. and Toulmé, J.-J. (2002) 2'-O-methyl-RNA hairpins generate loop–loop complexes and selectively inhibit HIV-1 Tat-mediated transcription. *Biochemistry* **41**, 12186–12192.
19. Darfeuille, F., Arzumanov, A., Gryaznov, S., Gait, M.J., Di Primo, C. and Toulmé, J.J. (2002) Loop–loop interaction of HIV-1 TAR RNA with N3'->P5' deoxyphosphoramidate aptamers inhibits *in vitro* Tat-mediated transcription. *Proc. Natl. Acad. Sci. U.S.A.* **99**, 9709–9714.
20. Darfeuille, F., Hansen, J.B., Orum, H., Di Primo, C. and Toulmé, J.J. (2004) LNA/DNA chimeric oligomers mimic RNA aptamers targeted to the TAR RNA element of HIV-1. *Nucleic Acids Res.* **32**, 3101–3107.
21. Kolb, G., Reigadas, S., Boiziau, C., Van Aerschot, A., Arzumanov, A., Gait, M., Herdewijn, P. and Toulmé, J.J. (2005) Hexitol nucleic acid-containing aptamers are efficient ligands of HIV-1 TAR RNA. *Biochemistry* **44**, 2926–2933.

22. Dausse, E., Cazenave, C., Rayner, B. and Toulmé, J.J. (2005) *In vitro* selection procedures for identifying DNA and RNA aptamers targeted to nucleic acids and proteins. *Methods Mol. Biol.* **288**, 391–410.
23. Corbett, P.T., Leclaire, J., Vial, L., West, K.R., Wietor, J.L., Sanders, J.K. and Otto, S. (2006) Dynamic combinatorial chemistry. *Chem. Rev.* **106**, 3652–3711.
24. Milligan, J.F., Groebe, D.R., Witherell, G.W. and Uhlenbeck, O.C. (1987) Oligoribonucleotide synthesis using T7 RNA polymerase and synthetic DNA templates. *Nucleic Acids Res.* **15**, 8783–8798.
25. Milligan, J.F., Groebe, D.R., Witherell, G.W. and Uhlenbeck O.C. (1989) Synthesis of small RNAs using T7 RNA polymerase. *Methods Enzymol.*, **180**, 51–62.
26. Wyatt, J.R., Chastain, M. and Puglisi, J.D. (1991) Synthesis and purification of large amounts of RNA oligonucleotides. *BioTechniques* **11**, 764–769.
27. Gold, L., Brown, D., He, Y., Shtatland, T., Singer, B.S. and Wu, Y. (1997) From oligonucleotide shapes to genomic SELEX: novel biological regulatory loops. *Proc. Natl. Acad. Sci. U.S.A.* **94**, 59–64.
28. Lorenz, C., von Pelchrzim, F. and Schroeder, R. (2006) Genomic systematic evolution of ligands by exponential enrichment (Genomic SELEX) for the identification of protein-binding RNAs independent of their expression levels. *Nat. Protoc.* **1**, 2204–2212.
29. Singer, B.S., Shtatland, T., Brown, D. and Gold, L. (1997) Libraries for genomic SELEX. *Nucleic Acids Res.* **25**, 781–786.
30. Bannwarth, S. and Gatignol, A. (2005) HIV-1 TAR RNA: the target of molecular interactions between the virus and its host. *Curr. HIV Res.* **3**, 61–71.
31. Gold, L., Brown, D., He, Y.Y., Shtatland, T., Singer, B.S. and Wu, Y. (1997) From oligonucleotide shapes to genomic SELEX: Novel biological regulatory loops. *Proc. Natl. Acad. Sci. U.S.A.* **94**, 59–64.
32. Darfeuille, F., Sekkai, D., Dausse, E., Kolb, G., Yurchenko, L., Boiziau, C. and Toulmé, J.-J. (2002) Driving *in vitro* selection of anti-HIV-1 TAR aptamers by magnesium concentration and temperature. *Combinat. Chem. High Throughput Screening* **5**, 315–327.
33. Pieles, U., Zurcher, W., Schar, M. and Moser, H.E. (1993) Matrix-assisted laser desorption ionization time-of-flight mass spectrometry: a powerful tool for the mass and sequence analysis of natural and modified oligonucleotides. *Nucleic Acids Res.* **21**, 3191–3196.
34. Bugaut, A., Bathany, K., Schmitter, J.M. and Rayner, B. (2005) Target-induced selection of ligands from a dynamic combinatorial library of mono- and bi-conjugated oligonucleotides. *Tetrahedron Lett.* **46**, 687–690.
35. Varani, G. and Tinoco, I. Jr. (1991) RNA structure and NMR spectroscopy. *Q. Rev. Biophys.*, **24**, 479–532.
36. Varani, G., Aboul-ela, F. and Allain, F.H.-T. (1996) NMR investigation of RNA structure. *Prog Nucl. Magn. Reso. Spectrosc.*, **29**, 51–127.
37. Fürtig, B., Richter, C., Wöhnert, J. and Schwalbe, H. (2003) NMR Spectroscopy of RNA. *ChemBioChem.*, **4**, 936–962.
38. Latham, M.P., Brown, D.J., McCallum, S.A. and Pardi A. (2005) NMR methods for studying the structure and dynamics of RNA. *ChemBioChem.*, **6**, 1492–1505.
39. Wüthrich, K. (1986) *NMR of Proteins and Nucleic Acids*. Wiley, New York.
40. Wüthrich, K. (1989) Protein structure determination in solution by nuclear magnetic resonance spectroscopy. *Science* **243**, 45–50.
41. Usman, N., Ogilvie, K.K., Jiang, M.-Y. and Cedergren, R.J. (1987) Automated chemical synthesis of long oligonucleotides using 2'-O-silylated ribonucleoside 3'-O-phosphoramidites on a controlled-pore glass support: Synthesis of a 43-nucleotide sequence similar to the 3'-half molecule of an Escherichia coli formylmethionine tRNA. *J.Am.Chem.Soc.* **109**, 7845–7854.
42. Ogilvie, K.K., Usman, N., Nicoghosian, K. and Cedergren, R.J. (1988) Total chemical synthesis of a 77-nucleotide-long RNA sequence having methionine acceptance activity. *Proc.Natl.Acad. Sci. U.S.A.* **85**, 5764–5768.
43. Wincott, F., DiRenzo, A., Shaffer, C., Grimm, S., Tracz, D., Workman, C., Sweedler, D., Gonzalez, C. et al. (1995) Synthesis, deprotection, analysis and purification of RNA and ribozymes. *Nucleic Acids Res.* **23**, 2677–2684.
44. Kao, C., Rüdisser, S. and Zheng, M. (2001) A simple and efficient method to transcribe RNAs with reduced 3' heterogeneity. *Methods* **23**, 201–205.

45. Lapham, J. and Crothers, D.M. (1996) RNase H cleavage fpr processing of *in vitro* transcribed RNA for NMR studies and RNA ligation. *RNA*, **2**, 289–296.
46. Grosshans, C.A. and Cech, T.R. (1991) A hammerhead ribozyme allows synthesis of a new form of the Tetrahymena ribozyme homogeneous in length with a 3' end blocked for transesterification. *Nucleic Acids Res.* **19**, 3875–3880.
47. Ferre-D'Amare, A.R. and Doudna, J.A. (1996) Use of cis- and trans-ribozymes to remove 5' and 3' heterogeneities from milligrams of *in vitro* transcribed RNA. *Nucleic Acids Res.* **24**, 977–978.
48. Bodenhausen, G. and Ruben, D.J. (1980) Natural abundance nitrogen-15 NMR by enhanced heteronuclear spectroscopy. *Chem. Phys. Lett.* **69**, 185.
49. Braunschweiler, L. and Ernst, R.R. (1983) Coherence transfer by isotropic mixing: application to proton correlation spectroscopy. *J. Magn. Reson.* **53**, 521.
50. Bax, A. and Davis, D.G. (1985) MLEV-17-based two-dimensional homonuclear magnetization transfer spectroscopy. *J. Magn. Reson.* **65**, 355–360.
51. Piantini, U., Sorensen, O.W. and Ernst, R.R. (1982) Multiple quantum filters for elucidating NMR coupling networks. *J. Am. Chem. Soc.*, **104**, 6800–6801.
52. Bax, A., Clore, G.M. and Gronenborn, A.M. (1990) ^{1}H-^{1}H correlation via isotropic mixing of ^{13}C magnetization: a new three-dimensional approach for assigning ^{1}H and ^{13}C spectra of ^{13}C-enriched proteins. *J. Magn. Reson.* **88**, 425–431.
53. Fesik, S.W., Eaton, H.L., Olejniczak, E.T. and Zuiderweg, E.R.P. (1990) 2D and 3D NMR Spectroscopy Employing ^{13}C-^{13}C Magnetization transfer by isotropic mixing. spin system identification in large proteins. *J. Am. Chem. Soc.* **112**, 886–888.
54. Kay, L.E., Marion, D. and Bax, A. (1989) Practical aspects of three-dimensional heteronuclear NMR of proteins. *J. Magn.Reson.* **84**, 72–84.
55. Sklenar, V., Miyashiro, H., Zon, G., Miles, H.T. and Bax, A. (1986) Assignment of ^{31}P and ^{1}H resonances in oligonucleotides by two-dimensional NMR spectroscopy. *FEBS Lett.*, **208**, 94–98.
56. Marino, J.P., Schwalbe, H., Anklin, C., Bermel, W., Crothers, D.M. and Griesinger, C. (1995) Sequential correlation of anomeric ribose protons and intervening phosphorus in RNA oligonucleotides by ^{1}H, ^{13}C, ^{31}P triple resonance experiment: HCP-CCH-TOCSY. *J. Biomol. NMR* **5**, 87–92.
57. Brünger, A.T. (1990) *X-PLOR Manual.* Yale University, New Haven, CT.
58. Brunger, A.T., Adams, P.D., Clore, G.M., DeLano, W.L., Gros, P., Grosse-Kunstleve, R.W., Jiang, J.S., Kuszewski, J., Nilges, M., Pannu, N.S., Read, R.J., Rice, L.M., Simonson, T. and Warren, G.L. (1998) Crystallography & NMR system: A new software suite for macromolecular structure determination. *Acta Crystallogr. D. Biol. Crystallogr.* **54**, 905–921.
59. Henning, M. and Williamson; J.R. (2000) Detection of N-H...N hydrogen bonding in RNA via scalar couplings in the absence of observable imino proton resonances. *Nucleic Acids Res.* **28**, 1585–1593.
60. Dingley, A.J. and Grzesiek, S. (1998) Direct observation of hydrogen bonds in nucleic acid base pairs by internucleotide 2JNN Couplings. *J. Am. Chem. Soc.* **120**, 8294–8297.
61. van Dongen, M.J.P., Wijmenga, S.S., Eritja, R., Azorin, F. and Hilbers, C.W. (1996) Through-bond correlation of adenine H2 and H8 protons in unlabeled DNA by HMBC spectroscopy. *J.Biomol.NMR* **8**, 207–212.

Chapter 7

Fluorescence Correlation Spectroscopy (FCS)-Based Characterisation of Aptamer Ligand Interaction

Arne Werner and Ulrich Hahn

Abstract

Fluorescence correlation spectroscopy (Bacia and Schwille (2007) *Nat. Protoc.* **2**, 2842–2856) reveals molecular mobilities, enabling to identify molecular interactions based on a change of diffusion times (Rigler and Elson, (2001) *Fluorescence Correlation Spectroscopy: Theory and Applications.* Springer, Berlin; Haustein, and Schwille, (2004) *Curr. Opin. Struct. Biol.* **14**, 531–540). This technique can be applied to determine the dissociation constant of a complex formed by a fluorescence-labelled target and its corresponding RNA aptamer selected via systematic evolution of ligands by exponential enrichment (SELEX) (Schürer, et al. (2001) *Biol. Chem.* **382**, 47948). Here, an FCS titration experiment is described in detail, where the binding properties of tetramethylrhodamine (TMR) labelled Moenomycin A to its corresponding RNA aptamer were determined (Schürer, et al. (2001) *Biol. Chem.* **382**, 47948).

Key words: Fluorescence correlation spectroscopy (FCS), RNA aptamer, Moenomycin A, tetramethylrhodamine (TMR).

1. Introduction

Fluorescence-based techniques provide a widely used alternative to isotope-related methods in quantifying intermolecular interactions. Single molecule spectroscopic methods reveal a high sensitivity, which minimises the consumption of sample materials and prevents self-quenching effects caused by high fluorophore concentrations. Fluorescence correlation spectroscopy (FCS) is a fluorescence-based method, which enables to identify molecular interactions based on a change of molecular mobilities at the single molecule level. For review please read the publications of Rudolf Rigler and Elliot S. Elson; Oleg Krichevsky and Gregoire Bonnet and of Petra Schwille and coworkers *(1–5)*. With the

Günter Mayer (ed.), *Nucleic Acid and Peptide Aptamers: Methods and Protocols, vol. 535*

DOI 10.1007/978-1-59745-557-2_7 Springerprotocols.com

SELEX (systematic evolution of ligands by exponential enrichment) technique a nuclease resistant 2'-aminopyrimidine RNA aptamer was selected, which specifically binds to the antibiotic Moenomycin A *(6)*. Moenomycin A is an inhibitor of the transglycosylation reaction, which is one of the last steps in peptidoglycan biosynthesis of the cell wall in gram-positive bacteria *(7)*. A dissociation constant value (K_d) of 437 nM was determined using tetramethylrhodamine (TMR)-labelled Moenomycin A in a FCS-based assay *(6)*.

With FCS diffusion coefficients can be derived in the range of 10^{-6}–10^{-9} cm^2 s^{-1}, determining the residence time of a fluorescent molecule in the accurately defined detection volume of a confocal microscope. By measuring fluorescence intensity fluctuations $\delta F(t)$ in the detection volume V_{eff} the entrance and exit of single particles can be identified. By the normalised autocorrelation function $G(\tau) = \langle \delta F(t+\tau)\delta F(t)\rangle / \langle F \rangle^2$, the average residence time τ_{Diff} of the fluorescent particles in the detection volume is defined. Due to $\tau_{Diff} \approx \sqrt[3]{m}$, between different diffusion species can be distinguished, if the masses of the species differ by a factor of 8 or more or the diffusion times differ by a factor of at least 1.6 *(8)*. The resolution also depends on the molecular brightness of the experiment.

From a model, describing three-dimensional diffusion and triplet state population, not only the translational diffusion time τ_{Diff} is defined but also the number of particles in the detection volume N and the fractional population f_i of n different diffusion species *(9–13)*:

$$G(\tau) = 1 + \frac{1}{N} \cdot \left(\frac{1 - T + Te^{-\frac{\tau}{\tau_T}}}{1 - T} \right) \cdot \left\{ \sum_{i=1}^{n} \frac{f_i}{(1 + \tau/\tau_{Di})\sqrt{1 + \tau(S^{-2}\tau_{Di})}} \right\} \tag{7.1}$$

The structure parameter S describes the ratio of the radial and axial distances from the centre of the laser beam focus to the $1/e^2$ fluorescence intensity, r_0 and z_0, with $S = z_0/r_0$. The fractional population and decay time T and τ_T of the triplet state are defined.

The ConfoCor set up performs the fit automatically. One has to choose the number of fluorescent species and is able to fix several parameters of Eq. (7.1), e.g. S. The reliability of the chosen model can be determined by X^2 test and the residual deviations of the fit results from the autocorrelation function *(8)*. The triplet state is a spin forbidden dark state, which can be populated by excited fluorophores. Triplet state contributes with a lifetime of 0.5–10 μs to the intensity fluctuations, identified by autocorrelation analysis.

To define the optical setup more accurately, the autocorrelation function of a standard fluorophore can be analysed using the same laser and filter system as in the experiment. The structure parameter S can be determined by measuring the diffusion time of a standard fluorophore, e.g. rhodamine 6 green (R6G), and should be kept fixed for the experiments. It should be at 5 for an excitation with 488 nm, about 6 for an excitation with 543 nm and about 8 for an excitation with 633 nm *(13)*. The radial distance to the centre of the laser beam focus ω_0 is calculated from

$$D_i = \varpi_0^2 / 4\tau_{Diff} \tag{7.2}$$

with the diffusion coefficient D_i (R6G: 2.8×10^{-10} m^2 s^{-1}[*13*]). The approximate size of V_{eff} can be calculated with

$$V_{eff} = \pi^{3/2} \cdot \omega_0^2 \cdot z_0 \tag{7.3}$$

To determine the concentration of the fluorescent particles in the detection volume [N], one uses Eqs. (7.2) and (7.3) and

$$[N] = \frac{N}{(N_A \cdot V_{eff})} \tag{7.4}$$

with the Avogadro constant N_A (6.023×10^{23} mol^{-1}).

To define the fraction of excitation light an acousto-optic tunable filter (AOTF) is used.

2. Materials

1. Binding buffer: 20 mM Na-Hepes pH 7.4, 150 mM NaCl, 5 mM $MgCl_2$.
2. ConfoCor1 (EVOTEC BIOSYSTEMS, Hamburg/CARL ZEISS, Jena, Germany) or ConfoCor2 (CARL ZEISS, Jena, Germany).
3. Tetramethylrhodamine isothiocyanate (TMR); rhodamine 6 green (R6G).
4. Chamber slides (0.1 nm borosilicate glas slides) (Nunc, Wiesbaden, Germany) or highly uniform borosilicate coverslips.
5. Objective with high numerical aperture and water immersion, f. e. C-Apochromat 40x, 1.2 NA, water immersion (Carl Zeiss, Jena, Germany).
6. FCS ACCESS 2.0 (EVOTEC BIOSYSTEMS GmbH, Hamburg, Germany).
7. GraFit (ERITHACEUS SOFTWARE Ltd., Staines, UK).

3. Methods

3.1. RNA Denaturation and Renaturation

1. Incubate RNA at 70°C for 10 min in selection buffer without $MgCl_2$ (*see* **Note 2**).
2. Add $MgCl_2$ to a final concentration of 5 mM.
3. Incubate at room temperature for 30 min to allow RNA folding.

3.2. FCS Measurement

1. Switch on confocal microscope and load FCS software (*see* **Note 3**).
2. Optimise absorption and signal detection with the beampath to prevent the detection of excitation light. Define the beampath by choosing the excitation filter (543 for 543 nm laser) and the emission filter (f.e. LP 580). For other dyes keep in mind that the excitation filter depends on the excitation maximum wavelength and the emission filter depends on the emission maximum wavelength.
3. Add water with a soft plastic tip to an objective, which is suited for FCS measurements.
4. Put a coverglass at the microscope table and scroll the microscope table until the water drop is in the near of the bottom glass edge, but do not destroy the objective surface!
5. Add 20 μl water at the glass surface, where the objective centre is placed to adjust the focus.
6. Switch on the laser at 543 nm wavelength for TMR and R6G (*see* **Note 1**). For other dyes use a suited laser, depending on the excitation maximum.
7. Focus into the water drop above the coverglass by scrolling up very slowly the microscope table. The focused light should not be observed directly by microscope, which would be harmful for eyes, but indirectly with the help of a CCD camera.
   Try to orient at the reflection points that appear at the bottom and upper edge of the coverglass, when they are in focus. The two reflection points should have a distance of 0.1 μm. When focussing the reflection point of the upper glass edge, scroll the focus of the confocal detection volume 200 μm in the probe solution.
8. Disconnect laser, when pipetting probes onto the coverglass. This can be done at ConfoCor2 by carefully tilting the microscope head back. Do not switch the laser on and off with rates below 15 min. Adjust the laser power to ~ zero. Pipet 10^{-7}–10^{-6} M R6G solution onto the coverglass. Raise laser power slowly until a count rate of 200–400 kHz is reached.

9. Adjust pinhole to define a Gaussian observation volume. The detection volume can be adjusted in *x*, *y* and *z* direction. The correct pinhole adjustment is a necessary prerequisite for reproducible diffusion time measurements. It can be tested by the structure parameter *S*, which should be 5–8. For 543 nm *S* should be ~6.
10. Dilute the ligand in titration buffer to 10 nM (in general 10^{-9}–10^{-8} M). Pipet the ligand solution and 10 nM R6G solution onto the coverglass.
11. Examine photobleaching and triplet state population and adjust laser power and measurement time (*see* **Notes 5 and 6**). Measure diffusion of R6G and the ligand.
12. Mix in tubes fluorescence-labelled ligand with aptamer RNA to reach RNA concentrations of 10 nM–10 μM at a ligand concentration of 10 nM in a volume of about 100 μl. Incubate at room temperature in darkness for about 15 min.
13. For 20 μl of each mixture perform an FCS measurement of 3 × 30 s. The diffusion time should increase with each titration step until a saturation point is reached. Then stop titration. Approximately 15 measurement points should be collected.

3.3. Data Analysis

1. Fix *S* in a fitting model for one diffusion species and apply the fit to all data. An increase in diffusion time should be observed after the addition of increasing amounts of aptamer RNA. Autocorrelation functions of free and bound Moenomycin A are shown in **Fig. 7.1**.
2. Examine X^2 test and the residual deviations of the fitting results from the autocorrelation curve, which may be calculated automatically by the FCS software. Delete data, which

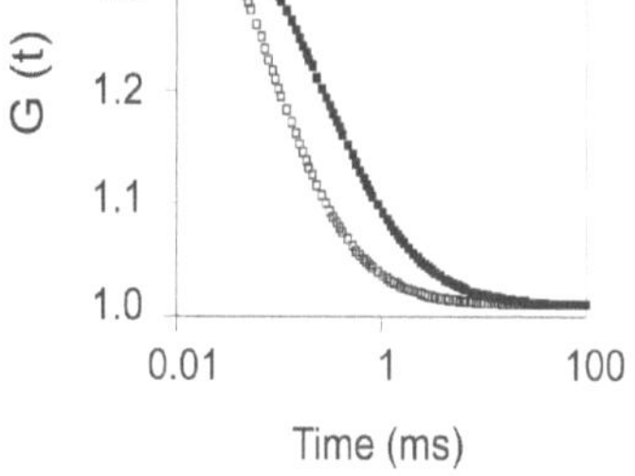

Fig. 7.1. Normalised autocorrelation functions of TMR-labelled Moenomycin A, free (*open squares*) and in complex with its corresponding RNA aptamer (*filled squares*).

show a significantly increased X^2 test result, an asymmetrically distributed residual curve, incoherent fluorescence traces or triplet fractions higher than 20% and triplet times higher than 20 μs.

3. Fix S and the diffusion time of the free fluorophore in a fitting model for two diffusion species and apply the fit to all data. Analyse the residual deviations of the autocorrelation curves, which were derived from the fit and from the measurement *(8)*. Analyse other statistical parameters, e.g. the X^2 test result. During the titration this fitting result should be more exact than in the fitting model assuming one diffusion species. At the beginning and the end of the titration the fit of the fitting model for one diffusion species should be more precisely.
4. Calculate from the fraction of complex and N the concentration of complex, using Eqs. (7.2), (7.3), and (7.4) and the R6G diffusion time. Divide complex concentration by the product of the concentrations of complex and free ligand to receive Υ. The fraction of aptamer/Moenomycin A complexes after addition of different amounts of RNA is shown in **Fig. 7.2**.
5. Fit Υ and the independent parameter RNA concentration to a hyperbolic equation, receiving the dissociation constant K_d.

$$\Upsilon = \frac{\Upsilon_{\max} \cdot [S]}{K_d + [S]} \tag{7.5}$$

The R^2 value should be at least 0.7 and may reach a maximum of 1, describing the exactness of the fit. The X^2 test result may also be used as a measure for the exactness of the fit. Compare with a fit

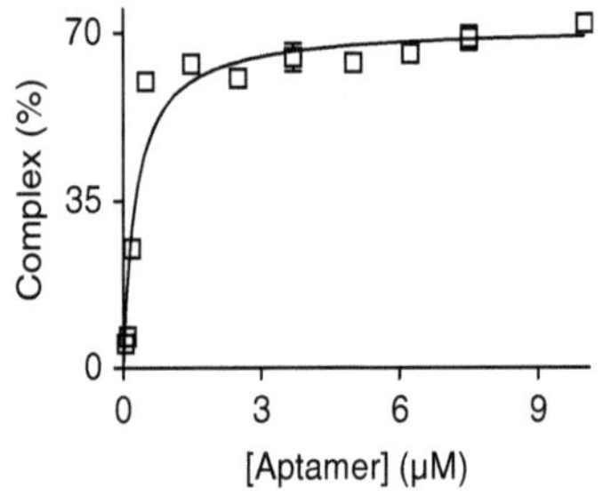

Fig. 7.2. Determination of the binding properties of aptamer/Moenomycin A interaction by FCS. The percentage of the complexes formed was calculated on the basis of the different diffusion times of free Moenomycin A (0.075 ms) and Moenomycin A/aptamer complexes (0.4 ms) using the FCS ACCESS 2.0 evaluation software (modified from *(6)*). A dissociation constant (K_d) of 437 nM can be determined.

to the Hill equation to identify whether more than one binding site is present by the hill coefficient n.

$$\Upsilon = \frac{\Upsilon_{\max} \cdot [S]^n}{[S]_{0,5}^n + [S]^n} \tag{7.6}$$

4. Notes

1. Avoid looking into the laser! Become familiar with laser safety.
2. Fluorescent probes should be stored at −20°C, protected against illumination and solubilised in the recommended medium, e.g. DMSO or water.
3. Be careful when raising the laser power, because a detector, suited for single photon counting, might be disturbed by high energy.
4. With FCS picomolar to nanomolar concentrations can be determined (N = 0.1–1,000).
5. The signal to noise ratio (S/N) is defined by the counts per molecule (cpm) *(14)*. Measurement times raise the S/N ratio, too, but to a lower extent than the cpm. Cpm and count rate should be at least 10–20 kHz. Optimal values depend on the dye and its concentration, reaching up to 100 kHz. By raising the laser power, the S/N can be optimised.
6. The limiting factors for laser power are photobleaching, saturation effects and triplet state population *(11, 15, 16)*. Triplet fraction should not be higher than 20%, triplet time should not exceed 20 μs. Photobleaching causes in a power series a significant reduction of diffusion time *(1)*. Fluorophores, which are suited for FCS, should have a high quantum yield and extinction coefficient, be photostable and show a low tendency to populate the triplet state. Many Alexa- and Atto-dyes, Cy5, tetramethylrhodamine (TMR) and rhodamine 6 green (R6G) can be used for FCS experiments. Laser power is typically between 1 and 10% AOTF for HeNe (543 and 633 nm) or Argon (488 nm) lasers. Measurement times are typically 10 × 10 s.
7. In contrast to fluorescence microscopy, for measurements with confocal microscopes, special glass coverslips are required, which show exactly a diameter of 100 nm *(16)*.
8. At low concentrations non-specific binding to the tube material may be a significant error source. For a titration, one should use special tubes with low affinity to DNA or protein. Non-specific binding of the sample to the coverglass might also occur. This could be examined by comparing the sample

concentration, which was determined in the FCS measurement with the concentration, determined by another method (Eqs. (7.2), (7.3) and (7.4)).

References

1. Bacia, K. and Schwille, P. (2007) Practical guidelines for dual-color fluorescence cross-correlation spectroscopy. *Nat. Protoc.* **2**, 2842–2856.
2. Rigler, R. and E.S. Elson, eds. (2001) *Fluorescence Correlation Spectroscopy: Theory and Applications.* Springer, Berlin.
3. Haustein, E. and Schwille, P. (2004) Single-molecule spectroscopic methods. *Curr. Opin. Struct. Biol.* **14**, 531–540.
4. Krichevsky, O. and Bonnet, G. (2002) Fluorescence correlation spectroscopy: the technique and its applications. *Rep. Prog. Phys.* **65**, 251–297.
5. Kim, S.A., Heinze, K.G. and Schwille, P. (2007) Fluorescence correlation spectroscopy in living cells. *Nat. Methods* **4**, 963–973.
6. Schürer, H., Buchynskyy, A., Korn, K., Famulok, M., Welzel, P. and Hahn, U. (2001) Fluorescence correlation spectroscopy as a new method for the investigation of aptamer/target interactions. *Biol. Chem.* **382**, 47948.
7. Vogel, S., Buchynskyy, A., Stembera, K., Richter, K., Hennig, L., Müller, D., Welzel, P., Maquin, F., Bonhomme, C. and Lampilas, M. (2000) Some selective reactions of Moenomycin A. *Bioorg. Med. Chem. Lett.* **10**, 1963–1965.
8. Meseth, U., Wohland, T., Rigler, R. and Vogel, H. (1999) Resolution of fluorescence correlation measurements. *Biophys. J.* **76**, 1619–1631.
9. Aragon, S.R. and Pecora, R. (1976) Fluorescence correlation spectroscopy as a probe of molecular dynamics. *J. Chem. Phys.* **64**, 1791–1803.
10. Rigler, R., Mets, Ü., Widengren, J. and Kask, P. (1993) Fluorescence correlation spectroscopy with high count rate and low-background: analysis of translational diffusion. *Eur. Biophy. J.* **22**, 169–175.
11. Widengren, J., Rigler, R. and Mets, Ü. (1994) Triplet-state monitoring by fluorescence correlation spectroscopy. J. Fluoresc. **4**, 255–258.
12. Widengren, J. Mets, Ü. and Rigler, R. (1995) Fluorescence correlation spectroscopy of triplet states in solution: a theoretical and experimental study. J. Phys. Chem. **99**: 13368–13379.
13. Weisshart, K., Jüngel, V. and Briddon, S.J. (2004) The LSM 510 META – ConfoCor 2 system: an integrated imaging and spectroscopic platform for single-molecule detection. Curr. Pharm. Biotech. **5**, 135–154.
14. Koppel, D.E. (1974) Statistical accuracy in fluorescence correlation spectroscopy. *Phys. Rev.* **10**, 1938–1945.
15. Eggeling, C., Widengren, J., Brand, L., Schaffer, J., Felekyan, S. and Seidel, C.A. (2006) Analysis of photobleaching in single molecule multicolor excitation and Förster resonance energy transfer measurements. *J. Phys. Chem. A* **110**, 2979–2995.
16. Enderlein, J., Gregor, I., Patra, D. and Fitter, J. (2004) Art and artefacts of fluorescence correlation spectroscopy. *Curr. Pharm. Biotech.* **5**, 155–161.

Chapter 8

Structural Probing Techniques on Natural Aptamers

Catherine A. Wakeman and Wade C. Winkler

Abstract

RNA sequences fold in a hierarchical manner to form complex structures. This folding pathway proceeds first with formation of secondary structure elements followed by the compilation of tertiary contacts. Although bioinformatics-based tools are commonly used to predict secondary structure models, it is notoriously difficult to achieve a high degree of accuracy via these approaches alone. Therefore, a diverse assortment of biochemical and biophysical techniques are regularly used to investigate the structural arrangements of biological RNAs. Among these different experimental techniques are structural probing methods, which are often times employed to determine which nucleotides for a given RNA polymer are paired or unpaired. Yet other probing methods assess whether certain RNA structures undergo dynamical structure changes. In this chapter we outline a general protocol for in-line probing, a method for analyzing secondary structure (and backbone flexibility) and describe a basic experimental protocol for hydroxyl radical footprinting as a method of investigating RNA folding.

Key words: In-line probing, riboswitch, hydroxyl radical footprinting, RNA folding, RNA secondary and tertiary structure.

1. Introduction

Riboswitches are RNA-based genetic control elements found in 5′ untranslated regions (UTRs) of the mRNA transcripts that they regulate *(1, 2)*. These RNA motifs function as direct sensors of specific metabolites in order to elicit control of gene expression. Specifically, binding of the target metabolite to the aptamer (ligand-binding) domain stabilizes an RNA conformation that either promotes or prevents downstream gene expression *(2–5)*. The three-dimensional structures of several riboswitch aptamer domains have been investigated by X-ray crystallography and NMR *(2–5)*. Most riboswitch RNA classes have also

Günter Mayer (ed.), *Nucleic Acid and Peptide Aptamers: Methods and Protocols, vol. 535*

DOI 10.1007/978-1-59745-557-2_8 Springerprotocols.com

been subjected to a variety of biochemical and biophysical experimentation. These different methods have been employed in order to: demonstrate association of metabolite ligands in the absence of accessory protein factors, provide evidence for secondary structure models, and to reveal portions of the RNA that undergo rearrangement upon ligand binding. One of the structural probing methods that has been utilized for these purposes is in-line probing. In-line probing assesses the relative flexibility of specific internucleotide linkages and is useful in demonstrating the formation of RNA secondary structure elements such as helices *(6)*.

Spontaneous cleavage of an RNA phosphodiester linkage occurs as the result of an internal nucleophilic attack by the 2′ oxygen on an adjacent phosphorus group *(6)*. The rate of this reaction depends upon the exact degree of "in-line" positioning of the 2′ oxygen, phosphorus, and 5′ leaving group oxygen atoms of a given RNA internucleotide linkage *(7–10)*, which is required for a productive nucleophilic attack by the 2′ oxygen. RNA linkages for nucleotides engaged in stable base-pairs exhibit rates of spontaneous cleavage that are substantially lower than for nucleotides that reside in relatively unstructured regions *(6, 9)*. Therefore, the rate at which spontaneous cleavage occurs for phosphodiester linkages is decidedly dependent upon the secondary and tertiary structure of the overall RNA. In-line probing of a given RNA can be used to provide evidence for RNA structural models that have been predicted by other means. Additionally, in-line probing of riboswitch RNAs in the presence and absence of metabolite ligands can reveal ligand-induced structural changes. When repeated at a range of ligand concentrations, this method can also be used to determine the apparent dissociation constants (K_D) for RNA–ligand interactions. Of course, application of in-line probing is not limited to metabolite-binding RNAs. Indeed, the probing method should be generally applicable for analysis of RNA structure and the study of RNA–ligand interactions.

In the past, experimentation similar to in-line probing (as described herein) has been interpreted as demonstrating that binding of magnesium to specific RNA sites leads to high rates of spontaneous cleavage for closely adjacent RNA backbone linkages (to Mg^{2+} binding sites). Indeed, other metals such as europium (Eu^{3+}) have been added to RNAs in vitro and positions of RNA cleavage have been interpreted as denoting specific sites of europium (Eu^{3+}) association (e.g., *11*). We recommend using caution when arriving at these conclusions given that cleavage patterns resulting from europium (Eu^{3+}) or by in-line probing in the absence of magnesium can closely resemble one another (Wakeman CA, Winkler WC, unpublished data). Also, sites of spontaneous cleavage observed by in-line probing reactions (conducted in the presence of magnesium) do not necessarily correlate well with highly occupied magnesium-binding pockets identified by structural methods such as X-ray crystallography. In contrast, the in-line character of RNA linkages in three-dimensional structures agrees well with the relative rates of RNA cleavages observed during in-line probing *(6)*. Additionally, magnesium can be removed from in-line probing reactions with a minor influence on spontaneous cleavages *(6)*. Indeed, in prior studies our laboratory has titrated magnesium into in-line probing reactions while maintaining high monovalent concentrations in order to characterize a particular magnesium-sensing riboswitch RNA *(12)*. A detailed discussion on any potential correlation between metal-binding sites and cleavages at specific RNA linkages, a subject that is not without debate, is beyond the scope of this article.

Every experimental technique exhibits particular strengths and weaknesses. Correspondingly, there are many different enzymatic and chemical probing techniques available to the interested RNA biochemist. For example, an RNA of interest can be subjected to partial digestion by enzymes that exhibit specific substrate preferences, such as cleavage of single-stranded nucleotides (e.g., RNase S1), cleavage of base-paired and stacked nucleotides (e.g., RNase V1) or cleavage of unpaired Gs (e.g., RNase T1) *(13)*. These enzymatic reactions can therefore assist in the identification of helical or unpaired nucleotides. Alternatively, chemical agents can be employed for structural probing purposes, such as cleavage of single-stranded regions during lead probing *(14, 15)*. Other chemical agents (e.g., methylation of guanine N1 and N2 positions by kethoxal) modify specific nucleobases only when they are unpaired or accessible. Yet another powerful method for analyzing RNA structure and function is through nucleotide analog interference mapping *(16)*, which is a method of rapidly investigating the effect(s) of substituting specific nucleotide functional groups.

Most of the abovementioned methods assist primarily in the determination of RNA secondary structure or in the mapping of individual tertiary contacts. However, one of the most frequently used methods for studying RNA folding is via hydroxyl radical footprinting *(17–19)*. For these reactions, hydroxyl radicals individually react with ribose sugars to trigger cleavage of the RNA backbone. Hydroxyl radical footprinting is not capable of detecting the formation of secondary structures because both single- and double-stranded RNA is susceptible to attack by hydroxyl radicals *(20)*; however the formation of solvent inaccessible regions found in tertiary RNA folds can be detected using this method *(21)*. In general, low concentration (~nanomolar) of DNA or RNA is labeled at either the 5' or 3' terminus with ^{32}P and the products from hydroxyl radical-mediated cleavage are resolved by denaturing gel electrophoresis. Hydroxyl radicals exhibit short lifetimes and travel small distances. Reactivity of the RNA backbone to hydroxyl radicals is thereby dependent upon accessibility to the bulk solvent. Therefore, conformational changes that promote formation of a solvent-protected core lead to decreased reactivity to hydroxyl radicals for RNA positions within the protected region (i.e., 'footprints'). Conversely, RNA regions that exhibit increased reactivity are likely to have become more solvent exposed during conformational changes, such as being moved to the outer surface of the overall global fold. Typically, hydroxyl radical footprints are observed for the interior portion of closely packed helical segments for structured RNAs. Therefore, in general, hydroxyl radical footprinting is an effective means of investigating RNA folding pathways and ligand-induced conformational changes.

2. Materials

2.1. Preparation of RNA by In Vitro Transcription

1. T7 RNA polymerase – store at –20°C (*see* **Note 1**).
2. DNA template. This template can be generated through PCR-amplification using a forward primer that incorporates the T7 promoter sequence (TAATACGACTCACTATA*GGG*).
3. 10X Transcription buffer: 300 mM Tris–HCl, pH 8.0, 100 mM DTT, 1% Triton X-100, 1 mM spermidine, 400 mM $MgCl_2$.
4. 25 mM NTP mix: 25 mM rATP, rCTP, rGTP, and rUTP (Roche).
5. Optional: Yeast inorganic pyrophosphatase (Sigma) (*see* **Note 2**).
6. 2X Urea loading buffer: 10 M urea, 1.5 mM EDTA, pH 8.0, 10 mM Tris–HCl, pH 8.0, 0.05% bromophenol blue, 0.05% xylene cyanol.
7. Crush-soak solution: 200 mM NaCl, 10 mM Tris–HCl, pH 7.5, and 1 mM EDTA. Filter-sterilize or autoclave the solution.
8. This section assumes the use of fluor-coated thin layer chromatography (TLC) plates and a hand held device for shortwave UV light (254 nm).

2.2. Radioactive-Labeling of RNA at the 5′ Terminus

1. Approximately 10–50 pmol synthetic RNA per ~13.3 pmol ATP [γ-^{32}P].
2. Calf intestinal alkaline phosphatase (CIP, New England Biolabs).

3. T4-polynucleotide kinase (PNK, New England Biolabs).
4. Adenosine 5′-triphosphate (ATP) [γ-^{32}P]: 6,000 Ci/mmole on reference date (Amersham).
5. 5X Kinase buffer: 25 mM $MgCl_2$, 125 mM CHES pH 9.0, 15 mM DTT (*see* **Note 3**).
6. Phenol/Chloroform/Isoamyl alcohol 25:24:1 (v/v).
7. Chloroform.
8. Glycogen 20 mg/ml.
9. 3 M Sodium acetate pH 5.2.
10. 2X Urea loading buffer: 10 M urea, 1.5 mM EDTA, pH 8.0, 10 mM Tris–HCl, pH 8.0, 0.05% bromophenol blue, 0.05% xylene cyanol.
11. Crush-soak solution: 200 mM NaCl, 10 mM Tris–HCl pH 7.5, and 1 mM EDTA.
12. Autoradiography film.

2.3. Hydroxyl Radical Footprinting

1. 5′-terminus radiolabeled RNA – store at –20°C. Note that 3′-terminus radiolabeled RNA could also be used as substrates for these reactions, although methods of 3′-labeling are not specifically discussed herein.
2. 10X Footprinting buffer: 210 mM HEPES, pH 7.4, 20 mM $MgCl_2$ (*see* **Notes 4 and 5**) – store at –20°C.
3. Yeast tRNA 2 μg/ml (Sigma) – store at –20°C.
4. Ammonium iron(II) sulfate hexahydrate (F3754, Sigma)
5. Fe(II)–EDTA solution: 20 mM EDTA pH 8.0, 10 mM ammonium iron(II) sulfate – prepare fresh.
6. (+)-sodium L-ascorbate (A7631, Sigma), dissolved in H_2O to 50 mM – store at –20°C.
7. 1% or 0.03% H_2O_2, diluted from commercially available stocks – make fresh. (*see* **Note 6**).
8. 3X Formamide loading buffer: 95% formamide, 20 mM EDTA pH 8.0, 0.05% bromophenol blue, 0.05% xylene cyanol – store at –20°C.
9. RNase T1 (Sigma) diluted to 4 units/μL in H_2O.
10. 10X T1 buffer: 0.25 M sodium citrate pH 5.0.
11. 10X OH buffer: 0.5 M Na_2CO_3 pH 9.0, 10 mM EDTA pH 8.0.
12. Glycogen 20 mg/ml.
13. 3 M Sodium acetate pH 5.2.
14. 2X Urea loading buffer: 10 M urea, 1.5 mM EDTA pH 8.0, 10 mM Tris–HCl pH 8.0, 0.05% bromophenol blue, 0.05% xylene cyanol.

2.4. In-Line Probing

1. 5′-terminus radiolabeled RNA – store at –20°C. Note that 3′-terminus radiolabeled RNA could also be used as substrates for these reactions, although methods of 3′-labeling are not discussed herein.
2. 2X in-line buffer: 100 mM Tris–HCl pH 8.3, 200 mM KCl, 40 mM $MgCl_2$ (*see* **Notes 4 and 7**).
3. 2X Urea loading buffer: 10 M urea, 1.5 mM EDTA pH 8.0, 10 mM Tris–HCl pH 8.0, 0.05% bromophenol blue, 0.05% xylene cyanol.
4. RNase T1 (Sigma) diluted to 4 units/μL in H_2O – store at 4°C.
5. 10X T1 buffer: 0.25 M sodium citrate pH 5.0.
6. 10X OH buffer: 0.5 M Na_2CO_3 pH 9.0, 10 mM EDTA pH 8.0.

2.5. Polyacrylamide Gel Electrophoresis (PAGE)

1. 37% Acrylamide/bis-acrylamide 29:1 (w/w) [add H_2O to 37% w/v], e.g., 89.5125 g acrylamide, 3.0875 g bis-acrylamide, and bring up to a final volume of 250 mL with H_2O. Dissolve completely and filter through Whatman paper. Store in amber bottles at 4°C. (*see* **Note 8**).
2. Urea.
3. 10X TBE: 108 g Tris, 55 g boric acid, and 3.725 g EDTA in 1 L H_2O.
4. Running buffer: 1X TBE (10X TBE diluted 1:10 in H_2O).
5. Ammonium persulfate (APS). Make a 10% solution (w/v) with H_2O – store at 4°C for up to a month.
6. TEMED (*N*,*N*,*N*′,*N*′-tetramethylethylenediamine, National Diagnostics) – store at 4°C.
7. This section assumes the use of glass plates that are 32.5 × 41 cm with 0.75 mm spacers and 24 well combs are used for hydroxyl radical footprinting, in-line probing, and SHAPE gels. We typically use 28 × 16.5 cm glass plates with 0.75 mm spacers and 4–8 well combs for RNA preparative techniques (*see* **Note 9**).
8. Vacuum pump and gel dryer.
9. Whatman paper – 3 mm chromatography paper.
10. 35 cm × 43 cm Phosphor screens (Amersham).
11. Phosphor imaging instrumentation (e.g., Typhoon 9200 Variable Mode Imager; Molecular Dynamics).

3. Methods

In hydroxyl radical footprinting, radioactively labeled RNA is exposed to hydroxyl radicals to induce strand breakage at riboses in a sequence-independent manner. Only regions of RNA located with a solvent inaccessible core will remain intact. Hydroxyl radicals attack the C4′ position of the sugar, resulting in cleavage of the RNA backbone. Hydroxyl radicals can be produced via radiolysis of water by high-energy radiation *(22)* or an X-ray synchrotron beam *(23)*. Hydroxyl radicals can also be generated by the disproportionation of peroxynitrous acid to •OH and nitrogen dioxide at neutral pH *(24)*. Finally, hydroxyl radicals can be generated by Fenton chemistry by disproportionation of hydrogen peroxide to hydroxyl radical, catalyzed by Fe–EDTA. Ferrous iron is chelated to EDTA to prevent direct binding to the negatively charged nucleic acids. Upon oxidation of H_2O_2, Fe(II)–EDTA is converted to Fe(III)–EDTA; therefore, ascorbate is provided in the reaction to regenerate the Fe(II)–EDTA *(21)*.

In-line probing is a method that takes advantage of the inherent instability of RNA. Due to the presence of the 2′-hydroxyl group, RNA molecules are susceptible to intramolecular transesterfication reactions which involve nucleophilic attack of the 2′-hydroxyl group on the nearby phosphorus center of the phosphodiester bond. This attack results in a 5′ cleavage fragment with a 2′,3′-cyclic phosphate and a 3′ cleavage fragment with a 5′ hydroxyl terminus. The likelihood of this reaction occurring is determined in part by the positioning of the chemical groups involved. Maximal rates are achieved when the 2′ oxygen, the phosphorus center, and the 5′ oxygen leaving group form a perfect 180°C. Flexible regions of RNA molecules are free to sample multiple conformations while structured regions are locked into place. Therefore flexible regions are more capable of randomly adopting near perfect in-line conformations. This results in higher rates of spontaneous cleavage for flexible regions of the RNA *(6)*.

3.1. Preparation of RNA by In Vitro Transcription

1. For a yield of >200 pmols of RNA, combine ~20–50 pmol DNA template, 2.5 μL 10X transcription buffer, 2.5 μL 25 mM NTP mix, ~50 μg ml^{-1} T7 RNA polymerase, and 0.0025–0.01 U inorganic pyrophosphatase (optional) in a final volume of 25 μL. These reactions can be scaled appropriately for recovery of the desired quantity of synthetic RNA The reaction should be incubated for 2–3 h at 37°C (*see* **Note 10**) and terminated with the addition of an equal volume of 2X urea loading buffer. (*see* **Note 11**) The transcription reaction can be stored at –20°C at this point.
2. The RNA should be purified on a denaturing 6–10% polyacrylamide gel (as detailed in **Section 3.5**) (*see* **Note 12**). The polyacrylamide percentage should be chosen as deemed appropriate for the length of the target RNAs.
3. Once the bromophenol blue has run ~2/3 the length of the plates, the gel can be turned off. At this point, the glass plates containing the gel can be removed from the gel rig. To remove the gel from the glass plates, slide out the spacers, lay the glass

plates flat on the bench top, and carefully pry them apart. The gel will typically preferentially stick to one plate. Flip the sandwich over so that the plate to which the gel is sticking is on the bottom and continue with the separation. Once one plate is removed, place plastic wrap over the exposed side of the gel. Flip the gel and peel off the remaining glass plate so that the gel sticks to the plastic wrap and then cover the other side of the gel with plastic wrap.

4. The synthetic RNA can then be visualized by UV shadowing. Place the gel sandwich over a TLC plate and expose to short-wave UV light. UV-absorbing material such as RNA polymers and free nucleotides will appear as dark shadows. Outline the top-most shadow with a fine-point marker as it should correspond to the target RNA. The lowest migrating band will likely represent the free nucleotides.
5. Excise the circled region with a razor blade, remove the outer layer of plastic wrap, and cut the gel slice into ~1 mm squares.
6. Place the gel bits into a 1.5 mL microcentrifuge tube and add approximately two volumes of crush-soak solution (typically ~400–600 μL). Incubate on a tube rotator at room temperature for 2 h or at 4°C overnight. Remove and save supernatant.
7. Ethanol precipitate the RNA by adding ~2.5 volumes cold 100% ethanol and incubating at -20°C for 30 min. Pellet the RNA by centrifuging at 20,000 × *g* for 15 min. Wash pellet with 200 μL 70% ethanol and centrifuging at 20,000 × *g* for 5 min. Carefully remove supernatant and dry the pellet via exposure to air for 1–5 min or by speedvac.
8. Resuspend the RNA in ~20–60 μL H_2O and quantify RNA yield via A_{260} measurements and calculation of extinction coefficient values. Store at -20°C until use.

3.2. Radioactive-Labeling of RNA at the 5′ Terminus

1. Prior to radioactive labeling, the RNA must first be dephosphorylated at the 5′ terminus. Combine ~10–40 pmols RNA and the commercial buffer for CIP in a 10 μL final volume and incubate at 50°C for 15 min. An alternative buffer for these reactions is 500 mM Tris–HCl, 1 mM EDTA, pH 8.5.
2. The CIP enzyme should then be removed by phenol/chloroform/isoamyl alcohol extraction. To accomplish this, bring the volume of the reaction up to 200 μL with H_2O. Add 200 μL phenol/chloroform/isoamyl alcohol and shake or vortex for ~5 s. to mix completely. Centrifuge at 20,000 × *g* for 5 min to separate into two phases. Remove the top, aqueous phase, which contains the RNA, and discard the bottom phase containing the protein.

Repeat this procedure with 200 μL of pure chloroform to remove traces of phenol.

3. Concentrate the RNA via ethanol precipitation. Add 1/10 volume (20 μL) 3 M sodium acetate and 1 μL glycogen to the RNA and mix (*see* **Note 13**). Add 2.5X volume (500 μL) 100% ethanol, mix by inversion, and incubate at -20°C for 30 min. Pellet the RNA by centrifuging at 20,000 × *g* for 15 min. The pellet should then be washed with the addition of 200 μL 70% ethanol and centrifuged at 20,000 × *g* for 5 min. Discard supernatant, air dry or speedvac the pellet for 2–5 min and resuspend the RNA in 10 μL H_2O.
4. Once the RNA has been dephosphorylated, it can be radioactively labeled using T4 PNK. For every 20 μL kinase reaction, use 5 μL of the CIP treated RNA, 4 μL 5X kinasation buffer, 4–12 μL ATP [γ-^{32}P], and 2 μL T4 polynucleotide kinase (PNK) at 10U/μL. Incubate the reaction for 35 min at 37°C.
5. The end-labeled RNA can be resolved on a 6% polyacrylamide gel for nucleic acids greater than 75 nucleotides and 10% polyacrylamide gel for nucleic acids less than 75 nucleotides (as detailed in **Section 3.5** and **Step 3** of **Section 3.1**). Prior to discarding, buffers in the upper and lower reservoirs of the gel rig should be checked for radioactivity.
6. Once the gel is wrapped in plastic wrap, the radioactively labeled RNA bands can be identified via exposure to autoradiography film. The gel should be secured inside the cassette so that it cannot move and can be reproducibly positioned against the autoradiography film. A sheet of autoradiography film should be exposed to the gel for ~1 min and developed. Outline the region of the gel sandwich that contains radiolabeled RNA as identified by dark band(s) on the autoradiographic film.
7. Excise the RNA from the gel using the procedure detailed in **Section 3.1 Steps 5–8**.

3.3. Hydroxyl Radical Footprinting

1. In a 7 μL final volume, combine ~1 nM radioactively labeled RNA (~150–200 kcpm), 1 μL 10X footprinting buffer, 1 μg carrier yeast tRNA, and the desired concentration of $MgCl_2$ or other substance to be tested in a 1.5 mL microcentrifuge tube (*see* **Note 14**). Incubate the reaction for 5 min at 37°C. This will allow the RNA to equilibrate with compounds that may affect global conformation (e.g., magnesium) and fold into the desired conformation.
2. Prior to initiating the hydroxyl radical footprinting reaction, place 1 μL of freshly prepared H_2O_2 solution inside the lid of the microcentrifuge tube (*see* **Note 6**). Additionally, prepare a 1:1 mixture of the Fe–EDTA solution and sodium ascorbate

solution and place 2 μL onto a separate spot on the inside of the microcentrifuge tube lid.

3. Initiate the production of hydroxyl radicals with a quick pulse spin to combine all of the separate solutions and incubate the reaction at 37°C for two additional minutes.
4. Precipitate the RNA with ethanol to stop the reaction. Add 170 μL H_2O, 2.5X (500 μL) of 100% ethanol, 1 μL glycogen, and 1/10 volume (20 μL) 3 M sodium acetate and incubate at –20°C for 30 min. Pellet the RNA by centrifugation at 20,000 × *g* for 15 min. The pellet should be washed in 200 μL 70% ethanol and centrifuged at 20,000 × *g* for 5 min. Discard supernatant and dry the pellet for 2–5 min.
5. The pellet should be resuspended in 10 μL H_2O and 5 μL 3X formamide loading buffer. At this point the reaction can be stored at –20°C for up to a week prior to resolving the reactions on a 10% polyacrylamide gel alongside size marker ladders. There are many different size markers that could be used. Our laboratory typically includes lanes containing RNAs that have been partially digested with RNase T1 to visualize guanosine residues and a lane that includes RNAs that were briefly exposed to increased pH and high temperature in order to generate partial cleavages at all nucleotide positions. We refer to these reactions herein as 'T1' and '–OH' ladders.
6. Prepare the T1 ladder by mixing 1 μL radioactively labeled RNA (~100 kcpm and similar in quantity to the experimental lanes) with 1 μL RNase T1 (1 U/μL) and 1 μL 10X T1 buffer and bring the volume up to 10 μL with 2X urea loading buffer. Incubate at 50°C for 20 min. Add 3 μL 2X urea loading buffer and 7 μL H_2O and store at –20°C prior to running on a gel.
7. Prepare the –OH ladder by mixing ~100 kcpm radioactively labeled RNA with 1 μL of 10X OH buffer and bring the volume up to 10 μL with H_2O. Incubate at 95°C for 3–8 min to induce scission after every base (the appropriate time interval required for these reactions will need to be optimized per target RNA). Stop reaction with 10 μL 2X urea loading buffer and immediately store at –20°C.
8. Resolve the hydroxyl radical footprinting reactions and size marker ladders by 6–10% polyacrylamide gel alongside a control lane containing ~150 kcpm radioactively labeled RNA that was not exposed to hydroxyl radicals (*see* **Section 3.5**).
9. Results should resemble those shown in **Fig. 8.1C**. When analyzing the data, consider bands and dark regions on the gel to be regions of RNA that are exposed to the solvent and therefore susceptible to backbone scission by hydroxyl radicals. Clearings on the gel are considered to be regions of RNA that are located within a solvent inaccessible core and therefore

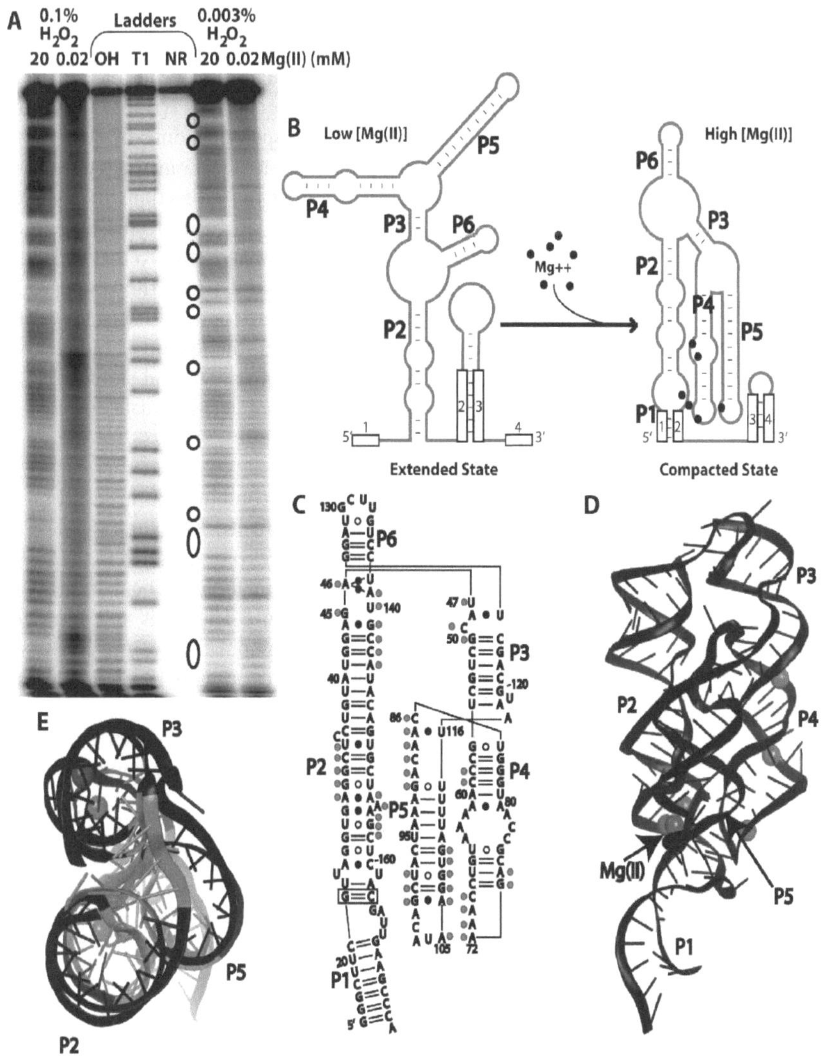

Fig. 8.1. Hydroxyl radical footprinting of the magnesium-binding M-box riboswitch found in the 5′ untranslated region (UTR) of the *Bacillus subtilis ykoK(mgtE)* gene. (**A**) Hydroxyl radical footprinting reactions resolved on a 10% denaturing polyacrylamide gel. The reactions were performed under magnesium concentrations both above and below the level required to induce compaction of the RNA. Sets of reactions using a final concentration of either 0.1% or 0.003% H_2O_2 have been included for comparison. Individual bands can be easily resolved and regions of protections and de-protections can be scored through the use of quantitative line trace comparisons of each individual lane. Protections have been marked on the gel by a series of *open circles*. T1 and OH lanes are ladders used for mapping probing changes onto the RNA sequence. The identity of each nucleotide in the footprinting reactions is shifted one band lower relative to the T1 ladder *(18)*. The NR lane is non-reacted RNA to demonstrate the quality of the RNA prior to incubation of the in-line reactions. (**B**) Line schematic showing the predicted structural rearrangement of the M-box RNA upon magnesium binding in which the RNA transitions from a secondary structure-dominated extended state in low magnesium concentrations to a compacted state containing extensive tertiary contacts under high magnesium conditions. *Black circles* represent magnesium ions. Helices within the aptamer domain are labeled P1–P6. The *numbered boxes* highlight a significant secondary structural rearrangement that

protected from cleavage by hydroxyl radicals. The location of these regions can be mapped onto the RNA sequence using the T1 and OH ladders as shown in **Fig. 8.1C**. The T1 ladder will reveal the location of all of the guanosine residues within the RNA while the OH ladder displays banding of every base, allowing the number of nucleotides separating each guanosine to be counted. When mapping the protections in hydroxyl radical footprinting, the identity of the protected nucleotide will be shifted one nucleotide lower relative to the T1 ladder because hydroxyl radical cleavage destroys the ribose while nuclease cleavage leaves the nucleotides intact *(25)*. The non-reacted RNA demonstrates the quality of the RNA prior to treatment with hydroxyl radicals and should display minimal to no banding. For more detail on data analysis, *see* **Section 3.6**.

3.4. In-Line Probing

1. In a final volume of 10 μL, combine 2X in-line buffer, ~75–200 kcpm radioactive 5′-labeled RNA, and the desired amount(s) of any other substance to be included in the assays (e.g., RNA-binding protein or metabolite). Incubate at room temperature for ~40 h (*see* **Note 15**). This will allow for spontaneous cleavage of a subpopulation of the RNAs via single-hit kinetics. Stop the reaction with 10 μL of 2X urea loading buffer and store at –20°C until resolution by denaturing gel electrophoresis alongside size marker ladders. There are many different size markers that could be used. Our laboratory typically includes lanes containing RNAs that have been partially digested with RNase T1 to visualize guanosine residues and a lane for RNAs that were briefly exposed to increased pH and high temperature in order to generate a ladder for cleavages at all nucleotide positions. We refer to these size marker control reactions herein as 'T1' and '–OH' ladders.
2. Follow **Steps 6 and 7** of **Section 3.3** to prepare the T1 and OH ladders.

Fig. 8.1 (continued) occurs upon the formation of magnesium-induced tertiary contacts. (**C**) Sequence of the *B. subtilis* M-box RNA aptamer arranged to reflect the magnesium-bound compact state of the RNA. Hydroxyl radical protections are denoted by *gray circles*. AU base pairs are shown as *single lines* connecting nucleotides while GC base pairs are shown as double lines. GU base pairs are denoted by *open circles* while all other non-canonical base pairs are denoted by the *black filled circles*. (**D**) Side view of the crystal structure of the *B. subtilis* M-box RNA aptamer in the magnesium-bound state supporting the magnesium-bound compacted state drawn in part B *(12)*. (**E**) *Top down* view of the magnesium-bound M-box RNA three-dimensional structure demonstrating that the protected nucleotides fall within a solvent inaccessible core formed by the packing of three parallel helices. Protected nucleotides are shown in *light gray*.

3. Resolve the in-line probing reactions and size marker ladders by denaturing 6–20% polyacrylamide gel electrophoresis (the most appropriate polyacrylamide percentage should depend on the size of the target RNAs) next to ~100 kcpm non-reacted radiolabeled RNA in 10 μL 2X urea loading buffer (NR, *see* **Section 3.5.**).
4. Results from these assays should resemble representative reactions shown in **Fig. 8.2**. When analyzing the data, consider bands to be regions of the RNA that are flexible/unstructured

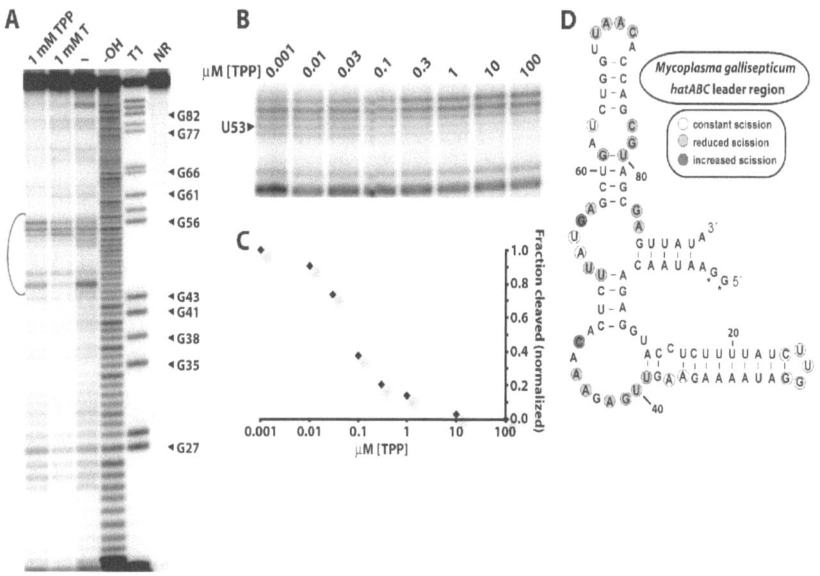

Fig. 8.2. In-line probing of the thiamine pyrophosphate (TPP)-binding riboswitch aptamer of the *Mycoplasma gallisepticum hatABC* leader region. (**A**) In-line probing reactions in the presence or absence of thiamine (T) or TPP, the preferred ligand, have been resolved on a 10% denaturing polyacrylamide gel. The RNA undergoes a significant conformational change upon ligand-binding as evidenced by a number of bands changing in intensity with the addition of substrate. The *gray curved line* denotes the region of this RNA that is quantified in subsequent parts of this figure. T1 and OH lanes are ladders used for mapping probing changes onto the RNA sequence. Bands in the T1 lane result from partial digestion with RNase T1 and represent guanosines within the RNA sequence. Bands in the OH lane represent cleavage after every nucleotide. Some of the guanosine bands have been labeled to assist in mapping banding changes onto the RNA sequence provided in part D. The NR lane is non-reacted RNA to demonstrate the quality of the RNA prior to incubation of the in-line reactions. (**B**) Enlarged image of the portion of an in-line probing gel, corresponding to the marked region in part A, in which increasing levels of TPP have been added to the reactions. The band chosen for quantification represents spontaneous scission at U53 of the *M. gallisepticum* TPP-binding aptamer. (**C**) A *line graph* showing the quantification of the intensity of the U53 position as increasing levels of TPP are titrated into the in-line probing reactions. The intensities of this band have been normalized to an unchanging band to account for loading variation in the gel. (**D**) The probing changes induced by the addition of 1 mM TPP as shown in part A of this figure have been mapped onto the RNA secondary structure and sequence. *Open circles* indicate regions of the RNA displaying a constant level of scission in all conditions, *light gray circles* indicate regions of the RNA displaying decreasing levels of scission upon addition of TPP, and *dark gray circles* indicate regions of the RNA displaying increasing levels of scission.

such as loops and bulges. Clearings on the gel can be considered regions of RNA, such as helices, that are structurally constrained (*see* **Note 16**). The location of these regions can be mapped onto the RNA sequence using the T1 and OH ladders as shown in **Fig. 8.2D**. The T1 ladder will reveal the location for guanosine residues within the RNA while the OH ladder displays bands for every nucleotide. The non-reacted RNA demonstrates the quality of the RNA prior to treatment with hydroxyl radicals and should display minimal to no banding. For more detail on data analysis, *see* **Section 3.6**.

3.5. Polyacrylamide Gel Electrophoresis

1. Prepare 10% gel solution. In a 500 mL final volume, combine 240 g urea, 50 mL 10X TBE, 135 mL 37% acrylamide/bis-acrylamide solution. Once the powder is dissolved, filter the solutions through Whatman paper. This solution can be stored at room temperature for up to a month. Adjust the volume of acrylamide/bis-acrylamide solution to achieve the desired polyacrylamide percentage.
2. To initiate polymerization of 100 mL of PAGE solution, gently add and mix 0.8 mL 10% APS and 0.04 mL TEMED, pour immediately, and slide in the comb. Allow the gel to polymerize for >30 min.
3. Gently remove the comb and rinse out the wells.
4. Assemble the gel electrophoresis rig and fill the upper and lower reservoirs with running buffer.
5. Pre-run the gel for 15 min. For glass plates that are 32.5 × 41 cm with 0.75 mm spacers we typically conduct electrophoresis at constant 60 W. For glass plates that are 28 × 16.5 cm with 0.75 mm spacers we typically electrophorese samples at constant 40 W.
6. Prior to loading the samples, rinse the wells thoroughly with running buffer.
7. For probing reactions, continue electrophoresis until the bromophenol blue indicator dye is ~1 in. from the bottom of the plate. For gels employed for preparative purposes we typically continue electrophoresis until the bromophenol blue dye has run ~2/3 the length of the plate, in order to retain free nucleotides within the gel.
8. For gels that are to be dried, remove one of the glass plates, allowing the gel to remain attached to the second plate. Press a sheet of Whatman paper on top of the exposed gel. The gel will adhere to the Whatman paper and can then be peeled away from the remaining glass plate. Cover the exposed side of the gel with plastic wrap and place in a gel dryer under vacuum pressure at 80°C for 2–3 h.

9. Expose the dried gel to a phosphor screen, which should then be scanned via phosphor imaging instrumentation.

3.6. Analysis

1. This section assumes use of software resembling ImageQuant (Molecular Dynamics).
2. Each lane of the gel should contain a distinctive banding pattern. Conformational changes in the RNA will be indicated by alterations in the overall banding pattern. Bands may darken, lighten, or disappear below detection. Dramatic banding changes can be easily observed by eye (*see* **Figs. 8.1** and **8.2**). Subtleties in these changes can be observed by examining the relative intensity of a line that cross-sections the lane. Specifically, a line trace can be drawn over the desired region of the gel and graphed, thereby generating a lane profile. Peaks and valleys correspond to bands and cleared regions, respectively. These data can be exported to spreadsheet analyses software such as Excel or SigmaPlot and carefully plotted and analyzed. Similar line traces copied onto control lanes, such as the T1 and –OH reactions, can be used to correlate line trace data to the overall RNA sequence. The line profile data can be normalized by converting the region of the each lane with the highest counts to 1 and the region with the lowest counts to 0.
3. Alternatively, a box can be drawn around individual bands and the relative intensity for the area within can be obtained by standard analyses by software such as ImageQuant. A similarly sized control box for background subtraction should be placed on an area of the gel that lacks obvious bands.
4. These assays can also be useful for characterization of ligand–RNA interactions by setting up reactions with a range of ligand concentrations. To account for subtle differences in loading, the relative intensity for the area within an individual box can be divided by the relative intensity for the area within a boxed region that encompasses the entire lane. If the lower ligand concentrations and upper ligand concentrations are below and at ligand saturation, respectively, these experimental data may be used for estimation of EC_{50}values or estimates of cooperativity. For this type of analysis, multiple individual bands that display increased and decreased intensity in response to ligand interactions or conformational changes should be directly compared with one another. The easiest and most rapid method is to normalize each box series to the boxed band with the highest and lowest intensity measurements. If these values are normalized to maximal and minimal values of 1 and 0, respectively, then multiple band series can be compared to one another despite potentially significant differences in their overall relative intensity. A rapidly accepted alternative to these types of data

analyses is a very useful software program called SAFA (Semi-Automated Footprinting Analysis) *(26)*. The latter software package significantly shortens the gel quantification process while reducing systematic error introduced during data analysis.

4. Notes

1. Other RNA polymerases can be used in the place of T7 RNA polymerase through incorporation of their individual promoter preferences onto DNA templates. For example, SP6 can also be used for the production of RNA by in vitro transcription.
2. While not necessary for in vitro transcription to occur, inorganic pyrophosphatase can be a useful addition to improve the yield of RNA. During transcription, pyrophosphates will be released into solution and chelate Mg^{2+}. When too much Mg^{2+} is bound by the pyrophosphates, function of RNA polymerase will be reduced.
3. The formation of structure near the 5′ portion of an RNA molecule can interfere with the ability of PNK to phosphorylate the 5′ termini. Use of CHES buffer (5X = 25 mM $MgCl_2$, 125 mM CHES pH 9, 15 mM DTT) with a higher pH can be useful for removing oligonucleotide structure. A short 2 min denaturation of the RNA substrate at 80–95°C can also be included prior to the kinase reaction to modestly improve end-labeling efficiency of RNAs with structured 5′ termini. The 10X buffer supplied by NEB can be used when structure is not a problem. One should also be aware that radiolysis of labeled RNAs can be a source of background noise in probing reactions. We recommend storing 5′-radiolabeled RNA at <100 kcpm/μl to reduce this problem.
4. It is often desired to vary $MgCl_2$ concentration in these experimentation in order to study Mg(II)-dependent folding pathways. This is particularly true for hydroxyl radical footprinting.
5. Tris buffer should not be used for hydroxyl radical footprinting because it reduces RNA cleavage by hydroxyl radicals likely by acting as a free-radical scavenger *(20)*. Sodium cacodylate buffer, pH 7, is another alternative to HEPES as described herein. Polyhydroxylated compounds such as glycerol, alcohols and sugars are also scavengers of hydroxyl radicals and should be avoided. Even glycerol concentrations as low as 0.5% have been observed to inhibit hydroxyl-mediated cleavage of DNA *(21, 27)*.

6. Either 1% or 0.03% H_2O_2 solution can be used. In our experience, if 1% H_2O_2 solution is used, clearings indicative of solvent inaccessible regions of RNA are more dramatically visualized. If 0.03% H_2O_2 solution is used, individual bands can be more easily resolved and regions of induced protection and deprotection can be distinguished (*see* **Fig. 8.2A**). While hydroxyl radical footprinting is typically used to report the induction of protections due to the formation of a solvent inaccessible core, deprotections have also been reported in the literature *(28)*.
7. If high monovalent ions do not disrupt the molecular interactions that one is studying, in-line probing reactions can be repeated in the presence of 2 M monovalents. Under these conditions, $MgCl_2$ concentration can be eliminated from the reactions or varied as desired with no effects upon the general efficiency of backbone self-cleavage. Therefore, in-line probing can be used for investigation of divalent metal-dependent folding pathways, as long as high concentrations of monovalents are included, which are included to outcompete the loosely associated 'atmosphere' of divalent ions *(29, 30)*. At lower monovalent concentrations, variations in $MgCl_2$ can lead to overall fluctuations in banding intensities for all bands within the individual lanes.
8. Acrylamide is neurotoxic until polymerized. A dust mask, gloves, and eye protection should be worn when handling powdered acrylamide.
9. Upon purchase, the outside of the glass plates should be permanently marked so that proper orientation of the plates can be maintained in all subsequent runs.
10. If a white precipitate forms, transcription efficiency is likely reduced due to accumulation of Mg^{2+}-chelated pyrophosphate, and the reaction can be stopped prior to the full 2–3 h incubation. The addition of inorganic pyrophosphatase should prevent the formation of this precipitate and increase the RNA yield.
11. Alternatively, the transcription reaction can be stopped by phenol/chloroform/isoamyl alcohol extraction – see methods from **Section 3.2 Steps 2 and 3**. This method is recommended if the transcription reaction has been scaled up in volumes greater than 100 μL to increase the yield of RNA.
12. Size exclusion chromatography can also be used to purify the RNA of interest if an FLPC and an appropriate gel filtration column is available. The Superdex 200 10/300 GL column (GE Healthcare) can be used successfully to purify RNA molecules ranging in size from 10,000 to 600,000 kilodaltons while a Superdex 75 10/300 GL column will resolve

RNAs sized between 3,000 and 70,000 kilodaltons. After the transcription reaction has been separated on the column, fractions containing the RNA can be concentrated by ethanol precipitation or on a centrifugation-based concentrator device (*see* **Section 3.2 Step 3**).

13. Glycogen is optional but can be included during ethanol precipitation. It is inert and its presence should not affect subsequent reactions. The glycogen precipitates into a pellet with the RNA and allows for easy visualization of the pellet. However, it should be noted that glycogen might reduce the efficiency of subsequent hydroxyl radical footprinting given that polyhydroxylated compounds can be •OH-scavenging molecules.
14. Although the volume at this point of the hydroxyl radical footprinting protocol is 7 μL, the final volume of the reaction will be 10 μL so this should be taken into account when calculating how much $MgCl_2$ or other potential ligand to add.
15. For in-line probing, shorter incubation times result in fainter bands but longer incubation times result in higher background. A highly recommended discussion on the rates of spontaneous cleavage of RNA linkages and effects of pH and metal ions can be found elsewhere *(6, 31)*.
16. While bands on an in-line probing gel are typically indicative of flexible RNA linkages, certain bands may result from nucleotides that are positioned such that the $2'$ hydroxyl and oxyanion leaving group most closely approximate the in-line conformation required for efficient phosphodiester scission *(6)*.

References

1. Winkler, W.C. and Breaker, R.R. (2005) Regulation of bacterial gene expression by riboswitches. *Annu. Rev. Microbiol.* **59**, 487–517.
2. Schwalbe, H., Buck, J., Furtig, B., Noeske, J., and Wohnert, J. (2007) Structures of RNA switches: insight into molecular recognition and tertiary structure. *Angew. Chem. Int. Ed. Engl.* **46**, 1212–1219.
3. Wakeman, C.A., Winkler, W.C. and Dann, C.E., III. (2007) Structural features of metabolite-sensing riboswitches. *Trends Biochem. Sci.* **32**, 415–424.
4. Edwards, T.E., Klein, D.J. and Ferré-D'Amaré, A.R. (2007) Riboswitches: small-molecule recognition by gene regulatory RNAs. *Curr. Opin. Struct. Biol.* **17**, 273–279.
5. Gilbert, S.D. and Batey, R.T. (2006) Riboswitches: fold and function. *Chem. Biol.* **13**, 805–807.
6. Soukup, G.A. and Breaker, R.R. (1999) Relationship between internucleotide linkage geometry and the stability of RNA. *RNA* **5**, 1308–1325.
7. Westheimer, F.H. (1968) Pseudo-rotation in the hydrolysis of phosphate esters. *Acc. Chem. Res.* **1**, 70–78.
8. Usher, D.A. (1969) On the mechanism of ribonuclease action. *Proc. Natl. Acad. Sci. U.S.A.* **62**, 661–627.

9. Usher, D.A. and McHale, A.H. (1976) Hydrolytic stability of helical RNA: a selective advantage for the natural 3′, 5′-bond. *Proc. Natl. Acad. Sci. U.S.A.* **73,** 1149–1153.
10. Dock-Bregeon, A.C. and Moras, D. (1987) Conformational changes and dynamics of tRNAs: evidence from hydrolysis patterns. *Cold Spring Harb. Symp. Quant. Biol.* **52,** 113–121.
11. Dorner, S. and Barta, A. (1999) Probing ribosome structure by europium-induced RNA cleavage. *Biol. Chem.* **380,** 243–251.
12. Dann, C.E. III, Wakeman, C.A., Sieling, C.L., Baker, S.C., Irnov, I. and Winkler, W.C. (2007) Structure and mechanism of a metal-sensing regulatory RNA. *Cell* **130,** 878–892.
13. Ehresmann, C., Baudin, F., Mougel, M., Romby, P., Ebel, J.P., and Ehresmann, B. (1987) Probing the structure of RNAs in solution. *Nucleic Acids Res.* **15,** 9109–9028.
14. Werner, C., Krebs, B., Keith, G. and Dirheimer, G. (1976) Specific cleavages of pure tRNAs by plumbous ions. *Biochim. Biophys. Acta.* **432,** 161–175.
15. Brunel, C. and Romby, P. (2000) Probing RNA structure and RNA-ligand complexes with chemical probes. *Methods Enzymol.* **318,**3–21.
16. Ryder, S.P. and Strobel, S.A. (1999) Nucleotide analog interference mapping. *Methods* **18,** 38–50.
17. Sclavi, B., Sullivan, M., Chance, M.R., Brenowitz, M. and Woodson, S.A. (1998) RNA folding at millisecond intervals by synchrotron hydroxyl radical footprinting. *Science* **279,** 1940–1943.
18. Latham, J.A. and Cech, T.R. (1989) Defining the inside and outside of a catalytic RNA molecule. *Science* **245,** 276–282.
19. Brenowitz, M., Chance, M.R., Dhavan, G. and Takamoto, K. (2002) Probing the structural dynamics of nucleic acids by quantitative time-resolved and equilibrium hydroxyl radical footprinting. *Curr. Opin. Struct. Biol.***12,** 648–653.
20. Celander, D.W. and Cech, T.R. (1990) Iron(II)-ethylenediamine tetraacetic acid-catalyzed cleavage of RNA and DNA oligonucleotides: similar reactivity toward single- and double-stranded forms. *Biochemistry* **29,** 1355–1361.
21. Hampel, K.D. and Burke, J.M. (2001) Time-resolved hydroxyl-radical footprinting of RNA using Fe (II)–EDTA. *Methods* **23,** 233–239.
22. Hayes, J.J., Kam, L. and Tullius, T.D. (1990) Footprinting protein-DNA complexes with γ-rays. *Methods Enzymol.* **186,** 545–549.
23. Sclavi, B., Woodson, S., Sullivan, M., Chance, M.R. and Brenowitz, M. (1997) Time-resolved synchrotron X-ray 'footprinting', a new approach to the study of nucleic acid structure and function: application to protein-DNA interactions and RNA folding. *J. Mol. Biol.* **266,** 144–159.
24. Swisher, J.F., Su, L.J., Brenowitz, M., Anderson, V.E. and Pyle, A.M. (2002) Productive folding to the native state by a group II intron ribozyme. *J. Mol. Biol.***315,** 297–310.
25. Hampel, K.J. and Tinsley, M.M. (2006) Evidence for preorganization of the *glmS* ribozyme ligand binding pocket. *Biochemistry* **45,** 7861–7871.
26. Das, R., Laederach, A., Pearlman, S.M., Herschlag, D. and Altman, R.B. (2005) SAFA: a semi-automated footprinting analysis software for high-throughput quantification of nucleic acid footprinting experiments. *RNA* **11,** 344–354.
27. Tullius, T.D., Dombrowski, B.A., Churchill, M.E.A. and Kam, L. (1987) Hydroxyl radical footprinting: a high-resolution method for mapping protein-DNA contacts. *Methods Enzymol.* **155,** 537–558.
28. Lipfert, J., Das, R., Chu, V.B., Kudaravalli, M., Boyd, N., Herschlag, D. and Doniach, S. (2007) Structural transitions and thermodynamics of a glycine-dependent riboswitch from *Vibrio cholerae. J. Mol. Bio.* **365,** 1393–1406.
29. Das, R., Travers, K.J., Bai, Y. and Herschlag, D. (2005) Determining the Mg^{2+} stoichiometry for folding an RNA metal ion core. *J. Am. Chem. Soc.***127,** 8272–8273.
30. Draper, D.E., Grilley, D. and Soto, A.M. (2005) Ions and RNA folding. *Annu. Rev. Biophys. Biomol. Struct.* **34,** 221–243.
31. Li, Y. and Breaker, R.R. (1999) Kinetics of RNA degradation by specific base catalysis of transesterification involving the 2'-hydroxyl group. *J. Am. Chem. Soc.* **121,** 5364–5372.

Chapter 9

Determining Structures of RNA Aptamers and Riboswitches by X-Ray Crystallography

Andrea L. Edwards, Andrew D. Garst, and Robert T. Batey

Abstract

Structural biology plays a central role in gaining a full understanding of the myriad roles of RNA in biology. In recent years, innovative approaches in RNA purification and crystallographic methods have lead to the visualization of an increasing number of unique structures, providing new insights into its function at the atomic level. This article presents general protocols which have streamlined the process of obtaining a homogeneous sample of properly folded and active RNA in high concentrations that crystallizes well in the presence of a suitable heavy-atom for phasing. Of particular importance are approaches toward RNA crystallography that include exploring "construct space" as opposed to "condition space". Moreover, development of a highly flexible method for experimentally phasing RNA crystals may open the door to a relatively simple means of solving these structures.

Key words: RNA purification, RNA synthesis, riboswitch, aptamer, RNA crystallography, heavy-atom derivative.

1. Introduction

The last decade has witnessed a rapid growth in the number of RNA structures determined by X-ray crystallography. Landmark structures such as the minimal hammerhead ribozyme *(1, 2)* and the P4–P6 domain of the *Tetrahymena thermophila* group I intron *(3)* substantially eroded the myth that RNA is extremely difficult to crystallize. Since then, the growing catalog of RNA structures has begun to reveal motifs such as the ribose zipper, the dinucleotide-platform, kink-turns, tetraloops, and the A-minor triple that serve as the foundation for establishing complex tertiary

---

Andrea L. Edwards and Andrew D. Garst contributed equally to this work.

Günter Mayer (ed.), *Nucleic Acid and Peptide Aptamers: Methods and Protocols, vol. 535*

DOI 10.1007/978-1-59745-557-2_9 Springerprotocols.com

architecture *(1, 3–5)*. Biological non-coding RNAs and artificial aptamers are being discovered at an accelerating pace, and their potential applications in the fields of medicine and biotechnology is increasing the demand for high-resolution structures to fully understand their function. This prompts a need for new and improved methods for RNA purification and crystallization to facilitate its structure determination.

The widespread use of labor-intensive denaturing purification techniques and the lack of a universal tool for obtaining phase information are among the most difficult issues faced in RNA structural biology *(6)*. Towards this end, a number of laboratories have developed techniques that address these problems and have generated strategies for engineering RNA that facilitate its crystallization. Choosing the correct set of approaches often determines the success of a crystallization effort. In this article, this process will be described with an emphasis on four steps specific for RNA (*see* **Fig. 9.1**): designing constructs for crystallization trials, RNA synthesis by T7 RNA polymerase using DNA templates generated by PCR, RNA purification under denaturing or native conditions, and initial screening for diffraction-quality crystals.

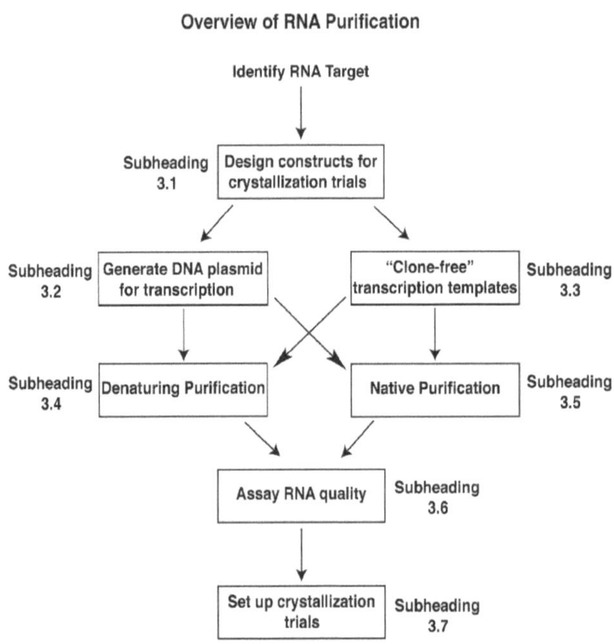

Fig. 9.1. Flowchart of RNA synthesis, purification, and initial crystallization trials as described in this chapter. Within this scheme, alternative protocols are presented for the synthesis of transcription templates (**Sections 3.2** and **3.3**) and for the purification of RNA (**Sections 3.4** and **3.5**). The *arrows* denote how each section can be linked to the next step in the process.

2. Materials

2.1. Designing a Library of RNA Variants

1. Access to Rfam database (rfam.sanger.ac.uk).
2. Access to GeneDesign (slam.bs.jhmi.edu/gd/index.html).

2.2. Construction of Plasmid Vectors for the Expression of RNA

2.2.1. PCR Construction of a DNA Gene

1. Milli-Q (18 mΩ) water.
2. 5 U/μL (working concentration) Taq DNA polymerase (New England Biolabs, Ipswich, MA).
3. 10x Thermophilic DNA polymerase buffer: 200 mM Tris–HCl, pH 8.8, 100 mM KCl, 100 mM $(NH_4)_2SO_4$, 20 mM $MgSO_4$, 1% Triton X-100.
4. 100 μM Stock concentration of DNA oligonucleotide primers (stored at −20°C).
5. 10 mM dNTPs mixture (stored at −20°C).
6. Thermocycling PCR machine.
7. QIAquick PCR purification kit (QIAGEN, Valencia, CA).
8. Restriction enzymes: *Eco*RI, *Nco*I, and *Kpn*I (New England Biolabs, Ipswich, MA).
9. Calf intestinal alkaline phosphatase (CIP), 0.2 U/μL (New England Biolabs).
10. Agarose gel electrophoresis equipment.
11. Shortwave ultraviolet (UV) illuminator.
12. QIAquick Gel Extraction Kit (QIAGEN).
13. 1x TE buffer: 10 mM Tris–HCl, pH 8.0, 1 mM EDTA.
14. Ethidium bromide (10 mg/mL aqueous stock).

2.2.2. Cloning into the pRAV Plasmid Vectors

1. DNA vectors (pRAV12 or pRAV23, *see* **Note 1**).
2. T4 DNA ligase (Invitrogen, Carlsbad, CA).
3. 10 mM ATP.
4. PCR thermocycler.
5. Luria broth (LB) agar plates containing ampicillin (50 μg/mL). Ampicillin stock solution is 50 mg/mL in 50% ethanol/50% water and stored at −20°C.
6. DH5α chemically competent *E. coli* cells (Stratagene, San Diego, CA).
7. Incubator at 37°C.
8. DNA Miniprep Kit (QIAGEN).

9. TE buffer: 10 mM Tris–HCl pH 8.0, 1 mM EDTA pH 8.0.
10. Sequencing primers: 10 μM stock concentration of M13 forward or reverse (New England Biolabs, Cat. # S1201 and S1212).

2.3. Clone-Free Generation of dsDNA Transcription Templates

1. Milli-Q (18 mΩ) water.
2. 2.5 U/μL (stock concentration) Pfu DNA polymerase (Stratagene).
3. 10x Thermophilic DNA polymerase buffer.
4. 100 μM Stock concentration of DNA oligonucleotide primers (stored at −20°C).
5. 10 mM dNTPs mixture (stored at −20°C).
6. Thermocycling PCR machine.
7. QIAquick PCR purification kit (QIAGEN).

2.4. Synthesis of RNA by In Vitro Transcription and Denaturing Purification

2.4.1. Large-Scale PCR Synthesis of Template

1. 10x Thermophilic DNA polymerase buffer.
2. 10 mM dNTP mixtures.
3. 5 U/μL Working concentration of Taq DNA polymerase (New England Biolabs).
4. 100 μM Stock concentration of DNA oligonucleotide primers.
5. Agarose gel electrophoresis equipment.

2.4.2. Preparation of rNTP Stocks

1. Ribonucleotide 5'-triphosphate disodium salt: ATP, CTP, GTP, and UTP (Sigma-Aldrich, St. Louis, MO).
2. Milli-Q water.
3. 5 M NaOH.
4. pH-indicator strips.

2.4.3. Synthesis of RNA by T7 RNA Polymerase for Denaturing Purification

1. 50 mL Disposable conical tube.
2. 10x Transcription buffer: 400 mM Tris–HCl, pH 8.0, 100 mM DTT, 20 mM spermidine, 0.1% Triton X-100.
3. 100 mM rNTP stocks (stored at −20°C and thawed immediately prior to application).
4. Inorganic pyrophosphatase (Sigma-Aldrich), lyophilized powder, suspended in storage buffer (20 mM KH_2PO_4 pH 7.0, 100 mM NaCl, 50% glycerol (v/v), 10 mM DTT, 0.1 mM EDTA pH 8.0, 0.2% (w/v) NaN_3) to a stock concentration of 20 U/μL. This is to be stored at −20°C and kept on ice while adding the appropriate aliquot to the reaction mixture.

5. T7 RNA polymerase (New England Biolabs or a 0.25 mg/mL working concentration of home-made). This enzyme should be stored at −20°C and kept on ice while adding the appropriate aliquot to the reaction mixture.
6. Incubator at 37°C.

2.4.4. Denaturing Polyacrylamide Gel Electrophoretic Purification of RNA

1. 100% Ethanol.
2. 8 M Urea.
3. 0.5 M Na_2EDTA, pH 8.0.
4. 40% Acrylamide/bisacrylamide solution (29:1).
5. 10% Ammonium persulfate solution.
6. *N*,*N*,*N'*,*N'*-tetramethylethylenediamine (TEMED).
7. 5x TBE buffer: 0.5 M Tris base, 0.42 M boric acid, and 5 mM Na_2EDTA.
8. Vertical polyacrylamide gel electrophoresis apparatus with 0.3 mm spacers.
9. Fluorescent thin layer chromatography plate (Sigma-Aldrich).
10. Shortwave ultraviolet (UV) lamp.
11. Electroelution apparatus (Elutrap) (Whatman, Florham Park, NJ).
12. 10,000 molecular weight cutoff centrifugal concentrators (Millipore, Billerica, MA).
13. 1x Exchange buffer: 10 mM Na-MES, pH 6.0, 2 mM $MgCl_2$, and 0.1% NaN_3 or a suitable alternative of choice.

2.5. Purification of RNA Using Native Ni-NTA Chromatography

2.5.1. Synthesis of RNA for Native Purification

1. 10x HEPES Thermophilic DNA polymerase buffer: 200 mM Na-HEPES, pH 8.6, 100 mM KCl, 100 mM $(NH_4)_2SO_4$, 20 mM $MgSO_4$, 1% Triton X-100.
2. 10 mM dNTP mixtures (stored at −20°C).
3. Taq DNA polymerase, 5 U/μL working concentration (New England Biolabs).
4. 100 μM Stock concentration of DNA oligonucleotide primers (stored at −20°C).
5. Agarose gel electrophoresis equipment.
6. 50 mL Disposable conical tube.
7. 10x HEPES Transcription buffer: 400 mM Na-HEPES, pH 8.0, 100 mM DTT, 20 mM spermidine, 0.1% Triton X-100.
8. 100 mM rNTP stocks.
9. Inorganic pyrophosphatase (Sigma-Aldrich).

10. T7 RNA polymerase (New England Biolabs or a ~10 mg/mL stock concentration of home-made).
11. 37°C Incubator or waterbath.

2.5.2. Preparation of HMM Tagging Protein

1. Chemically competent BL21(DE3) *E. coli* cells (Novagen, Madison, WI).
2. pHMM expression vector (*see* **Note 1**).
3. Luria broth (LB) agar plates containing kanamycin (30 μg/mL). Kanamycin stock solution is 30 mg/mL in water and stored at −20°C.
4. Isopropyl-thio-ß-galactopyranoside (IPTG), molecular biology grade (Sigma-Aldrich, St. Louis, MO). Prepare a working 1,000x stock of 1 M in ddH_2O, filter sterilize, and store at −20°C.
5. *E. coli* protease inhibitor solution (#P8849, Sigma-Aldrich, St. Louis, MO)
6. Lysis buffer: 50 mM sodium phosphate, pH 8.0, 300 mM NaCl, 10% glycerol, 0.5% Tween-20, 10 mM imidazole.
7. Cell sonic disruptor.
8. Ni-NTA agarose (QIAGEN).
9. SP-Sepharose (GE Healthcare, Piscataway, NJ).

2.5.3. Ni-NTA Purification of RNA

1. Ni-NTA agarose (QIAGEN).
2. Econo-Pac 20 mL columns (BioRad, Hercules, CA).
3. RNA wash buffer: 50 mM K^+-HEPES, pH 7.5, 150 mM NaCl, 10 mM $MgCl_2$, 10 mM imidazole.
4. Glucosamine-6-phosphate (GlcN6P), 0.1 M working concentration in ddH_2O (Sigma-Aldrich).
5. Stripping buffer: 50 mM K^+-HEPES, pH 7.5, 150 mM NaCl, 10 mM $MgCl_2$, 250 mM imidazole.
6. 10,000 molecular weight cutoff centrifugal concentrators (Millipore).

2.6. Assaying RNA Quality

1. Vertical polyacrylamide gel electrophoresis apparatus with 1 mm spacers.
2. 5x TBE buffer.
3. 40% Acrylamide/bisacrylamide solution (29:1) stored at 4°C.
4. 10% Ammonium persulfate solution stored at 4°C.
5. *N,N,N′,N′*-tetramethylethylenediamine (TEMED) stored at 4°C.
6. Ethidium bromide solution (10 mg/mL in H_2O).
7. Ultraviolet (UV) transilluminator box.

2.7. Crystallization of RNA

2.7.1. Screening of RNA for Crystallizability

1. 24-well Linbro trays (Hampton Research, Aliso Viejo, CA).
2. Siliconized glass cover slips (Hampton Research).
3. Sparse Matrix screening kits (Hampton Research).
4. 0.22 μm Cellulose acetate microcentrifuge tube filters (Corning Inc., Corning, NY)
5. Incubator at 30°C.

2.7.2. A Simple Synthesis of Iridium(III) Hexamine

1. Heavy-walled Ace pressure tube (35 mL volume) (Sigma-Aldrich).
2. Silicone oil bath.
3. Teflon tape.
4. Iridium (III) chloride (Sigma-Aldrich).
5. Ammonium hydroxide.
6. Rotary evaporator.
7. Concentrated hydrochloric acid.
8. Absolute (100%) ethanol.

3. Methods

3.1. Designing a Library of RNA Variants

Before undertaking a crystallization effort, it is highly advisable to devise a strategy for systematically surveying a number of RNA constructs containing the motif or activity of interest, as well as a means of obtaining phase information. It is now generally appreciated that small changes in an RNA, particularly in peripheral helices and terminal loops, have a significant effect on the formation of productive lattice contacts. Therefore, it is useful to identify nucleotide positions or regions in the RNA that are highly variable by sequence comparison of phylogenetic variants; this information is often available in a database such as Rfam (rfam.sanger.ac.uk) *(7)*. These regions are often not required for function, and thus are prime candidates for sequence alteration. Combined with biochemical and genetic information, often one can be very confident that alterations to an RNA sequence will not be deleterious to its function or structure.

Systematic variation of the length and sequence composition of peripheral elements is the primary means of generating a library of constructs for crystallization trials. Each variant in the library is synthesized and purified, refolded if necessary, and subjected to several commercially available sparse matrix screens (*see* **Section 3.7.1**) at a single temperature (30°C) to assess whether it is inherently crystallizable. Only those individuals that yield crystals in a number of conditions within the initial survey are further examined for suitably diffracting crystals via further condition screening. In this fashion, our crystallization strategy focuses initially upon "construct space" with only those RNAs showing an inherent tendency to crystallize subjected to refinement in "condition space" (*see* **Note 2**). The following are two recent examples from our laboratory as to how we specifically approached this problem. Other examples in the literature can be found detailing how crystals were found for the U1A–RNA complex *(8, 9)*, signal recognition particle *(10)*, hammerhead ribozyme *(11)*, hepatitis delta ribozyme *(12)*, and the guanine riboswitch *(13)*.

3.1.1. The SAM-I Riboswitch: A Case Study in Exploring Construct Space

The SAM-I riboswitch is an mRNA element that binds *S*-adenosylmethionine to regulate gene expression in a variety of bacteria *(14, 15)*. In our strategy for this RNA, we decided to use a tactic often employed in protein crystallography – thermophilic variants *(16)*. As riboswitches are found in thermophilic species (*Thermotoga maritima* and *Thermus thermophilus*, for example), we identified in Rfam a promising RNA that controlled the *metF-H2* operon in *Thermoanaerobacter tengcongensis* (*see* **Fig. 9.2A**). Phylogenetic alignment of this RNA motif revealed several highly variable peripheral regions: the lengths of the P1, P3, and P4 helices and the composition of the terminal loops of P3 and P4.

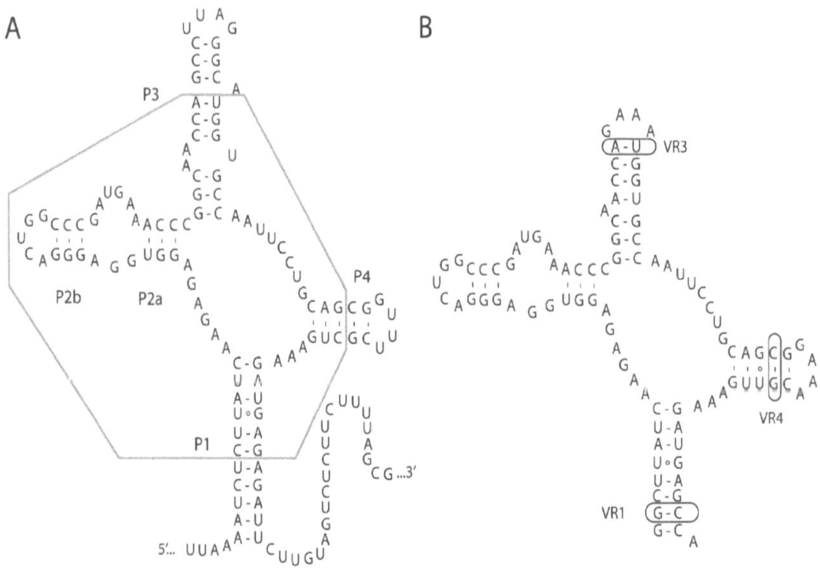

Fig. 9.2. Conversion of the SAM-I riboswitch to a sequence that was successfully crystallized. **(A)** Raw sequence of the SAM-I riboswitch aptamer domain that controls the *metF-H2* operon in *T. tencongensis*. The box encloses all sequence elements that are >90% conserved across phylogeny and implicated in ligand binding. **(B)** Sequence of the RNA that was crystallized complexed with *S*-adenosylmethionine. VR1, VR2, and VR3 denote the three regions of the P1, P3, and P4 helices that contained expansions to create a library of different variants of the SAM-I aptamer domain.

Based upon the available phylogenetic and biochemical data, we employed a directed engineering strategy involving two approaches. First, the terminal tetraloops in the natural sequence were changed to GAAA tetraloops (*see* **Fig. 9.2B**). The GAAA tetraloop has been observed in a large number of crystal structures to mediate formation of lattice contacts *(10, 17)*, often via the use of A-minor triples *(18, 19)*. Second, the lengths of the P1, P3, and P4 helices were systematically varied such that an array of RNAs was produced *(13)*. Subsequent screening of these variants against three commercially available sparse matrices (Crystal Screen I, Natrix, and Nucleic Acid Mini Screen; Hampton Research) showed a strong trend towards the smaller SAM-I RNAs being more crystallizable. The 8/3/5 variant (eight base pairs in P1, three base pairs above a conserved internal loop in P3, and five base pairs in P4) yielded crystals in a large number of conditions in the initial survey, and with a little fine tuning of the mother liquor, crystals were reproducibly obtained that diffracted X-rays to 2.9 Å resolution *(20)*.

3.1.2. The SAM-II Riboswitch: A Case Study in Exploring Phylogenetic Variants

The SAM-II riboswitch *(21)* presented a significant problem toward implementing the directed engineering strategy described above. The pseudoknot that forms the conserved functional core of this RNA (*see* **Fig. 9.3**) did not allow us to simply take one species variant and alter its peripheral elements. Instead, we initially screened phylogenetic variants in which the lengths of the P1 and P2 stems naturally differed (*see* **Note 3**). Using a published phylogenetic

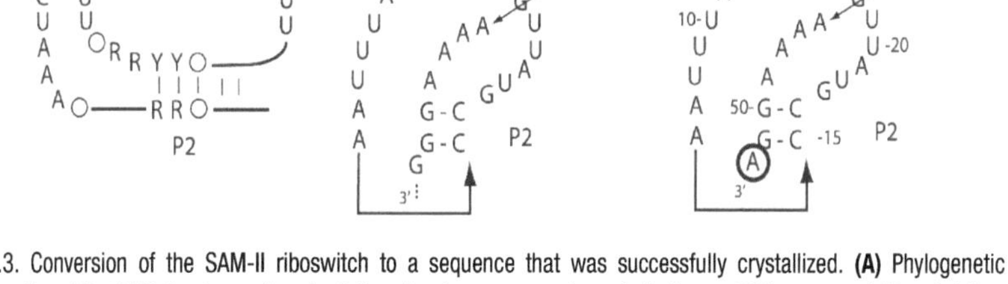

Fig. 9.3. Conversion of the SAM-II riboswitch to a sequence that was successfully crystallized. **(A)** Phylogenetic conservation of the SAM-II aptamer domain; *letters* denote sequence elements that are >90% conserved (Y, pyrimidine; R, purine) and *circles* denote the presence of a base pair at that position in >90% of sequences. Adapted from *(21)* **(B)** Sequence and secondary structure of the SAM-II aptamer domain that controls the *metX* gene in a sequence from the Sargasso Sea metagenome. **(C)** Sequence of the RNA that was successfully crystallized. *Circled* nucleotides are sequences that were changed in order to facilitate synthesis (G at the 5'-end), processing by the H∂V ribozyme (A at the 3'-end), or phasing (pairs in the P1 helix).

alignment of RNAs bearing the SAM-II sequence signature (*see* **Fig. 9.3A**), we targeted 13 variants for initial crystallization trials *(21)*. From this series, two RNAs (6/2 and 7/6) yielded diffraction-quality crystals from initial screens. Characterization of each crystal revealed that the 7/6 variant crystallized in a P1 space group, making data collection and phasing more difficult. Therefore it was abandoned in favor of the 6/2 variant (*see* **Fig. 9.3B**) that yielded a more favorable C2 space group. After further engineering (*see* **Fig. 9.3C**, *see* **Section 3.1.3**), this RNA yielded a 2.8 Å structure *(22)*.

3.1.3. The Phasing Module

It is never too early to think about how to solve the phase problem. For RNA, this is a significant issue because there is no generally accepted and simple means of creating a heavy-atom derivative as there is for proteins via bioincorporation of selenomethionine. Over the last decade, a number of approaches have been employed including the use of an RNA binding protein to provide selenium sites *(12)*, incorporation of 5-bromouracil *(23, 24)* or 2'-selenoribose *(25, 26)*, and standard "soak-and-pray" methods using multivalent cations *(27)*. A more recent technique that we have pioneered is the use of a small sequence motif that can be placed into virtually any A-form helix based upon the observation that a single G•U wobble pair often has cations bound to its major groove face and does not significantly alter the helical geometry *(28)*. A systematic survey of single G•U pairs with differing flanking sequences (Watson–Crick A–U or G–C pairs) reveals a striking trend: the nucleotide on the 5'-side of the G•U pair has a strong influence on the strength of a metal's interaction with the RNA. If a guanine or uracil occupies each of these two positions (*see* **Fig. 9.4**), the G•U pair has a very high likelihood of binding a hexammine ion or cesium ion with both high occupancy and low B-factor, making it a favorable derivative for phasing. This "phasing module" was placed into the P1 helix of the SAM-II 6/2 variant RNA (*see* **Fig. 9.3C**) to yield a cesium derivative that was used for phasing by the SIRAS (single isomorphous replacement with anomalous scattering) method.

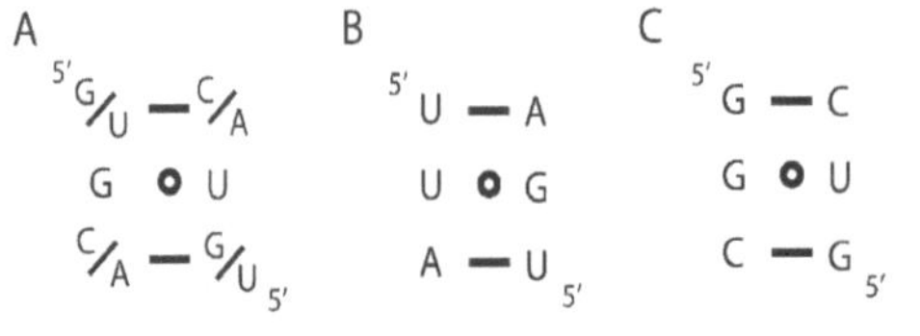

Fig. 9.4. Sequence of the iridium(III) hexammine binding motif (the "phasing module"). **(A)** The consensus sequence of the module consisting of a single wobble G•U pair flanked by two Watson–Crick pairs. **(B)** The phasing module containing two A–U pairs. **(C)** The phasing module containing two G–C pairs.

3.2. Construction of Plasmid Vectors for the Expression of RNA

For each RNA in the library, the corresponding DNA sequence is created by PCR with overlapping DNA oligonucleotides and placed into one of two plasmid vectors. The choice of vector is based upon the purification approach, which in turn depends upon the RNA being studied. In general, the majority of small RNA species (<50 nucleotides in length) can be folded into an active conformation following denaturation. Thus, a purification scheme involving classical denaturing gel electrophoresis is the most straightforward method. For this approach, we clone into the pRAV12 vector (*see* **Note 1**) *(6)*, placing the RNA of interest upstream of the H∂V ribozyme for processing of the 3'-end *(29)*.

Larger RNAs, on the other hand, are susceptible to kinetic folding traps, presenting the researcher with the dilemma of testing a number of refolding protocols before a homogeneously folded and active RNA population is achieved. In this case, a native purification scheme may be the best route. Therefore, we clone into the pRAV23 vector (*see* **Note 1**) containing a glmS ribozyme/MS2 coat protein affinity tag for Ni-NTA chromatography *(30)*. In our experience, assaying different phylogenetic variants of the RNA of interest by native gel electrophoresis (*see* **Section 3.6.2**) allows us to assess the ability of each RNA to refold and thus make an informed decision about the purification strategy. As these types of initial decisions about the experimental strategy are often crucial to success, it is advisable to try a number of approaches proven to generate diffraction-quality crystals of RNA and its complexes with small molecules and proteins.

Once sequences and a purification strategy have been chosen, it is a relatively straightforward process to convert the sequence into a series of DNA primers to construct a transcription template. For example, to clone the SAM-I RNA (*see* **Figs. 9.2B** and **9.5A**) into pRAV12, an *Eco*R1 site and the T7 promoter sequence is added upstream of the desired RNA sequence and a H∂V/*Nco*I site on the 3' side (*see* **Fig. 9.5B**). This sequence is then input into the GeneDesign website (slam.bs.jhmi.edu/gd/index.html, "oligo design" component) which breaks it up into a series of overlapping DNA oligonucleotides of approximately the same length (*see* **Fig. 9.5C**). These are output as a set of oligonucleotides (*see* **Fig. 9.5D**) to be synthesized for use as inner primers in a PCR reaction that will reconstruct the intact double-stranded DNA template (*see* **Section 3.2.1**).

3.2.1. PCR Construction of a DNA Gene

1. The PCR amplification reaction is assembled as follows: 10 μL of 10x Taq DNA polymerase buffer, 2 μL dNTP solution (10 mM each dNTP), 2 μL of a 100 nM solution of each inner DNA oligonucleotide (*see* **Note 4**), 1 μL 100 μM 5'-

A. Desired RNA sequence (SAM-I riboswitch, Figure 2B):

```
5'-ggcuuaucaagagagguggagggacuggcccgaugaaacccggcaaccagaaauggugccaauuccugcagcggaaacguugaaagaugagcca-3'
```

B. DNA sequence needed (for cloning into pRAV12):

```
GGCCGGGAATTC TAATACGACTCACTATA GGCTTATCAAGAGAGGTGGAGGGACTGGCCCGATGAAACCCGGCAACCAGAAATGGTGCCAATTCCTGCAGCGGAAACGTTGAAAGATGAGCCA GCCGGCCATGGTCCCAGC
      EcoR1    T7 promoter                    RNA of interest                                                                 HδV/NcoI
```

C. Gene design output:

```
5'GGCCGGGAATTCTAATACGACTCACTATAGGCTTATCAAGAGAGGTGGAGGGACTGGCCCGATGAAACCCGGCAACCAGAAATGGTGCCAATTCCTGCAGCGGAAACGTTGAAAGATGAGCCAGCCGGCCATGGTCCCAGC
5'GGCCGGGAATTCTAATACGACTCACTATAGGCTTATCAAGAGAGGTGGAGGGACTGG          AAACCCGGCAACCAGAAATGGTGCCAATTCCTGCAGCGGAAACGTTGAAAG
                                    AGTTCTCTCCACCTCCCTGACCGGGCTACTTTGGGCCGTTGGTCTTTACCACG      GGACGTCGCCTTTGCAACTTTCTACTCGGTCGGCCGGTACCAGGGTCG-5'
  CCGGCCCTTAAGATTATGCTGAGTGATATCCGAATAGTTCTCTCCACCTCCCTGACCGGGCTACTTTGGGCCGTTGGTCTTTACCACGGTTAAGGACGTCGCCTTTGCAACTTTCTACTCGGTCGGCCGGTACCAGGGTCG-5'
```

D. Primers to order:

```
A    5'-GGCCGGGAATTCTAATACGACTCACTATAGGCTTATCAAGAGAGGTGGAGGGACTGG
B    5'-GCACCATTTCTGGTTGCCGGGTTTCATCGGGCCAGTCCCTCCACCTCTCTTGA
C    5'-AAACCCGGCAACCAGAAATGGTGCCAATTCCTGCAGCGGAAACGTTGAAAG
D    5'-GCTGGGACCATGGCCGGCTGGCTCATCTTTCAACGTTTCCGCTGCAGG
```

Fig. 9.5. Conversion of an RNA sequence into a DNA sequence suitable for cloning into the pRAV12 expression vector. **(A)** Sequence of the RNA used to crystallize the SAM-I riboswitch in complex with *S*-adenosylmethionine. **(B)** DNA sequence encoding this RNA (*third block of letters*) with the addition at the 5'-end of an *Eco*RI restriction site (*bold, block 1*) and T7 RNA polymerase promoter (*block 2*) and the H∂V ribozyme/*Nco*I restriction sequence at the 3'-end (*block 4*). **(C)** Output from the GeneDesign website, in which the sequence of **(B)** was input with target primer length of 60 nucleotides, 18 base pair overlap, and 56°C overlap melting temperature. **(D)** The four DNA oligonucleotides that would be synthesized for use in a PCR reaction to create the full length DNA insert.

Table 9.1
General DNA primers for use with pRAV12 and –23 vector systems

Name	Sequence
5'-GEN	5'-GCGCGCGAATTCTAATACGACTCACTATAG
5' H∂V	5'-GCCGGCCATGGTCCCAGCCTCCTCG
5' glmS	5'-GCGCCCGAACACCGGTACC
3'-glmS	5'-TACCGGTACCGGTAGTTCGGGCGCT
3'-GEN12	5'-AGAGGTCCCATTCGCCATGCCGAAGCATGTTG
3'-MS2	5'-CAGACCCTGATGGTGTCTGAA
3'-H∂V	5'-CTGGGACCATGGCCGGC

GEN outer primer (*see* **Table 9.1**), 1 μL 100 μM outer primer 3'-H∂V (*see* **Table 9.1**), 1 μL Taq DNA polymerase, and 82 μL of ddH_2O (100 μL total volume).

2. The reaction is amplified in a PCR machine with the program: initial melt for 1 min at 95°C; 30 cycles of 94°C for 30 s, 50°C for 30 s, and 72°C for 60 s; 5 min final extension at 72°C; hold at 4°C.
3. The PCR amplification product is purified using the QIAquick PCR Purification Kit (QIAGEN) according to the manufacturer's instructions and recovered in 50 μL of TE buffer (or ddH_2O, which works equally well).
4. Restriction digestion of the purified PCR product is performed by adding the following constituents to the 50 μL of purified DNA from the previous step: 6 μL of the appropriate 10x restriction buffer provided from New England Biolabs, 1 μL of *Eco*R1, and 1 μL of either *Nco*I (pRAV12) or *Kpn*I (pRAV23) and incubating at 37°C for 1 h. 20 μg of Plasmid vector is also digested using the same reaction except that 1 μL of calf intestinal alkaline phosphatase (CIAP) is added to the reaction to reduce ligation background.
5. The product is purified using a 2% (PCR inserts) or 0.5% (plasmid) agarose gel in TAE buffer. The correct band is visualized by staining with ethidium bromide (1:1,000 dilution of 10 mg/mL stock), illuminated by short-wave UV, and rapidly excised from the gel with a clean razor blade.
6. The DNA fragment is removed from the gel matrix and purified using the QIAquick Gel Extraction Kit (QIAGEN)

according to manufacturer's instructions and recovered in 50 μL of TE buffer.

3.2.2. Cloning into the pRAV Plasmid Vectors

1. Ligation of the PCR insert into the DNA vector is performed in the following reaction: 4 μL 5x ligase buffer, 2 μL 10 mM ATP, 1 μL 0.1 μg/μL vector, 1 μL 1:5 dilution of PCR insert, and 1 μL T4 DNA ligase (*see* **Note 5**). Incubate the reaction at 16°C for 2 h or overnight (16 h).
2. Transform 2 μL into a standard *E. coli* strain used for plasmid propagation (DH5α or XL-10) and plate on LB-agar plates containing 50 μg/mL ampicillin for resistance selection.
3. Pick individual colonies and inoculate 5 mL LB broth supplemented with 50 μg/mL ampicillin. Use a DNA miniprep kit (QIAGEN) to purify plasmid DNA.
4. Verify the sequence of the resulting plasmid using the M13 reverse sequencing primer. While the M13 forward primer also works, artifacts may arise from sequencing in this direction because the RNA structure is reflected in the DNA being sequenced.

3.3. Clone-Free Generation of dsDNA Transcription Templates

In the initial stages of high-throughput screening of RNA constructs, the rate-limiting factor is often the cloning process. While cloning provides a reliable and sequence-verified DNA template, during initial screening stages we prefer to circumvent this step in favor of rapidly identifying crystallizable constructs. This system allows the use of simple PCR protocols to generate a sequence of interest conjugated to the 3' processing ribozyme of choice (*see* **Fig. 9.6**).

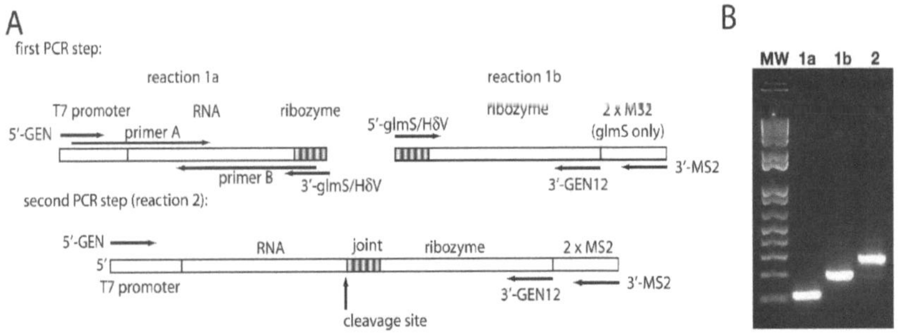

Fig. 9.6. Recombinant PCR as a means of generating a DNA fragment for use as a transcription template without the need for cloning. **(A)** Schematic of the PCR reactions needed to generate a full length fragment for transcription. Reactions 1a, 1b, and 1c are analogous to the reactions outlined in **Sections 3.3.1**, **3.3.2**, and **3.3.3**, respectively. **(B)** Ethidium bromide stained gel showing fragments generated in **Step 1** (*lane a, b*) and **Step 2** (*lane c*) of the scheme.

It should be noted that cloning is still recommended based on aforementioned criteria, particularly once promising leads have been identified.

3.3.1. Design of Primers and PCR Synthesis of RNA of Interest

1. Design a set of DNA oligonucleotides that contain the RNA gene of interest with the sequence of the 5'-GEN outer primer (*see* **Table 9.1**) appended to the 5' end and the sequence of the 3' ribozyme of choice at the 3'-end (3'-H∂V or 3'-glmS, *see* **Table 9.1**).
2. Amplify the RNA of interest using primers described above in the following reaction: 10 μL 10X Thermophilic DNA polymerase buffer, 2 μL 10 mM dNTPs mix, 1 μL of 100 μM 5' outer primer, 1 μL of 100 μM 3' outer primer, 1 μL of inner DNA primer mix (*see* **Fig. 9.6A**, "reaction 1a" this corresponds to the inner primers A and B; *see* **Note 4**), 1 μL Pfu polymerase (*see* **Note 6**) and 84 μL ddH_2O.
3. Amplify in a PCR thermocycler using the following protocol: initial melt for 1 min at 95°C; 30 cycles of 94°C for 30 s, 50°C for 30 s, and 72°C for 60 s; 5 min final extension at 72°C; hold at 4°C.
4. Load the entire 100 μL PCR reaction aliquot in large wells of a 2% electrophoresis grade agarose gel containing 1 μg/mL ethidium bromide and run at 100 V for 45 min. Visualize the band using a short-wave UV lamp, and cut out the band corresponding to the expected size.
5. Extract DNA for gel slice with the QIAquick gel extraction kit using the protocol provided by the manufacturer.
6. Make a 1:100 dilution of this stock into TE buffer. This dilution will be used in the subsequent recombination reaction (*see* **Section 3.3.4**).

3.3.2. PCR Synthesis of H • V Ribozyme (Denaturing Purification Protocol).

1. Set up a PCR reaction with the following components: 10 μL 10X Thermophilic DNA polymerase buffer, 2 μL 10 mM dNTPs mix, 1 μL of 100 μM 5'-H∂V primer (*see* **Table 9.1**), 1 μL of 100 μM 3'-GEN12 primer (*see* **Table 9.1**), 1 μL pRAV12 vector, 1 μL Pfu enzyme (*see* **Note 6**) and 84 μL ddH_2O (*see* **Fig. 9.6A**, "reaction 1b").
2. Follow **Steps 3–5** outlined in **Section 3.3.1**.

3.3.3. PCR Synthesis of glmS Ribozyme (Native Purification Protocol)

1. Set up a PCR reaction with the following components: 10 μL 10X Thermophilic DNA polymerase buffer, 2 μL 10 mM dNTPs mix, 1 μL of 100 μM 5'-glmS primer (*see* **Table 9.1**), 1 μL of 100 μM 3'-MS2 primer (*see* **Table 9.1**), 1 μL pRAV23 vector, 1 μL Pfu polymerase (*see* **Note 6**) and 84 μL ddH_2O (*see* **Fig. 9.6A**, "reaction 1b").
2. Follow **Steps 3–5** outlined in **Section 3.3.1**.

3.3.4. Recombination Reaction

1. Set up following PCR reaction: 10 μL of 10X Thermophilic DNA polymerase buffer, 2 μL 10 mM dNTPs mix, 1 μL of 100 μM 5'-GEN, 1 μL of 100 μM 3'-GEN12 or 3'-MS2 (depending on the ribozyme choice), 1 μL of a 1:100 dilution of your RNA gene (from **Section 3.3.1**), 1 μL of 1:100 dilution of ribozyme DNA piece (from **Section 3.3.2** or **3.3.3**), 1 μL Pfu enzyme and 84 μL ddH_2O (*see* **Fig. 9.6A**, "reaction 2").
2. Follow **Steps 2 and 3** from **Section 3.3.1**. Note that an apparent shift in the overall length of the product by about 90 or 200 base pairs, as judged by a 2% agarose gel, indicates successful conjugation of the H∂V or glmS ribozymes, respectively (*see* **Fig. 9.6B**).
3. Gel purify the product according to **Step 4** in **Section 3.3.1**. This will then serve as the template for large-scale PCR reactions to generate template for RNA synthesis by T7 in vitro transcription.

3.4. Synthesis of RNA by In Vitro Transcription and Denaturing Purification

3.4.1. Large-Scale PCR Synthesis of Template

1. Assemble in a 1.5 mL Eppendorf tube the following 1 mL reaction: 847 μL ddH_2O, 100 μL 10x Thermophilic DNA polymerase buffer, 10 μL of 100 μM 5'-GEN (*see* **Table 9.1**), 10 μL of 100 μM 3'-GEN12 (*see* **Table 9.1**), 20 μL 10 mM mixture of dNTPs, 3 μL of template plasmid (*see* **Section 3.2**) or product from clone-free PCR reaction (*see* **Section 3.3**) and 10 μL of Taq DNA polymerase (5 U/μL).
2. After vortexing the reaction to mix it thoroughly, make eight 125 μL aliquots into 200 μL thin-walled PCR tubes.
3. Amplify the target gene in a PCR thermocycler using the following protocol: initial melt for 1 min at 95°C; 30 cycles of 94°C for 30 s, 50°C for 30 s, and 72°C for 30 s; 5 min final extension at 72°C; hold at 4°C.
4. Pool the eight 125 μL reaction aliquots into a single tube. Assay 5 μL on a 2% agarose gel next to an appropriate size marker (1 Kb Plus DNA ladder, Invitrogen) to ensure strong amplification has occurred.
5. Freeze DNA solution at –20°C until use for transcription. No other purification of the reaction is necessary as residual dNTPs, Taq polymerase, and primers do not interfere with RNA synthesis by T7 RNA polymerase.

3.4.2. Preparation of rNTP Stocks

For large-scale RNA synthesis, the most economical option for obtaining ribonucleotide 5'-triphosphate precursors is to purchase dry powder from Sigma or a comparable supplier and dissolve each in a sufficient amount of ddH_2O to yield a 100 mM working stock. Most importantly, the solution's pH must be adjusted such that it is approximately 7.5. The amount of sodium hydroxide needed to adjust the pH will differ depending upon manufacturer of the rNTPs, so it is prudent to add the base slowly and monitor the pH by spotting 1–2 μL on pH strips. Thus, the recipes below will likely need to be slightly modified. All of the working stocks are stored in 1 mL aliquots at −20°C and thawed immediately prior to application.

1. 1 g of ATP (Sigma, A3377) is dissolved in ddH_2O to a final volume of 10 mL. Then the pH is adjusted to 7.5 by adding 580 μL 5 N NaOH and adjusting the final volume to 16 mL with ddH_2O to yield a working stock of 100 mM.
2. 1 g of CTP (Sigma, C1506) is dissolved in ddH_2O to a final volume of 10 mL. Then the pH is adjusted to 7.5 by adding 600 μL 5 N NaOH and adjusting the final volume to 17.1 mL with ddH_2O to yield a working stock of 100 mM.
3. 1 g of GTP (Sigma, G8877) dissolved in ddH_2O to a final volume of 10 mL. Then the pH is adjusted to 7.5 by adding 310 μL 5 N NaOH and adjusting the final volume to 16.5 mL with ddH_2O to yield a working stock of 100 mM.
4. 1 g of UTP (Sigma, U6570) dissolved in ddH_2O to a final volume of 10 mL ddH_2O. Then the pH is adjusted to 7.5 by adding 270 μL 5 N NaOH and adjusting the final volume to 17 mL with ddH_2O to yield a working stock of 100 mM.

3.4.3. Preparation of T7 RNA Polymerase

Synthesis of crystallographic quantities (5–20 mg) of RNA requires a substantial amount of T7 RNA polymerase, such that it is impractical to buy it from a commercial source. Most laboratories studying RNA structure over-express and purify their own polymerase using one of several published protocols *(31–33)*. Our laboratory uses the expression vector pT7-911Q transformed into BL21 *E. coli* cells and purified using a single immobilized nickel ion affinity column *(34)*. The protein obtained from this procedure is stored in a buffer containing 20 mM potassium phosphate, pH 7.5, 100 mM NaCl, 10 mM DTT, 0.1 mM Na_2EDTA, 0.2% NaN_3, and 50% glycerol at a stock concentration of ~10 mg/mL.

3.4.4. Synthesis of RNA by T7 RNA Polymerase for Denaturing Purification

1. Assemble in a disposable 50 mL conical tube the following 12.5 mL reaction: 7.25 mL of ddH_2O, 1.25 mL 10x transcription buffer, 300 μL 1 $MgCl_2$, 100 μL 1 M DTT, 500 μL 100 mM ATP, 500 μL 100 mM CTP, 500 μL 100 mM GTP, 500 μL 100 mM UTP, 1 mL PCR reaction containing the appropriate template, 100 μL inorganic pyrophosphate (20 U/μL), 250 μL of 10 mg/mL T7 RNA polymerase. Vortex the reaction to ensure complete mixing.
2. Incubate the reaction for 1–2 h at 37°C. While further reaction may result in minor increases in overall yield, the reaction is

generally complete after 1 h and additional synthesis tends to initiate with 2 or 3 nucleotide abortive products, yielding 5'-end heterogeneity *(35)*.

3.4.5. Denaturing Polyacrylamide Gel Electrophoretic Purification of RNA

1. Add 37.5 mL of –20°C ethanol to the 12.5 mL transcription reaction and incubate for at least 1–2 h at –20°C (incubation overnight is also permissible).
2. Centrifuge the reaction for 30 min at 4°C and 4,000*g* to collect the precipitate. Decant off the supernatant and dry the pellet under a vacuum for 10–15 min.
3. Suspend the pellet in 2 mL of 8 M urea, 500 μL 0.5 M EDTA pH 8.0, and 1 mL of formamide load dye. Vortex vigorously to resuspend all of the precipitate and heat to 65°C for 5 min to ensure a complete denaturation of the RNA.
4. Apply solution to a denaturing polyacrylamide gel (8–15% 29:1 acrylamide/bisacrylamide, 1x TBE buffer, 8 M urea). The dimensions of the gel we typically use are 35 cm wide, 23 cm long and 3 mm thick. The gel is electrophoresed in 1x TBE with constant power at 30 W until the desired RNA has migrated approximately 80% down the gel. The amount of time required for this step will depend on the size of the RNA. An analytical denaturing gel (described in **Section 3.6.1**) can be run prior to purification in order to gauge the extent of RNA migration relative to load dye migration.
5. Remove the spacers and the top plate from the gel. Then transfer the gel to a piece of plastic wrap. Gently, remove the other glass plate and place another piece of plastic wrap over the other side. In a dark room, place the gel on a fluorescent TLC plate and shadow the RNA with a short-wave UV lamp.
6. Working quickly to minimize UV-induced damage, outline the appropriate band on the plastic wrap with a marker.
7. Remove the other piece of plastic wrap and excise the band using a clean razor blade. Cut gel into equal sized pieces.
8. Place the gel slices into an electroeluter cell (three pieces side-by-side until the elutrap is full) and elute into 1x TBE at 100 V, constant voltage, at 4°C overnight. The elution of RNA from the gel slices can be monitored by removing them and shadowing them in a dark room using a short-wave UV lamp; they will be clear when the RNA is completely gone from the gel. An alternative protocol for extracting RNA out of a polyacrylamide gel is the "crush and soak" method as described by Golden *(36)*.
9. Remove the buffer from the trap and place in a 15 mL 10,000 molecular weight cutoff centrifugal concentrator. Centrifuge

the sample at 4,000*g* for 20 min at 4°C in order to concentrate the sample.

10. Add 15 mL of an appropriate exchange buffer (such as 10 mM K-HEPES, pH 7.5 or 10 mM Na-MES, pH 6.0). Centrifuge for 20 min at 4°C and 4,000*g*.
11. Repeat **Step 10** twice more.
12. After the final round of buffer exchange, concentrate the RNA to approximately 200 μL.
13. Calculate the concentration of the RNA by measuring the absorbance of an appropriately diluted RNA sample (1:1,000) at a wavelength of 260 nm. The extinction coefficient of the RNA can be calculated by summing the individual extinction coefficients for each nucleotide in the RNA.
14. Store the RNA at –20°C until use.

3.5. Purification of RNA Using Native Ni-NTA Chromatography

This method describes a means of native purification using an affinity-immobilization column, requiring the glmS-MS2 coat 3'-tag on the RNA of interest *(30)*. Another means to accomplish an entirely native purification of RNA on the milligram scale has been developed by Puglisi and coworkers that involves the use of size exclusion chromatography *(37)*. Each of these techniques has certain advantages that will influence their adoption for a specific RNA.

3.5.1. Synthesis of RNA for Native Purification

1. Amplify DNA template using the same protocol outlined in **Section 3.4.1**. However, the 10x Thermophilic DNA polymerase PCR buffer must be replaced by 10x HEPES Thermophilic DNA polymerase buffer (*see* **Note** 7).
2. Synthesis of RNA is performed according to the protocol of **Section 3.4.4**. However, 10x Transcription buffer is replaced by 10x HEPES Transcription buffer (*see* **Note** 7).

3.5.2. Preparation of HMM Tagging Protein

1. Transform pHMM expression vector into chemically competent BL21(DE3) cells and plate on LB-Agar supplemented with 30 μg/mL kanamycin. Incubate overnight at 37°C.
2. Grow a 50 mL of starter culture (10 aliquots of 5 mL culture) in LB supplemented with 30 μg/mL kanamycin at 37°C and 220 rpm until the OD_{600} reaches saturation.
3. Inoculate 1 L of LB broth containing 30 μg/mL kanamycin with 10 mL of starter culture and incubate at 37°C and 220 rpm while monitoring turbidity at OD_{600}.
4. When the OD_{600} reaches ~0.6, add IPTG to a final concentration of 0.5 mM. Continue incubation at 37°C and 220 rpm for 4 h.
5. Pellet cells by centrifugation at 4°C and 6,000*g* for 10 min. Suspend cell pellet in Lysis Buffer (37.5 mL of buffer per liter

of cell culture). The cell suspension can be stored at –80°C until further use.

6. Into a 600 mL stainless steel beaker, combine cell suspension aliquots to a total of 75 mL and add 500 μL of *E. coli* protease inhibitor solution.
7. Lyse the cells on ice with sonication for a total of 5 min (15 s bursts with a 45 s rest between pulses) at 75% total power.
8. Centrifuge the cell lysate at 30,000*g* at 4°C for 20 min. Immediately remove the supernatant and discard the pellet.
9. Apply the supernatant to a clean column containing a 10 mL bed volume of Ni-NTA agarose equilibrated in Lysis Buffer. Collect the flow-through in a single fraction.
10. Wash column with 20 column volumes of Lysis Buffer containing 20 mM imidazole.
11. Elute the protein with Lysis Buffer supplemented with 250 mM imidazole. Collect 10 mL fractions; the majority of the protein usually elutes in fractions 2 and 3. The concentration of protein can be obtained by measuring the A_{280} (the extinction coefficient for HMM protein is 83,310 $M^{-1}cm^{-1}$ and has a molecular weight of 59,050 g/mol).
12. Dialyze the protein solution exhaustively against three 1 L exchanges of a buffer containing 10 mM Na-MES, pH 6.0 and 10 mM NaCl at 4°C.
13. Apply protein to a Sepharose-SP column, and wash with ten column volumes of 10 mM Na-MES, pH 6.0 or until the absorption of the eluate returns to baseline, as monitored by the A_{280}. Elute the protein using a NaCl concentration gradient from 0 to 1 M in 10 mM Na-MES, pH 6.0.
14. Pool fractions containing protein and dialyze into Storage Buffer overnight at 4°C.
15. Determine the final protein concentration by measuring the A_{280} and store the solution at –20°C until needed.

3.5.3. Ni NTA Purification of RNA

1. Add 20 mg of HMM protein directly to the transcription reaction (*see* **Section 3.5.1**) and incubate the reaction at 37°C for 15 min to allow the protein to bind to full length RNA transcript.
2. Apply the reaction to a 5 mL bed volume of Ni-NTA agarose in a disposable 20 mL gravity flow column and collect the flow-through fraction.
3. Wash the column three times with 15 mL of ice-cold RNA Wash Buffer.
4. Initiate cleavage of the glmS ribozyme by adding 10 mL of room temperature RNA Wash Buffer supplemented with

1 μM glucosamine-6-phosphate (Gln6P) and collect this fraction.

5. Close the column at the bottom and apply 500 μL of RNA Wash Buffer to the top of the column in order to prevent the resin from drying, and allow the column to sit at room temperature for 15 min to allow the ribozyme to fully cleave.
6. Apply a second 10 mL aliquot of room temperature RNA elution buffer and collect the elution fraction.
7. Remove the bound protein using Stripping buffer. The recovered protein can be saved and purified again using the protocol described in **Section 3.5.2**.
8. Pool elution fractions containing RNA (to ensure fractions contain RNA check the absorption at 260 nm). Add these fractions to a 10,000 molecular weight cutoff centrifugal concentrator.
9. Concentrate the RNA sample by centrifugation at 4,000*g* and 4°C for 20 min.
10. Add 15 mL of an appropriate exchange buffer such as 10 mM K-HEPES, pH 7.5 or 10 mM Na-MES, pH 6.0. Centrifuge for 20 min at 4,000*g* and 4°C.
11. Repeat **Step 10** twice more.
12. After the final exchange, the RNA should be recovered in 200 μL volume.
13. Determine the concentration of the RNA by measuring the absorption at 260 nm in the fashion described in **Section 3.4.5**.
14. Store the RNA sample at −20°C. Alternatively, to avoid potential aggregation of the RNA upon freezing, it can be stored at 4°C if the RNA is to be used within 1 week.

3.6. Assaying RNA Quality

The simplest method for assaying overall RNA quality is to examine the homogeneity of the sample preparation by running denaturing and native polyacrylamide gels followed by staining with ethidium bromide or SYBR Green (Invitrogen). The denaturing gel reveals whether the purification was successful and if any significant RNA degradation occurred. A native gel is used to assess whether the RNA species is folded uniformly, multimerizing, or aggregating allowing the researcher to optimize a folding protocol. The gels are typically 6–12% acrylamide (29:1 acrylamide/bisacrylamide), depending on the RNA size and desired resolution (*see* **Fig. 9.7**). The analytical gel dimensions are 16.0 cm high, 20.0 cm wide, and 2.0 cm thick; this size of gel is run at 15 W, constant power.

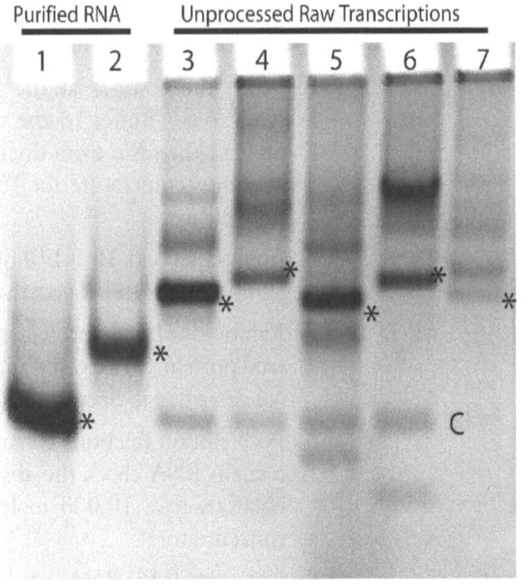

Fig. 9.7. Assessing RNA quality using an 8% native polyacrylamide gel electrophoresis as described in **Section 3.6.1** to identify well-folded phylogenetic variants of an RNA currently under investigation. *Asterisks* denote natively folded species and "C" highlights the 3'-H∂V tag that cleaved off during transcription (all of the RNAs were cloned into pRAV12). *Lanes 1 and 2* are the adenine and SAM-I riboswitches, respectively, purified using denaturing gel electrophoresis and refolded. *Lanes 3–7* represent samples from small scale (100 μL) trial transcriptions of lysine riboswitch variants from *Bacillus subtilis*, *Escherichia coli*, *Haemophilus influenzae*, *Shigella flexneri*, and *Thermotoga maritima*, respectively. The *B. subtilis* and *H. influenzae* appear to be mostly monomeric, suggesting that these may provide better starting points in the pursuit of diffraction-quality crystals. Also the relative intensities of the banding patterns suggest that the efficiency of transcription can be effected by the starting template as observed with *T. maritima*. Unlabeled bands above the monomer are likely multimeric, aggregate, or uncleaved species.

3.6.1. Analytical Polyacrylamide Gel Electrophoresis Under Denaturing Conditions

1. Remove the purified RNA sample from storage at –20°C and allow it to thaw at room temperature.
2. Mix 5 μL of RNA sample with 5 μL of formamide load dye, and load the RNA sample onto a denaturing acrylamide gel (6–12% (29:1) acrylamide/bisacrylamide, 8 M urea, 1x TBE buffer).
3. Run the gel at 15 W, constant power until the desired resolution is achieved.
4. Remove the spacers and gel from the plates. Place the gel in a bath of 1x TBE containing 1 μg/mL of ethidium bromide. Alternatively, SYBR Green may be used as per the manufacturer's instructions.
5. Visualize and assay the RNA quality by exposing the gel to short-wave UV light.

6. If there appears to be a significant number of extra bands on the gel, then it is recommended to repeat the purification procedure (*see* **Note 8**), as this RNA may not be suitable for crystallization trials (*see* **Section 3.7.1**).

3.6.2. Analytical Polyacrylamide Gel Electrophoresis Under Native Conditions

Prior to using the purified RNA sample for structural or biochemical studies, it is useful to assay the overall folded state of the sample preparation. This is especially important if the RNA of interest was purified under denaturing conditions. For small RNA species (<50 nucleotides in length) a heating step at 90°C for 3 min followed by cooling for 5 min on slushy ice is usually sufficient to refold the sample. Keep in mind that each RNA species is unique. Therefore, one folding protocol will not necessarily translate well from one RNA species to the next, even for very closely related RNAs in sequence.

1. Fold the purified RNA by starting with the simplest folding protocol and working your way toward more difficult strategies.
2. Mix 1 μL of folded RNA with 5 μL of type III load dye (30% (w/v) glycerol in H_2O, 0.05% xylene cyanol and 0.05% bromophenol blue) and load the sample onto the gel.
3. Run the native gel using a native gel running buffer (0.5x TBE, 2 mM $MgCl_2$) at 7 W of constant power (the gel should remain cool to the touch during electrophoresis) until the desired resolution is achieved.
4. Visualize and assay the RNA quality as described in **Steps 4 and 5** of **Section 3.4.1**.

3.7. Crystallization of RNA.

3.7.1. Screening of RNA for Crystallizability

The most important aspect of crystallography at this stage is to adopt techniques that are capable of yielding crystals in a reproducible fashion. An excellent resource that describes in detail many techniques pertaining to screening macromolecules for crystallization can be found in the Hampton Research Catalog (www.hamptonresearch.com). What follows is a brief overview of the initial screening step. Commercial screens including Natrix, Crystal Screen 1, Nucleic Acids Mini Screen and PEG-Ion (Hampton Research) are typically employed in early trials to identify good crystallization conditions. Further trials can be attempted using variation of drop size, screening temperature, and the evaporation time allowed before inversion of drops in order to identify suitable techniques. Initial conditions that yield crystals as identified from sparse matrix screening are then used as a starting point to optimize crystal growth *(38)*.

1. Add 500 μL of each crystallization solution from a sparse matrix screen to individual well in a Linbro tray (Hampton). Line the well edges with petroleum jelly to act as a seal when the glass cover slip is applied. Incubate the trays at room temperature for 30 min to allow the crystallization solutions to come to room temperature.
2. Remove an aliquot of RNA sample and pass through a 0.22 μm centrifugal filter to remove particulate matter that may interfere with the crystallization process. At this point, for our riboswitch RNAs, we add sufficient ligand to the solution to yield a 2:1 molar ratio of ligand/RNA.
3. Incubate the ligand/RNA complex at room temperature for 30 min.
4. Using a clean pressurized air source, blow off any particulate matter from individual cover slips and place them along the edge of the lid of the Linbro tray.
5. Pipette 2.0 μL RNA sample onto the center of a cover slip and add an equal volume of the mother liquor (well buffer). Invert the cover slip over the source well and press down lightly until the petroleum forms an air-tight seal.
6. Repeat **Step 5** until tray is complete.
7. Incubate the tray for 3–4 days at 30°C (our preferred temperature, but room temperature can suffice as long as there is not significant temperature fluctuations).
8. Examine drops under a stereomicroscope to determine if each drop remains clear, contains precipitate, or some form of crystalline material. RNA constructs that show crystallization (microcrystalline precipitate through discrete single crystals) under a number of conditions (usually >5% of conditions tested) become the focus of further screening and refinement of crystallization conditions.

3.7.2. A Simple Synthesis of Iridium(III) Hexammine

One of the primary reagents used for phase determination is iridium hexammine, a compound that is not commercially available. Nonetheless, its synthesis is very easy to accomplish using the following protocol to yield sufficient material for heavy-atom soaks or co-crystallization.

1. In a heavy-walled Ace pressure tube (Aldrich), add 2 g of iridium chloride ($IrCl_3$) and 35 mL (fill tube) of ammonium hydroxide. Seal the tube well with teflon tape, particularly around the O-ring seal. Screw the top on tightly and place it halfway into a silicone oil bath that is set at 150°C. Incubate for 4 days. (If volume decreases, remove, cool down, and refill with fresh ammonium hydroxide, but this means that the seal is leaking and you need to reseal with teflon tape).

2. Solution should be light-brown and clear with most of the solid dissolved at the end of the reaction period. Allow it to completely cool and then incubate it in an ice-water bath.
3. Pass solution through a sintered-glass filter to remove solid material.
4. In a 50°C waterbath, use a rotovap in order to distill the solution to dryness.
5. Suspend solid in 5 mL of water. Solid material will not completely go into solution; do not worry about this.
6. Remove two 500 μL aliquots and place each in a separate Eppendorf tube. Add 200 μL conc. HCl to each tube; a significant amount of white precipitate should come out of solution. Spin down for 1 min at 10,000*g* at room temperature.
7. Remove supernatant (usually light yellow) and wash pellet with a 1 mL solution of 2:1 (v/v) water/conc. HCl by vortexing. Centrifuge for 1 min at 10,000*g* at room temperature and remove supernatant.
8. Repeat **Step** 7 twice more.
9. Wash pellet three times with 10 mL of ice-cold absolute ethanol.
10. Air dry for 1 h and resuspend in ddH_2O in order to bring most of the white solid into solution. Spin down insoluble material and transfer supernatant to fresh Eppendorf tubes.
11. Take an absorbance spectrum of the material. There should be a clear peak at 251 nm. Calculate the concentration of iridium(III) hexammine using the extinction coefficient of 92 $M^{-1}cm^{-1}$.
12. Store solution at −20°C.

4. Notes

1. These plasmids are available from the laboratory of the authors upon request.
2. This approach is based upon the idea that a macromolecule's ability to crystallize is primarily determined by inherent properties such as the shape, the presence of functional groups on the molecule's surface, and conformational homogeneity in the sample. In this sense, the solution conditions and temperature act secondarily to promote a controlled precipitation event. Thus, we refer to "construct space" as the range

of RNA sequences that contain a particular motif of interest that can be tested for crystallization and "condition space" as the range of solution conditions (mother liquors) that can be tested.

3. The 13 sequences used in an initial screen for an RNA sequence that was crystallizable was: Env17 (5/2), Env15 (5/3), Bth (5/4), Env01 (5/5), Env12 (6/2), Bja3 (6/3), Mma2 (6/4), Atu2 (6/5), Rpa2 (6/6), Env49 (7/2), Env03 (7/6), Env13 (8/2), and Env24 (8/5) *(21)*. The numbers in parentheses refer to the length of the P1 and P2 helices in the pseudoknot, respectively. These sequences were chosen to maximize the diversity of the size of the pseudoknot within the library of RNAs. Conversion of the mRNA sequence into a SAM-binding RNA for crystallization involved using only the sequence between the 5'-side of the P1 helix to the 3'-side of the P2 helix. An additional guanine residue (*see* **Fig. 9.3C**, *circled*) was added to the 5'-end to ensure efficient transcription by T7 RNA polymerase and an adenosine residue was added to the 3'-end to facilitate cleavage of a downstream H∂V ribozyme.
4. The concentration of the inner oligonucleotides should be much less than the outer primers in order to limit the number of incorrect products and favor the full-length PCR product. A range between 500 and 100,000-fold excess of outer primers to inner oligonucleotides works well, such that in our typical PCR reaction the outer primers at 1 μM concentration while the inner primers are at 0.1 nM.
5. A molar ratio of between 3 and 6 of PCR insert to vector works well for ligation reactions. Therefore, to ensure a successful ligation it is advisable to first estimate the ratio of cut PCR product to cut vector. This can be done by simply running a 1% agarose gel and comparing the band intensities to a known amount of DNA, perhaps from a ladder. Then the molar quantity of PCR insert and vector can be estimated based on the molecular weight of each species.
6. It is important to use Pfu polymerase in the clone-free reactions until a full length template has been generated. This is due to Taq polymerase adding a non-templated adenosine to the 3' end of each strand resulting in insertions within the ribozyme during the recombination reaction. Conversely, Pfu polymerase creates clean 3' ends and has higher inherent fidelity yielding a template that is of the desired sequence.
7. The use of a HEPES buffer system for PCR and transcription reactions associated with the native purification method is necessitated by the use of the glmS ribozyme. It has been shown by several laboratories that this ribozyme is weakly

activated by Tris buffer, causing cleavage before the addition of the activating ligand GlcN6P *(39, 40)*.

8. The nature of contaminants will depend on the chosen purification method and the presence or absence of a 3'-end processing ribozyme. For RNA purification under denaturing conditions, an impure RNA sample most often is the result of a failure to resolve the sample RNA band from degradation products, the cleaved HδV ribozyme, or the uncleaved RNA–ribozyme sample depending on the RNA size. Either decreasing the percentage of acrylamide in the gel for large RNA species or increasing the percentage of acrylamide for small RNA species may alleviate this problem.

Acknowledgments

The contents of this work comprise much of the collected wisdom of a number of colleagues and members of the Batey laboratory. In particular, we would like to thank Jeffrey Kieft of the C.U. Health Sciences Center who has been instrumental in developing many of the ideas presented in this article. This work was made possible by a Research Scholar Grant from the American Cancer Society and support from the National Institutes of Health to R.T.B.

References

1. Pley, H.W., Flaherty, K.M. and McKay, D.B. (1994) Three-dimensional structure of a hammerhead ribozyme. *Nature* **372**, 68–74.
2. Scott, W.G., Finch, J.T. and Klug, A. (1995) The crystal structure of an all-RNA hammerhead ribozyme: a proposed mechanism for RNA catalytic cleavage. *Cell* **81**, 991–1002.
3. Cate, J.H., Gooding, A.R., Podell, E., Zhou, K., Golden, B.L., Kundrot, C.E., Cech, T.R. and Doudna, J.A. (1996) Crystal structure of a group I ribozyme domain: principles of RNA packing. *Science* **273**, 1678–1685.
4. Cate, J.H., Gooding, A.R., Podell, E., Zhou, K., Golden, B.L., Szewczak, A.A., Kundrot, C.E., Cech, T.R. and Doudna, J.A. (1996) RNA tertiary structure mediation by adenosine platforms. *Science* **273**, 1696–1699.
5. Klein, D.J., Schmeing, T.M., Moore, P.B. and Steitz, T.A. (2001) The kink-turn: a new RNA secondary structure motif. *EMBO J.* **20**, 4214–4221.
6. Kieft, J.S. and Batey, R.T. (2004) A general method for rapid and nondenaturing purification of RNAs. *RNA* **10**, 988–995.
7. Griffiths-Jones, S., Moxon, S., Marshall, M., Khanna, A., Eddy, S.R. and Bateman, A. (2005) Rfam: annotating non-coding RNAs in complete genomes. *Nucleic Acids Res.* **33**, D121–124.
8. Oubridge, C., Ito, N., Teo, C.H., Fearnley, I. and Nagai, K. (1995) Crystallisation of RNA-protein complexes. II. The application of protein engineering for crystallisation of the U1A protein-RNA complex. *J. Mol. Biol.* **249**, 409–423.
9. Price, S.R., Ito, N., Oubridge, C., Avis, J.M. and Nagai, K. (1995) Crystallization of RNA-protein complexes. I. Methods for the large-scale preparation of RNA suitable for crystallographic studies. *J. Mol. Biol.* **249**, 398–408.
10. Batey, R.T., Sagar, M.B. and Doudna, J.A. (2001) Structural and energetic analysis of

RNA recognition by a universally conserved protein from the signal recognition particle. *J. Mol. Biol.* **307**, 229–246.

11. Scott, W.G., Finch, J.T., Grenfell, R., Fogg, J., Smith, T., Gait, M.J. and Klug, A. (1995) Rapid crystallization of chemically synthesized hammerhead RNAs using a double screening procedure. *J. Mol. Biol.* **250**, 327–332.
12. Ferre-D'Amare, A.R. and Doudna, J.A. (2000) Crystallization and structure determination of a hepatitis delta virus ribozyme: use of the RNA-binding protein U1A as a crystallization module. *J. Mol. Biol.* **295**, 541–556.
13. Gilbert, S.D., Montange, R.K., Stoddard, C.D. and Batey, R.T. (2006) Structural studies of the purine and SAM binding riboswitches. *Cold Spring Harb Symp. Quant. Biol.* **71**, 259–268.
14. Grundy, F.J. and Henkin, T.M. (1998) The S box regulon: a new global transcription termination control system for methionine and cysteine biosynthesis genes in gram-positive bacteria. *Mol. Microbiol.* **30**, 737–749.
15. Winkler, W.C., Nahvi, A., Sudarsan, N., Barrick, J.E. and Breaker, R.R. (2003) An mRNA structure that controls gene expression by binding S-adenosylmethionine. *Nat. Struct. Biol.* **10**, 701–707.
16. McPherson, A. (1999) *Crystallization of Biological Macromolecules.* Cold Spring Harbor Press, Cold Spring Harbor, NY.
17. Pley, H.W., Flaherty, K.M. and McKay, D.B. (1994) Model for an RNA tertiary interaction from the structure of an intermolecular complex between a GAAA tetraloop and an RNA helix. *Nature* **372**, 111–113.
18. Doherty, E.A., Batey, R.T., Masquida, B. and Doudna, J.A. (2001) A universal mode of helix packing in RNA. *Nat. Struct. Biol.* **8**, 339–343.
19. Nissen, P., Ippolito, J.A., Ban, N., Moore, P.B. and Steitz, T.A. (2001) RNA tertiary interactions in the large ribosomal subunit: the A-minor motif. *Proc. Natl. Acad. Sci. U.S.A.* **98**, 4899–4903.
20. Montange, R.K. and Batey, R.T. (2006) Structure of the S-adenosylmethionine riboswitch regulatory mRNA element. *Nature* **441**, 1172–1175.
21. Corbino, K.A., Barrick, J.E., Lim, J., Welz, R., Tucker, B.J., Puskarz, I., Mandal, M., Rudnick, N.D. and Breaker, R.R. (2005) Evidence for a second class of S-adenosylmethionine riboswitches and other regulatory RNA motifs in alpha-proteobacteria. *Genome. Biol.* **6**, R70.
22. Gilbert, S.D., Rambo, R.P., Van Tyne, D., and Batey, R.T. (2008) Structure of the SAM-II riboswitch bound to S-adenosylmethionine. *Nat. Struct. Biol.* **15**, 177–182.
23. Baugh, C., Grate, D. and Wilson, C. (2000) 2.8 A crystal structure of the malachite green aptamer. *J. Mol. Biol.* **301**, 117–128.
24. Kieft, J.S., Zhou, K., Grech, A., Jubin, R. and Doudna, J.A. (2002) Crystal structure of an RNA tertiary domain essential to HCV IRES-mediated translation initiation. *Nat. Struct. Biol.* **9**, 370–374.
25. Carrasco, N., Buzin, Y., Tyson, E., Halpert, E. and Huang, Z. (2004) Selenium derivatization and crystallization of DNA and RNA oligonucleotides for X-ray crystallography using multiple anomalous dispersion. *Nucleic Acids Res.* **32**, 1638–1646.
26. Hobartner, C., Rieder, R., Kreutz, C., Puffer, B., Lang, K., Polonskaia, A., Serganov, A. and Micura, R. (2005) Syntheses of RNAs with up to 100 nucleotides containing site-specific 2'-methylseleno labels for use in X-ray crystallography. *J. Am. Chem. Soc.* **127**, 12035–12045.
27. Golden, B.L. and Kundrot, C.E. (2003) RNA crystallization. *J. Struct. Biol.* **142**, 98–107.
28. Keel, A.Y., Rambo, R.P., Batey, R.T. and Kieft, J.S. (2007) A general strategy to solve the phase problem in RNA crystallography. *Structure* **15**, 761–772.
29. Ferre-D'Amare, A.R. and Doudna, J.A. (1996) Use of cis- and trans-ribozymes to remove 5' and 3' heterogeneities from milligrams of *in vitro* transcribed RNA. *Nucleic Acids Res.* **24**, 977–978.
30. Batey, R.T. and Kieft, J.S. (2007) Improved native affinity purification of RNA. *RNA* **13**, 1384–1389.
31. Grodberg, J. and Dunn, J.J. (1988) ompT encodes the Escherichia coli outer membrane protease that cleaves T7 RNA polymerase during purification. *J. Bacteriol.* **170**, 1245–1253.
32. Zawadzki, V. and Gross, H.J. (1991) Rapid and simple purification of T7 RNA polymerase. *Nucleic Acids Res.* **19**, 1948.
33. Li, Y., Wang, E. and Wang, Y. (1999) A modified procedure for fast purification of T7 RNA polymerase. *Protein. Expr. Purif.* **16**, 355–358.

34. Ichetovkin, I.E., Abramochkin, G. and Shrader, T.E. (1997) Substrate recognition by the leucyl/phenylalanyl-tRNA-protein transferase. Conservation within the enzyme family and localization to the trypsin-resistant domain. *J. Biol. Chem.* **272**, 33009–33014.
35. Pleiss, J.A., Derrick, M.L. and Uhlenbeck, O.C. (1998) T7 RNA polymerase produces 5' end heterogeneity during *in vitro* transcription from certain templates. *RNA* **4**, 1313–1317.
36. Golden, B. L. (2007) Preparation and crystallization of RNA. *Methods. Mol. Biol.* **363**, 239–257.
37. Kim, I., McKenna, S.A., Viani Puglisi, E. and Puglisi, J.D. (2007) Rapid purification of RNAs using fast performance liquid chromatography (FPLC). *RNA* **13**, 289–294.
38. Bergfors, T. (2007) Screening and optimization methods for nonautomated crystallization laboratories. *Methods Mol. Biol.* **363**, 131–151.
39. McCarthy, T.J., Plog, M.A., Floy, S.A., Jansen, J.A., Soukup, J.K. and Soukup, G.A. (2005) Ligand requirements for glmS ribozyme self-cleavage. *Chem. Biol.* **12**, 1221–1226.
40. Roth, A., Nahvi, A., Lee, M., Jona, I. and Breaker, R.R. (2006) Characteristics of the glmS ribozyme suggest only structural roles for divalent metal ions. *RNA* **12**, 607–619.

Chapter 10

Locked Nucleic Acid Aptamers

Jan Barciszewski, Michael Medgaard, Troels Koch, Jens Kurreck, and Volker A. Erdmann

Abstract

The aptamer technology has been introduced in the early 1990s. With this technique ligands for organic dyes and proteins have been identified in many research field, providing various inhibitory molecules that allow functional interference in biological systems. Aptamers can therefore be employed for various applications ranging from diagnostic to therapeutic assay formats. Locked nucleic acid aptamers (LNA-Aps) are oligonucleotides containing one or more LNA nucleotide monomers with a bicyclic furanose unit locked in an RNA mimicking sugar conformation, evolved in vitro to bind target ligands with high affinity and specificity. LNA-Aps are attractive alternatives to antibody- and small-molecule-based therapeutics due to their stability, low toxicity and immunogenecity.

Key words: LNA – locked nucleic acid, LNA-As – LNA antisense, LNA-Ag – LNA antigene, LNA-Ap – LNA aptamer.

1. Introduction

LNA (locked nucleic acid) comprises a new class of bicyclic high-affinity RNA analogues (2′-*O*, 4′-*C*-methylene-β-D-ribofuranosyl nucleotide) in which the furanose ring of the ribose sugar is chemically locked in an RNA-mimicking conformation by the introduction of a 2′-*O*, 4′-*C*-methylene bridge (*see* **Fig. 10.1**). The flexibility of the ribofuranose ring is restricted and the structure is locked into a rigid C3'-*endo* conformation. LNAs are among the most promising candidates of the third generation of nucleic acids derivatives (for reviews, *see (1–5)*).

LNA-modified oligonucleotides exhibit high thermal stability when hybridized with their DNA and RNA target molecules *(3)*. Introduced LNAs increase melting temperature of +1 to +8°C and of +2 to +10°C per monomer against complementary DNA and

Günter Mayer (ed.), *Nucleic Acid and Peptide Aptamers: Methods and Protocols, vol. 535*

DOI 10.1007/978-1-59745-557-2_10 Springerprotocols.com

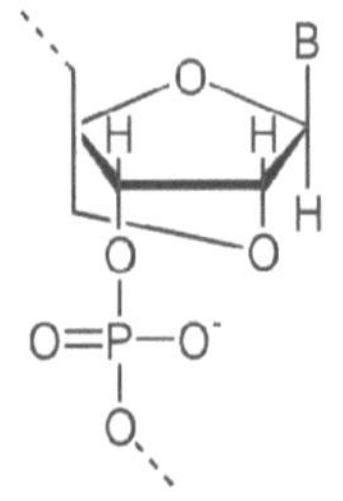

Fig. 10.1. Structure of locked nucleic acid (LNA).

RNA, respectively, compared to unmodified duplexes *(1, 2, 6, 7)*. Structural studies have shown interesting properties of oligonucleotides containing LNA and DNA residues. LNA nucleotides induce neighbouring DNA to adopt an overall A-type conformation in a DNA–RNA duplex *(8, 9)*.

Aptamers are single-stranded nucleic acid molecules that possess properties comparable to those of protein monoclonal antibodies, and thus are clear alternatives to antibody-based diagnostic and therapy *(10)*. They offer a valuable complement to loss-of-function phenotypic knockdown approaches and the assignment of novel activities to members of highly homologous protein families *(11)*. Like any other type of oligonucleotides, aptamers are highly susceptible to nucleolytic degradation and have therefore to be protected by the introduction of modified nucleotides. Conceptually, two different strategies can be followed: Modified nucleotides can be introduced during the SELEX procedure by their incorporation into the initial library or post-SELEX by site-specific engineering of the selected aptamer. Since LNAs are not compatible with the standard enzymes being required for the SELEX process, only the post-SELEX strategies has been followed to date.

The introduction of LNA monomers into an aptamer renders the molecule highly resistant toward degradation by nucleases. The additional functional groups can furthermore lead to ligands with novel physical and chemical properties or can provide additional handles to be utilized for functional improvement. Significant stabilization of aptamers has been achieved by introducing LNAs at various positions of the oligonucleotides. LNA building blocks can be incorporated into oligonucleotides by standard phosphoramidite chemistry; optimized protocols for efficient synthesis of LNA oligonucleotides will be outlined below. It should be noted that commercially provided LNA building blocks are thymine for uracil and 5-methylcytosine instead of cytosine *(12, 13)*.

LNA oligonucleotides can be divided into different classes based on the content of LNA monomers:

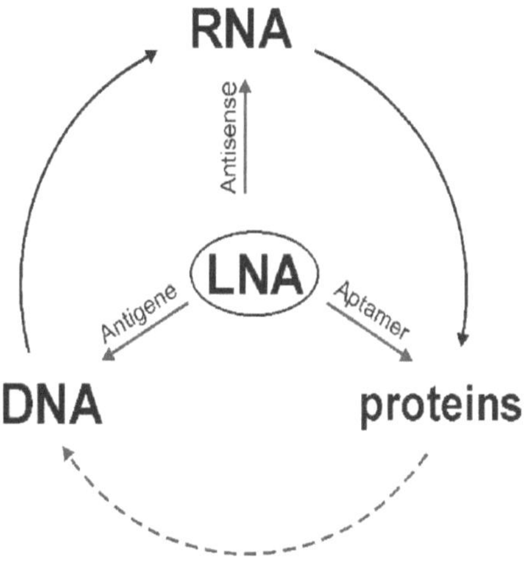

Fig. 10.2. Scheme presenting three different types of LNA oligonucleotides: antigene (Ag), antisense (As) and aptamers (Ap).

a. all-LNA,

b. LNA mixmers containing various combination of LNA and DNA residues,

c. LNA gapmers with LNA units at both ends of the oligonucleotide,

On the other hand LNA oligonucleotides can be divided into three different groups with respect to their application (*see* **Fig. 10.2**):

a. Antigenic LNA,

b. Antisense LNA,

c. LNA aptamers

The first two groups have exhaustively been reviewed recently *(4, 5)* and will therefore not be dealt with in the present chapter. In recent years, a number aptamers containing LNA building blocks have been described in the literature. Some representative examples will be described in the following sections. Finally protocols for efficient synthesis and purification of LNA oligonucleotides will be given.

1.1. LNA Aptamer to Tenascin-C

Tenascin-C is a large hexameric extracellular matrix protein that is newly expressed during tissue remodelling processes including angiogenesis, embryonic development, wound healing and tumour growth. Tenascin-C is abundantly and continuously expressed in neovasculature and tumour stroma. It is overexpressed in carcinomas of the lung, breast, prostate, and colon as well as lymphomas, sarcomas, glioblastomas and melanomas.

The RNA aptamer (TTA1) is an in vitro selected 39mer oligonucleotide (M_r 13.4 kDa) that structurally recognizes human tenascin C with very high affinity of a nanomoles range *(14)*. TTA1 labelled with technetium-99 m is a promising candidate for imaging of tumours expressing tenascin-C that has already been tested in clinical trials. A significant increase in stability of TTA1 against nucleolytic degradation has been achieved by replacement of pyrimidine nucleotides with 2'-deoxy-2'-fluoro derivatives and most purine nucleotides by 2'-deoxy-2'-*O*-methylnucleotides. TTA1 is a three-stem junction with the binding domain being located on the junction, while stem I is important for the structural stability of the whole aptamer and maintenance of binding activity. To achieve maximal thermal stabilization of stem I LNA building blocks were introduced. TTA1 derivatives with LNA modifications in the non-binding stem exhibited significant stem stabilization and markedly improved plasma stability while maintaining high binding affinity to the target *(15)*. In addition, a higher tumour uptake and longer blood retention was found in tumour-bearing nude mice. We consider the introduction of LNA modifications after the selection procedure to be generally applicable to improve the in vivo stability of aptamers without compromising their binding properties. The LNA variant of stem I of TTA1 was the first LNA/LNA duplex to be crystallized *(16)* and will provide first crystallographic insight into this type of modified hybrid.

1.2. TAR RNA Aptamers

The TAR RNA element contributes to the regulation of the HIV-1 genome transcription through the interaction with several proteins, including the retroviral *trans*-activator protein Tat. RNA aptamers have been developed that recognize the TAR RNA hairpin through loop–loop interactions. The strongest ligand termed R06 is characterized by a dissociation equilibrium constant of about 20 nM. The key features of such a so-called kissing loop interaction reside in the hairpin structure of the aptamer with a stem of at least 3 bp, in a loop complementary to the TAR apical one and in a G–A combination of residues at the positions closing the aptamer loop.

The RNA hairpin R06 selected against the TAR RNA element of HIV-1 was truncated to generate the aptamer $R06_{16}$, which retains the originally selected eight nucleotide loop and a 4 bp stem as well as the binding properties of the parent aptamer. The modified analogue $LD06_{16}$ contains two base pairs at the bottom of the stem and LNA residues were incorporated at positions T6, C8, A10 and G11 of the loop. The chimeric oligonucleotide adopts an A-type helix geometry and mimics selected RNA aptamers. $LD06_{16}$ forms a complex with the RNA hairpin corresponding to the apical loop of the TAR element through a loop–loop (kissing loops) interaction. In contrast, the antisense sequence corresponding to the apical loop of the chimera $LD06_{16}$ binds as strongly to TAR as to the LNA/DNA aptamer. The fully LNA-modified octamer is an even stronger ligand than the antisense chimera. This means that the loop–loop helix in the kissing complex cannot assimilate to a canonical A-type double helix *(17)*.

The LNA/DNA aptamer discriminates between two TAR variants, differing only by a single mutation in the loop, whereas the LNA/DNA antisense sequence does not. The modified aptamer competes with a peptide derived from the viral protein Tat for binding to TAR, while no such competition was observed for the antisense LNA/DNA octamer *(18)*.

The TAR RNA binding sites of the aptamer and the Tat peptide, the apical loop and a bulge, 4 bp below the loop, respectively, do not overlap. This suggests a teleo-specific effect. The aptamer, but not the antisense octamer, switches the conformation of the TAR bulge region to a structure that is no longer recognized by the Tat peptide.

A fully modified version of the aptamer R06 with LNA is a poor TAR ligand suggesting that the geometry of the LNA loop is probably too constrained for optimal loop–loop interaction. This demonstrates that a post-selection modification of an aptamer is a challenging task. Chimeric LNA/DNA derivatives with an LNA loop and a DNA stem were also less efficient TAR ligands, similarly to the DNA version of the RNA aptamer, which did not recognize the TAR hairpin. One can conclude that the loop geometry plays a crucial role for target recognition *(19)*.

It is known that the antisense RNA octamer corresponding to the aptamer loop is not a good TAR ligand. On the other hand, the LNA/DNA antisense octamer corresponding to the loop of the chimeric aptamer showed the same affinity for the viral target as the aptamer, however, the two variants did not behave kinetically similarly *(20)*. Although the stability of interactions of the aptamer with its target is increased by the introduction of LNA units it also induces structural modifications which interfere with the loop–loop interaction at the same time. This means that there is a competition between intrinsic stability caused by LNA and unfavourable structural constraint involved by LNA inclusion in the junction regions. Since the aptamer recognizes a site (the apical loop) that does not overlap the Tat binding site (the bulge region), this supports the above mentioned assumption that the aptamer switches the TAR element to a conformation no longer appropriate for Tat binding *(20)*.

1.3. LNA Decoys for the Transcriptional Factor NF-κB

Decoys are a class of oligonucleotides that is conceptually related to aptamers. They are double-stranded oligonucleotides designed to possess binding sites for transcription factors and to compete with the native DNA sequence for available transcription factors.

The Rel/NF-κB family of inducible transcription factors mediates cellular responses to a variety of stimuli by regulating the expression of hundreds of genes with distinct functions. These cellular activities include immune and inflammatory responses, programmed cell death, growth, proliferation, and development. NF-κB dimers bind transcriptional regulatory DNA elements that are κB-DNA binding sites.

LNA nucleotides were incorporated into the termini of a decoy molecule, outside of the NF-κB binding sequence, to generate LNA–DNA–LNA copolymers resistant to nucleolytic hydrolysis without interfering with transcription factor binding. Significant stabilization was achieved when LNA monomers were included within the NF-κB site. On the other hand, modification of the κB binding site with LNA units increases thermal stability and decreases the affinity of NF-κB for the target sequence. This effect was dependent on positioning of the LNA substitutions *(21)*. Although less efficient in binding, this decoy molecule has been shown to compete with a DNA probe in the nanomolar range of concentrations *(22)*. It seems that LNAs induce changes in the molecular structure of the κB-binding sequence, which are proportional to the extent of modification, leading to a different degree of perturbation of the interactions with NF-κB. This is due to the C3'-*endo* conformation of LNA, which introduces a higher population of the N-type conformations of the neighbouring unmodified nucleotides on the same strand, resulting in a local change in the phosphate backbone geometry. When LNA nucleotides are incorporated into several positions in both strands of the duplex, a cooperative effect of the local conformational changes introduced by each mutation has been observed. It results in changes of the helix structure towards A-type geometry, which in contrast to the DNA B form is not optimal to sustain interactions between the transcription factor and LNA–DNA mix-mers. The B-type structure can be retained after substitutions with α-L-RNA *(22)*. Further details about the improvement of decoy oligonucleotides by the introduction of LNAs are given in *(9)*.

1.4. LNA Thrombin Aptamer

Recently, a thrombin binding aptamer was described that acts as a strong anticoagulant in vitro and inhibits thrombin-catalysed activation of fibrinogen and thrombin induced platelet aggregation. The aptamer is 15mer DNA molecule. Its solution structure is characterized by a chair-like quadruplex structure consisting of two G-tetrads connected by two TT loops and a single TGT loop. The aptamer is a thrombin inhibitor with a K_D of 2 nM that is considered to be useful as an anticoagulant during coronary artery bypass procedures. In order to improve the properties of the thrombin aptamer, LNA residues can be incorporated in the antiparallel quadruplex forming oligonucleotide. It seems that the overall structure of the LNA-modified aptamer is similar to its parental oligonucleotide. Base stacking of the two quadruplexes is quite similar, with the five-membered rings of the top guanines overlapped to the five-membered rings of the underneath guanine bases *(23)*. The biological data for the thrombin aptamer containing a G-LNA residue imply the modification of the original structure to influence the specific activity. However, the analysis of the overall folding of the modified aptamer does not argue for a significant reduction of its biological activity. It suggests that the mode of action of the thrombin aptamer requires a wider recognition process that involves even locally a single residue.

1.5. LNA Synthesis

LNA oligonucleotides are synthesized by the phosphoramidite approach. They are protected with a 4,4'-dimethoxytrityl (DMT) group on the 5'-OH and functionalized with the 2-cyanethyl, *N,N*-diisopropyl-phosphoramidite moiety on the 3'-OH. As a consequence of this classic amidite structure chimeric LNA, DNA and RNA oligonucleotides can easily be synthesized on a common DNA-synthesizer *(24–26)*. Synthesis of LNA comprising both phosphordiester and phosphorothioate internucleoside linkages has been performed in small scales (0.2 and 1 μmol) on an Expedite DNA synthesizer and has previously been published *(26)*. Small-scale synthesis of phosphordiester LNA has also been performed on an ABI 3900 high-throughput synthesizer (www.exiqon.com). Larger scale synthesis of LNA has been reported on the Äkta system platform (i.e. Oligo Pilot 10, 100 or 400) in scales ranging from 1 μmol to 60 mmol *(27)*. Here we report for the first time synthesis of LNA on the MerMade 12 DNA synthesizer (*see* **Note 1**). On this platform LNA with phosphordiester and phosphorothioate internucleoside linkages and mixtures thereof can be synthesized in scales ranging from 1 to 200 μmol, i.e. the MerMade12 DNA synthesizer can synthesize up to 12 different oligonucleotides (1 μmol scale) in a single run. The instrument can be reconfigured for synthesis in larger scales while maintaining the capability to perform small-scale synthesis and hence four 1 and four 100 μmol syntheses can be performed in a single run. In this chapter we will describe in detail LNA synthesis on the MerMade12 DNA synthesizer. We will provide the synthesis protocols, materials and methods, work-up, purification, analytical procedures for LNA synthesis and finally show chromatograms of the crude and purified material.

2. Materials

The materials and reagents mentioned here are for LNA phosphorothioate synthesis. The materials and reagents are listed here as required for the individual steps in the entire process: Synthesis, cleavage from solid support, deprotection, purification, fraction analysis, desalting, and final QC-analysis of the purified oligonucleotide.

2.1. Synthesis

1. MerMade 12 DNA synthesizer configured for 100 μmol synthesis.
2. Reagents: deblock (3% dichloroacetic acid in dichloromethane), anhydrous acetonitrile, Cap A, Cap B (commercially available, e.g. from Sigma-Aldrich), DEA (20% diethylamine and 1% 2,6-lutidine in acetonitrile), sulfurizing reagent (0.0225 M xanthan hydride in 20% pyridine/acetonitrile), Activator (0.5 M 4,5-dicyanoimidazole in anhydrous acetonitrile).
3. LNA phosphoramidites: LNA-A^{Bz}, LNA-T, LNA-G^{DMF}, LNA-mC^{Bz} (Glen, Exiqon) (*see* **Note 2**).
4. DNA phosphoramidites: A^{Bz}, T, G^{iBu}, C^{Bz} .
5. Suitable support, e.g. primer support 200 (GE Healthcare) and an empty synthesis column (www.bioautomation.com, product no. MM12-6-50).
6. Anhydrous acetonitrile and dichloromethane for dissolving amidites.
7. Argon.

2.2. Cleavage from Solid Support and Deprotection

1. 50 ml or 100 ml Blue-cap bottle.
2. 2 ml Acetonitrile.
3. 40 ml Concentrated aqueous ammonia.
4. Oven.
5. Büchner funnel (pore size 3), 250 or 500 ml round-bottomed flask, adaptor and vacuum pump.
6. Ethanol/water solution (1:1).
7. Rotavapor.
8. Dry ice and acetone.
9. Lyophilization equipment.

2.3. Purification

1. Anion exchange column (e.g. custom-packed column with Source15Q (GE Healthcare) with a column volume of approximately 130 ml).
2. Suitable HPLC system, e.g. Äkta Pilot HPLC. Optionally, the HPLC can be equipped with fraction collector set up for prep-mode chromatography.
3. Centrifuge and centrifuge tube (50 ml).
4. NaOH and NaCl for Buffer A: 10 mM NaOH, Buffer B: 2 M NaCl in 10 mM NaOH.

2.4. Fraction Analysis (see Note 3)

1. RP-UPLC column (e.g. Waters Acquity, BEH C18, 1.7 μm, 2.1 × 50 mm).
2. UPLC system (e.g. Waters Acquity UPLC connected to an electrospray mass spectrometer).
3. Buffer A: 0.2 M hexafluoroisopropanol, 8.4 mM triethylamine, 5% MeOH/MilliQ water.
4. Buffer B: 0.2 M hexafluoroisopropanol, 8.4 mM triethylamine, 60% MeOH/MilliQ water.

2.5. Desalting

1. Suitable HPLC system, e.g. Äkta explorer HPLC with UV and conductivity monitor Reservoir that can be incorporated into the flowpath of the HPLC system (*see* **Note 4**).
2. Pellicon 2 Mini Ultrafiltration Module and a Pellicon Mini Cassette Holder (Millipore).
3. 0.1 M HCl.
4. Milli Q water.
5. 500 ml Round bottomed flask.
6. Dry ice and acetone.
7. Lyophilization equipment.

2.6. Analysis of LNA (see Note 3)

1. RP-UPLC column (e.g. Waters Acquity, BEH C18, 1.7 μm, 2.1 × 50 mm).
2. UPLC system (e.g. Waters Acquity UPLC connected to an electrospray mass spectrometer).
3. Buffer A: 0.2 M hexafluoroisopropanol, 8.4 mM triethylamine, 5% MeOH/MilliQ water.
4. Buffer B: 0.2 M hexafluoroisopropanol, 8.4 mM triethylamine, 60% MeOH/MilliQ water.

3. Methods

Synthesis of LNA oligonucleotides can be performed in many different synthesis scales on the MerMade DNA synthesizer. The cycle consumption for the 100 μmol synthesis – described here in detail – as for other synthesis scales are listed in **Table 10.1**.

The oxidizer position on the instrument is used for synthesizing phosphordiester oligonucleotides, but for the synthesis of LNA phosphorothioates in a 100 μmol scale, the bottle position is used for a solution of a weak base used to remove the cyanoethyl protective group prior to the cleavage and deprotection step. The presence of the cyanoethyl protective group in the cleavage and

Table 10.1
Overview of the amount of reagents used on the MerMade DNA synthesizer for various synthesis scales. Amidites are used at a concentration of 0.1 M. Activator is 0.5 M DCI, Deblock is 3% dichloroacetic acid in dichloromethane, Sulphur is 0.0225 M xanthan hydride in 20% pyridine/acetonitrile. Oxidize is 20% diethylamine, 1% 2,6-lutidine in ACN or commercially available oxidize solution. Cap A and B are commercially available capping reagents

Synthesis scale (μmol)	DNA (ml)	Activator (ml)	LNA (ml)	Activator (ml)	Deblock (ml)	Sulphur (ml)	Capping (Cap A + B, ml)	ACN wash (complete cycle, ml)	Oxidize (ml)
1	0.16	0.22	0,2	0.3	0.5	0.25	0.3	2.1	0.2
10	0.8	1.2	0,8	1.2	3	1	1	7	0.75
50	2	3	2	3	9	6	3	21	6
100	3	4.5	3	4.5	21	12	6	45	12
200	6	9	6	9	30	20	12	91	20

deprotection step will give rise to an acrylonitrile adduct, which will appear as an impurity after the purification.

The instrument dispenses the reagents on top of the support and the reagent flows through the support by applying vacuum. Reagents are kept in bottles under Ar pressure and dispensing of a reagent is performed when a valve is opened, thus allowing a flow. The Ar pressure and vacuum can be regulated and hence the "open-times" of valves is dependent on these settings and must be configured individually.

Synthesis of an LNA oligonucleotide on the MerMade can be divided into three parts: The run, the cycle and the module

The run comprises the entire synthesis of the oligonucleotide and is composed of a number of cycles.

The cycle comprises a sequence of modules that each defines a unit operation. The cycles are shown in **Table 10.2** and the modules that the cycles are composed of are shown in **Table 10.3**.

Before synthesizing LNA oligonucleotides on a MerMade instrument, the cycles and modules must be programmed according to the values from **Tables 10.2** and **10.3**, but it is recommended that the user calibrates the valves and ensures that applied vacuum pulses are sufficient to drain the columns properly.

As can be seen from **Table 10.1**, the 100 μmol synthesis method uses fewer equivalents of amidites as compared to, e.g. a 1 μmol synthesis. In order to compensate for this reduced amount rather long coupling times are used. Compared to DNA coupling cycles, LNA coupling cycles have a prolonged reaction time for the coupling reaction and for the thiolation of the phosphit triester.

3.1. Synthesis of LNA Phosphorothioates

Oligonucleotides synthesized in a 100 μmol scale are synthesized without the final DMT group and are purified by anion-exchange chromatography.

1. Dissolve DNA and LNA amidites in dry bottles according to **Table 10.4**. LNA-mC is dissolved in anhydrous dichloromethane (*see* **Note 5**). The remaining LNA and DNA amidites are dissolved in anhydrous acetonitrile.
2. Add molecular sieves to the solution of the amidites.
3. Install the reagents and amidites on the instrument and prime thoroughly.
4. Prepare a synthesis column by placing a bottom frit and add 0.5 g primer support.
5. Add acetonitrile (3 ml) to the column to swell the support.
6. Let the acetonitrile drip through the support and place a top frit in the column.
7. Install the column on the instrument.

Table 10.2
The steps in the pre-, synthesis-, and post-cycle. Each step in the cycle is described in detail in Table 10.3. The pre-cycle specifies the steps performed before the synthesis is initiated, the synthesis cycle specifies the actual steps performed in a synthesis cycle and the post-cycle specifies the steps performed after the synthesis is complete. The Deblock steps in the post-cycle are skipped if a DMT-ON synthesis is selected. The solution of DEA/2,6-lutidine is placed on the oxidize position and hence these three oxidize steps in the post-cycle are wash steps with a solution of these weak bases in acetonitrile in order to remove the cyanoethyl protective group

Pre-cycle	Synthesis cycle	Post-cycle
ACN wash	Deblock	Deblock
ACN wash	Deblock	Deblock
ACN wash	Deblock	Deblock
	ACN wash	ACN wash
	ACN wash	ACN wash
	ACN wash	Oxidize
	Coupling	Oxidize
	ACN wash	Oxidize
	Sulphur	ACN wash
	Sulphur	ACN wash
	ACN wash	ACN wash
	Capping	
	ACN wash	
	ACN wash	
	Drain	

8. Setup a synthesis using the instruments software. Select DMT-OFF when setting up the instrument for the synthesis. The parameters for programming the synthesis cycles in the 100 μmol scale are described in **Tables 10.2** and **10.3**.
9. Run the synthesis and follow the progression in the log and by monitoring the trityl monitor.

Table 10.3
The instrument settings for each step in a synthesis cycle. The settings are applicable for an instrument with the following settings: vacuum: 16 in Hg (406 torr), amidite pressure: 6 psi, modifier pressure: 6 psi, reagent pressure: 7 psi. The tubing from the column to the vacuum valve is a 1/8" tubing.

	Deblock	Cap A	Cap B	Sulphur	ACN wash	DNA	LNA	Activator	Oxidize
Injection time (ms)	23,000	8,000	8,300	14,500	15,000	8,800	8,800	10,700	14,500
Injection volume (μl)	7,000	3,000	3,000	6,000	6,400	3,000	3,000	4,500	6,000
No. of subinjections	1	6	–*	1	1	8	8	–‡	1
Belay before injection (ms)	30	30	30	30	30	30	30	30	30
Priming	Before use	Before use	Before use	Before use	Before use	Before use	Before use	Before use	Before use
No. of primes	1	1	1	1	1	1	1	1	1
Prime injection time (ms)	50	50	50	50	50	50	50	50	50
Wait time (s)	40	120	0	150 (DNA) 300 (LNA)	30	360	480	0	240
Length of vacuum pulse (ms)	530	430	0	300	700	400	400	0	400

(continued)

Table 10.3 (continued)

	Deblock	Cap A	Cap B	Sulphur	ACN wash	DNA	LNA	Activator	Oxidize
Wait to vacuum pulse (s)	4	12	0	15 (DNA) 30 (LNA)	5	20	30	0	10
No. of vacuum pulses	10	10	0	10	6	12	12	0	12
Drain time (s)	15	25	0	25	15	30	30	0	25
Equalize time	2	2	0	2	2	2	2	0	2

*The number of subinjections for Cap B is specified by the number of subinjection for Cap A. Hence, six subinjections are used for Cap B. ‡The number of subinjections for the activator is specified by the number of subinjections for the amidite. Hence, eight subinjections are used for the activator.

Table 10.4
The amounts of amidite and solvent for preparing 0.1 M solutions of amidites in anhydrous acetonitrile/dichloromethane

Monomer	Concentration (M)	Amount (g)	DNA Volume (ml)	LNA Volume (ml)
A	0.1	5	58.3	56.4
T	0.1	5	67.1	64.7
G(DMF)	0.1	5	60.6	58.6
G(iBu)	0.1	5	59.5	57.6
C	0.1	5	60.7	57.1

3.2. Deprotection of LNA

1. Empty the synthesis column into a blue-cap bottle.
2. Swell the solid support in 2 ml acetonitrile.
3. Add 40 ml concentrated aqueous ammonia.
4. Seal the bottle and place it in the oven at 55°C for at least 5 h or overnight.
5. Allow the bottle to cool to room temperature before opening.
6. Filter the suspension on a Büchner funnel.
7. Wash the support with ethanol/water (min 150 ml).
8. Remove excess ammonia on a rotavapor.
9. Freeze the sample and lyophilize.

3.3. Purification of LNA (see Note 6)

1. Prepare the buffers for preparative HPLC.
2. Prepare the HPLC system by equilibrating the column in 10 mM NaOH (Solvent A) with at least four column volumes (CV) with a flow rate of 65 ml/min. The HPLC column is kept at room temperature.
3. Dissolve the oligonucleotide in 40 ml buffer A (10 mM NaOH) and transfer the solution to a centrifuge tube.
4. Spin the solution in a centrifuge for 15 min at 4,000 rpm.
5. Load the solution onto the column and leave the pellet (if any) in the centrifuge tube.
6. Purify the LNA oligonucleotide by running a gradient of 10 mM NaOH (A) and 2 M NaCl in 10 mM NaOH (B): 0% B in 3CV, 0–27.5% B in 2CV, 27.5–32.5% B in 5CV, 32.5–80% B in 17.5CV, 80–100% B in 2CV with a flow of 65 ml/min.
7. Monitor the absorbance at 260 nm and collect 150 ml fractions as the peak eludes from the column (*see* **Fig. 10.3**).

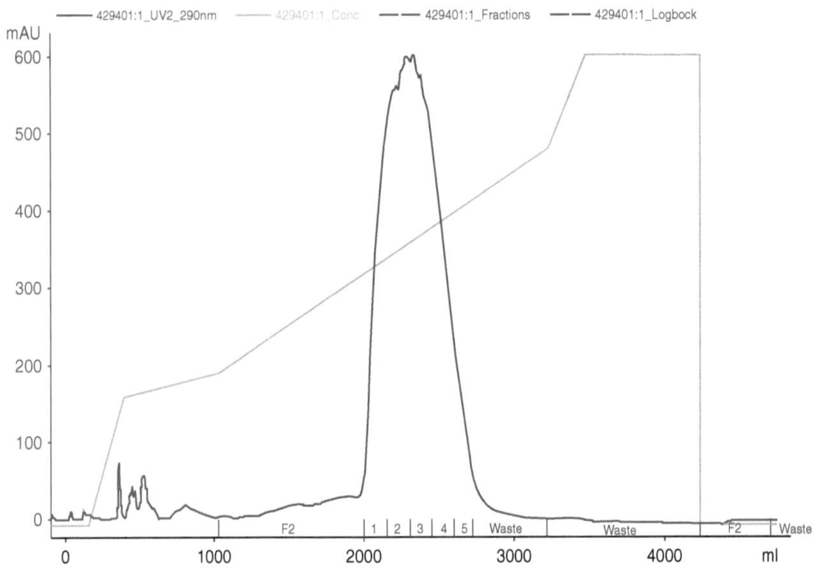

Fig. 10.3. HPLC chromatogram of the anion exchange HPLC purification (**Sections 2.3** and **3.3**). The UV absorbance is plotted at 290 nm to avoid saturation of the signal. The gradient as percentage B is shown and the collected fractions are depicted by numbers.

3.4. Fraction Analysis (see Note 3)

1. Prepare the buffers described in **Section 2.4**. Install the column and buffers and prime the buffers.
2. Equilibrate the column with at least four column volumes of starting conditions: 4%B
3. Inject 2 μl of the sample and analyse by running the following gradient with a flow rate of 0.5 ml/min with a column temperature of 65°C: 0–5.3 min: 4–50%B; 5.3–5.4 min: 50–100%B; 5.4–6.4 min: 100%B; 6.4–6.5 min: 100–4%B; 6.5–10,0 min: 4%B.
4. Analyse the chromatograms at 260 nm (*see* **Fig. 10.4**). Select the fractions with a purity $\geq$ 80%.

3.5. Desalting

1. Pool the selected fractions and adjust the pH to 7–8 with diluted HCl.
2. Install the pellicon filter in the holder and connect it to the system according to the manufacturer's protocol. Create a closed circuit, i.e. the flow must go from a reservoir through the pump to the filter, to the detector and back to the reservoir. In the flowpath, a pressure regulator can be inserted at the exit from the filter to adjust the pressure during the diafiltration (*see* **Note 4**).
3. Start the HPLC's pumps with a flow rate of 65 ml/min and adjust the pressure over the membrane to approximately 1 bar.

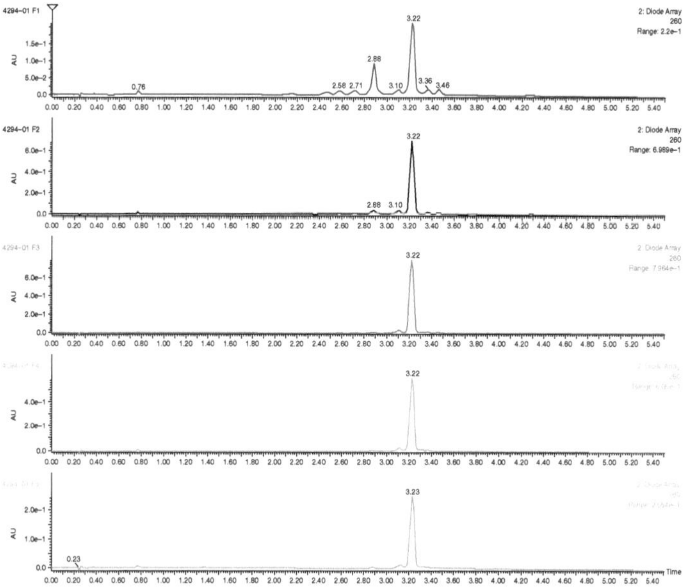

Fig. 10.4. Analysis of fractions obtained from the anion exchange HPLC purification. The fractions were analysed according to **Sections 2.4** and **3.4**.

Load the sample by pouring it into the reservoir. As the desalting progresses, a flow of water goes from the filter to the waste and the sample is concentrated which can be monitored by the UV detector.

4. When the sample has been loaded, water is added stepwise until a uniform conductivity level over 2 or more water additions has been reached. The typical time for the desalting of a 100 μmol synthesis is 120 min.
5. When the sample has been desalted, stop the flow and change the flow path from filter to reservoir to a flow from the filter to a round bottomed flask. Release the pressure over the membrane and start the pump. Wash the reservoir until the UV detector reaches the baseline.
6. Change the flow path from filter to flask to filter to waste. Wash with minimum 300 ml Milli Q water and stop the pump.
7. Freeze the desalted sample in a dry-ice/acetone bath and lyophilize.

3.6. Analysis of LNA (see Note 3)

1. Prepare the buffers described in **Section 2.6**. Install the column and buffers and prime the buffers.
2. Equilibrate the column with at least four column volumes of starting conditions 4%B.

3. Dissolve the purified and desalted oligonucleotide in Milli Q water (1 mg/ml).
4. Inject 2 μl of the sample and analyse by running the following gradient with a flow rate of 0.5 ml/min with a column temperature of 65°C: 0–5.3 min: 4–50%B; 5.3–5.4 min: 50–100%B; 5.4–6.4 min: 100%B; 6.4–6.5 min: 100–4%B; 6.5–10.0 min: 4%B.
5. Analyse the chromatogram at 260 nm (*see* **Fig. 10.5**).
6. The mass of the oligonucleotide can be found by deconvoluting the mass spectrum of the main peak.

LNA phosphoramidites are commercially available and LNA synthesis has been demonstrated on several synthesis platforms *(26, 27)*. Here, we have described the synthesis of LNA on a MerMade instrument in a 100 micromol synthesis scale. Under routine circumstances a 100 micromol synthesis scale will provide ca. 200 mg purified material. As a representative for a routine synthesis we have chosen synthesis of SPC4294 a fully thiolated 16-mer LNA gapmer having three LNAs in both the 5'- and the 3'- end of the molecule. The amount of crude material was 540 mg, and of this the target molecule represented 75% as

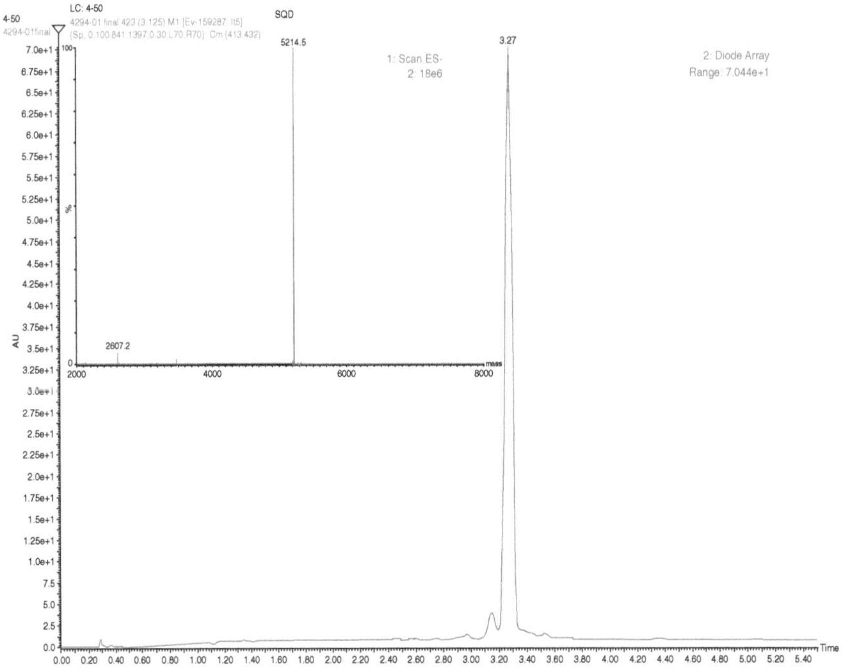

Fig. 10.5. Chromatogram from the analysis of the finished product according to **Sections 2.6** and **3.6**. The mass

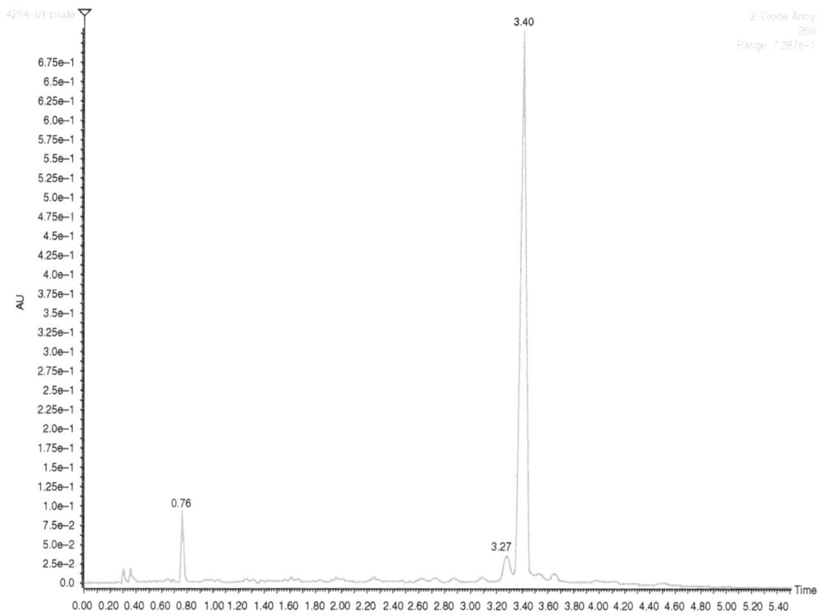

Fig. 10.6. Crude analysis of an oligonucleotide synthesized using the described protocol. The oligonucleotide has been analysed according to **Sections 2.4** and **3.4**.

determined by UPLC (*see* **Fig. 10.6**). After purification and desalting, the total yield was 240 mg with the target molecule integrating 88% on UPLC (*see* **Fig. 10.5**). In conclusion, synthesis of LNA by this protocol provides full length products in the range 40–45% of the theoretical yield with purities of normally in the range of 85–95% of the target molecule.

4. Conclusion

Although the first therapeutic aptamers are currently emerging, a potential shortcoming that narrows their therapeutic utility to targets on the cell surface or secreted proteins lies perhaps in their nucleic acid nature. Being negatively charged aptamers will not easily cross cellular membranes. However, a wide range of delivery systems have been developed to facilitate intracellular delivery of therapeutic nucleic acids in general, many of which are likely to work for aptamers as well. Taken together, LNA aptamers represent versatile tools for the functional characterization of biomolecules, their detection, therapeutic intervention, and the development of small molecules that serve as pharmaceutical lead compounds. LNA aptamers are now established as key players on these stages, and it can be expected that they will continue to open up entirely new possibilities in the future. In addition to that LNA aptamers offer an exciting novel interface between target validation and drug screening as a biologically active aptamer can be used to identify functionally equivalent small molecules directly in competitive high-throughput screening assays.

5. Notes

1. Different MerMade synthesizers are available with the emphasis on small scale synthesis of oligonucleotides, e.g. the MerMade 384 synthesizes oligonucleotides in 96-well plates and can synthesize up to 384 oligonucleotides in one run.
2. LNA amidites are commercially available at e.g. Glen and Exiqon.
3. For analysis on LNA using ion-exchange HPLC or LC-MS *(26)*.
4. Commercially available instruments specialized for a desalting step are available – e.g. Crossflow (GE Healthcare) and Cogent M (Millipore).
5. The amidite can also be dissolved in anhydrous THF and diluted with anhydrous acetonitrile – www.exiqon.com.
6. It might be advantageous to analyse the crude material before initiating the purification process to confirm the mass of the oligonucleotide and ensure that crude purity is at least 60% as determined by the area% at 260 nm. Deletion sequences, e.g. *n*-1 and *n*-2 are difficult to remove by preparative anion-exchange chromatography.

References

1. Jepsen, J.S. and Wengel, J. (2004) LNA-antisense rivals siRNA for gene silencing. *Curr. Opin. Drug Discov. Devel.* 7, 188–194.
2. Vester, B. and Wengel, J. (2004) LNA (locked nucleic acid): high-affinity targeting of complementary RNA and DNA. *Biochemistry* **43**, 13233–13241.
3. Braasch, D.A. and Corey, D.R. (2001) Locked nucleic acid (LNA): fine-tuning the recognition of DNA and RNA. *Chem. Biol.* **8**, 1–7.
4. Grunweller, A. and Hartmann, R.K. (2007) Locked nucleic Acid oligonucleotides: the next generation of antisense agents? *BioDrugs* **21**, 235–243.
5. Kaur, H., Babu, B.R. and Maiti, S. (2007) Perspectives on chemistry and therapeutic applications of Locked Nucleic Acid (LNA). *Chem. Rev.* **107**, 4672–4697.
6. Petersen, M. and Wengel, J. (2003) LNA: a versatile tool for therapeutics and genomics. *Trends Biotechnol.* **21**, 74–81.

7. Kurreck, J., Wyszko, E., Gillen, C. and Erdmann, V.A. (2002) Design of antisense oligonucleotides stabilized by locked nucleic acids. *Nucleic Acids Res.* **30**, 1911–1918.
8. Ivanova, A. and Rosch, N. (2007) The structure of LNA:DNA hybrids from molecular dynamics simulations: the effect of locked nucleotides. *J. Phys. Chem.* **111**, 9307–9319.
9. Crinelli, R., Bianchi, M., Gentilini, L., Palma, L. and Magnani, M. (2004) Locked nucleic acids (LNA): versatile tools for designing oligonucleotide decoys with high stability and affinity. *Curr. Drug Targets* **5**, 745–752.
10. Famulok, M., Hartig, J.S. and Mayer, G. (2007) Functional aptamers and aptazymes in biotechnology, diagnostics, and therapy. *Chem. Rev.* **107**, 3715–3743.
11. Famulok, M. and Mayer, G. (2005) Intramers and aptamers: applications in protein-function analyses and potential for drug screening. *ChemBioChem* **6**, 19–26.
12. Srivastava, P., Barman, J., Pathmasiri, W., Plashkevych, O., Wenska, M. and Chattopadhyaya, J. (2007) Five- and six-membered conformationally locked 2′,4′-carbocyclic ribo-thymidines: synthesis, structure, and biochemical studies. *J. Am. Chem. Soc.* **129**, 8362–8379.
13. Pasternak, A., Kierzek, E., Pasternak, K., Turner, D.H. and Kierzek, R. (2007) A chemical synthesis of LNA-2,6-diaminopurine riboside, and the influence of 2'-O-methyl-2,6-diaminopurine and LNA-2,6-diaminopurine ribosides on the thermodynamic properties of 2′-O-methyl RNA/RNA heteroduplexes. *Nucleic Acids Res.* **35**, 4055–4063.
14. Hicke, B.J., Stephens, A.W., Gould, T., Chang, Y.F., Lynott, C.K., Heil, J., Borkowski, S., Hilger, C.S., Cook, G., Warren, S. and Schmidt, P.G. (2006) Tumor targeting by an aptamer. *J. Nucl. Med.* **47**, 668–678.
15. Schmidt, K.S., Borkowski, S., Kurreck, J., Stephens, A.W., Bald, R., Hecht, M., Friebe, M., Dinkelborg, L. and Erdmann, V.A. (2004) Application of locked nucleic acids to improve aptamer *in vivo* stability and targeting function. *Nucleic Acids Res.* **32**, 5757–5765.
16. Forster, C., Brauer, A.B., Brode, S., Schmidt, K.S., Perbandt, M., Meyer, A., Rypniewski, W., Betzel, C., Kurreck, J., Fürste, J.P. and Erdmann, V.A. (2006) Comparative crystallization and preliminary X-ray diffraction studies of locked nucleic acid and RNA stems of a tenascin C-binding aptamer. *Acta Crystallogr. F* **62**, 665–668.
17. Lebars, I., Richard, T., Di Primo, C. and Toulmé, J.J. (2007) NMR structure of a kissing complex formed between the TAR RNA element of HIV-1 and a LNA-modified aptamer. *Nucleic Acids Res.* **35**, 6103–6114.
18. Di Primo, C., Rudloff, I., Reigadas, S., Arzumanov, A.A., Gait, M.J. and Toulmé, J.J. (2007) Systematic screening of LNA/2'-O-methyl chimeric derivatives of a TAR RNA aptamer. *FEBS Lett.* **581**, 771–774.
19. Lebars, I., Richard, T., Di Primo, C. and Toulmé, J.J. (2007) LNA derivatives of a kissing aptamer targeted to the trans-activating responsive RNA element of HIV-1. *Blood Cells Mol. Dis.* **38**, 204–209.
20. Darfeuille, F., Reigadas, S., Hansen, J.B., Orum, H., Di Primo, C. and Toulmé, J.J. (2006) Aptamers targeted to an RNA hairpin show improved specificity compared to that of complementary oligonucleotides. *Biochemistry* **45**, 12076–12082.
21. Crinelli, R., Bianchi, M., Gentilini, L. and Magnani, M. (2002) Design and characterization of decoy oligonucleotides containing locked nucleic acids. *Nucleic Acids Res.* **30**, 2435–2443.
22. Crinelli, R., Bianchi, M., Gentilini, L., Palma, L., Sørensen, M.D., Bryld, T., Babu, R.B., Arar, K., Wengel, J. and Magnani, M. (2004) Transcription factor decoy oligonucleotides modified with locked nucleic acids: an *in vitro* study to reconcile biostability with binding affinity. *Nucleic Acids Res.* **32**, 1874–1885.
23. Virno, A., Randazzo, A., Giancola, C., Bucci, M., Cirino, G. and Mayol, L. (2007) A novel thrombin binding aptamer containing a G-LNA residue. *Bioorg. Med. Chem.* **15**, 5710–5718.
24. Brown, T. and Brown, D. J. S. (1991) Oligonucleotides and analogues. *A Practical Approach*, Eckstein, F. (Ed.), IRL, Oxford, 1–24.
25. Beaucage, S.L. and Caruthers, M.H. (2000) *Current protocols in nucleic acid chemistry*, Beaucage, S.L., Bergstrom, D.E., Herdewijn, P., and Matsuda, A. (Eds.). John Wiley, New York, pp. 3.3.1–3.3.20.

26. Pfundheller, H.M., Sørensen, A.M., Lomholt, C., Johansen, A.M., Koch, T. and Wengel, J. (2005) Locked nucleic acid synthesis. *Methods Mol. Biol.* **288**, 127–146.

27. Hansen, H.F., Olsen, O. and Koch, T. (2003) New standards in LNA synthesis. *Nucleosides Nucleotides Nucleic Acids* **22**, 1273–1275.

Chapter 11

Screening of Novel Inhibitors of HIV-1 Reverse Transcriptase with a Reporter Ribozyme Assay

Satoko Yamazaki and Michael Famulok

Abstract

"Highly active anti-retroviral therapy (HAART)" is currently the standard treatment for human immunodeficiency virus (HIV). This treatment consists of a cocktail of two reverse transcriptase (RT) inhibitors and a protease inhibitor. Despite the success of this regimen, there is a continuing need for innovative drug to overcome problems with tolerability and the emergence of viral resistance. The present protocol describes a novel strategy to rapidly screening a new class of small molecule HIV-1 RT inhibitors, which bind to the primer/template binding site of RT, as yet an unexplored site for small molecule interference on this target. The assay is based on aptamer-displacement which is visualized by applying a rationally designed HIV-1 RT responsive ribozyme. The handiness of the assay procedure permits automation, compatible with high-throughput screening (HTS). Subsequently, the identified hit compounds have been evaluated by an in vitro enzymatic assay to test the inhibitory potential. The strategy provides a powerful and efficient screening format for site-directed inhibitors with biological activity.

Key words: HTS, aptamer, allosteric ribozyme, reporter ribozyme, HIV-1 RT, FRET.

1. Introduction

We have recently reported the construction of an allosteric hammerhead ribozyme (or aptazyme) fused to an HIV-1 reverse transcriptase (RT) binding aptamer *(1)*. This aptamer recognizes the primer/template binding site of HIV-1 RT *(2, 3)* and inhibits viral replication *(4, 5)*. Binding of the protein to the aptamer part of the allosteric ribozyme leads to the formation of a pseudoknot structure that allosterically interferes with the cleavage activity of the catalytic site of the ribozyme moiety. In the absence of HIV-1 RT, the aptazyme remains active. When

Günter Mayer (ed.), *Nucleic Acid and Peptide Aptamers: Methods and Protocols, vol. 535*

DOI 10.1007/978-1-59745-557-2_11 Springerprotocols.com

using a substrate oligonucleotide that is labeled with a fluorescence emitting group on one end, and a fluorescence quencher group on the other end, cleavage of the substrate can be monitored by an increase in fluorescence. This setting is compatible with high-throughput screening and we describe here a protocol for the application of this HIV-1 RT-regulated allosteric hammerhead ribozyme for the search of novel small molecule inhibitors of HIV-1 RT.

HIV-1 RT is one of the major drug targets for anti-HIV therapy. However, the primer/template binding site has not been exploited so far by the currently available anti-RT drugs, including nucleoside analogue RT inhibitors (NRTIs) and non-nucleoside RT inhibitors (NNRTIs) *(6)*. Because the administration of these inhibitors often leads to the emergence of multidrug resistant strains of HIV-1 *(7, 8)* we sought to explore regions in this protein as new drugable interfaces that are not targeted by any of the currently available anti-HIV-1 RT inhibitors. We therefore utilized the anti-HIV-1 RT aptamer with its unique inhibitory mechanism of interference with primer/template binding as a screening tool to find novel types of anti-retroviral small-molecule inhibitors that target the same site *(9)*.

Although oligonucleotide-based therapeutics are currently attracting considerable attention, an example being the FDA-approved aptamer drug Macugen^{TM} that treats age-related macular degeneration (AMD) *(10)*, aptamers are currently not routinely used as drugs, mainly due to their insufficient lifetime in vivo and complex delivery. In contrast, these drawbacks of nucleic acid-based drugs could be overcome if an aptamer could be 'converted' into a small molecule. Therefore, we have developed strategies that enable the functional conversion of aptamer binding and inhibitory properties into a small organic inhibitor by screening small-molecule libraries for small molecules that displace the aptamer by directly competing with its binding to the same target. Protein-dependent allosteric ribozymes (or reporter ribozymes) are powerful tools to monitor molecular interactions *(11)*, and have been utilized in screening assay formats also for other targets. Here, the anti-RT aptamer sequence was fused to the ribozyme sequence in a rational design approach, allowing highly specific detection of HIV-1 RT *(12)* (*see* **Fig.** 11.1).

We have employed the HIV-1 RT-regulated reporter ribozyme in an assay in which we screened 2,500 small molecules for their ability to displace the aptazyme/HIV-1 RT complex and have identified potential inhibitors that exhibited this activity. These molecules were confirmed in biological and structural analyses as HIV-1 RT inhibitors that presumably target the same site as the aptamer, inhibited the DNA-dependent but not

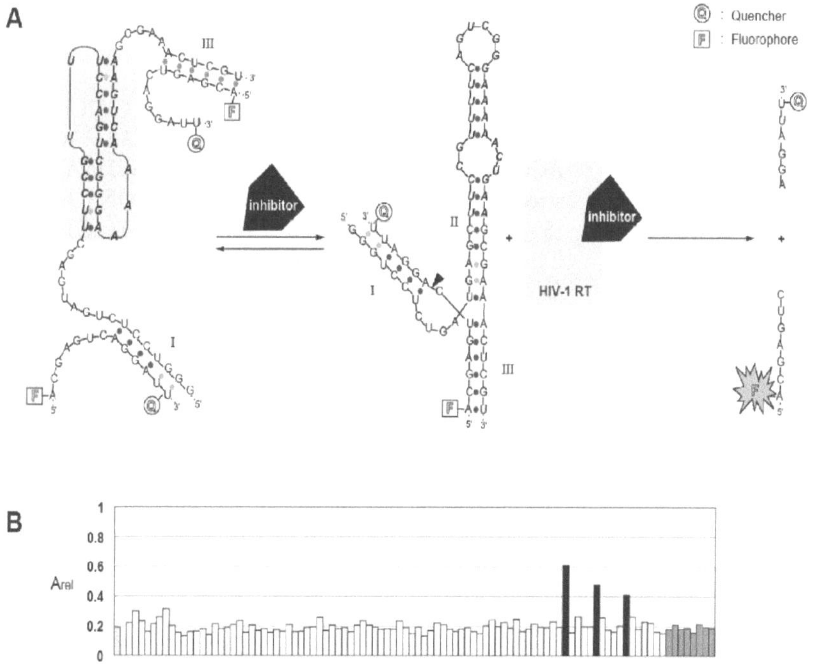

Fig. 11.1. Schematic of the reporter ribozyme-based screening assay and a plot of hit validation. (**A**) Aptamer sequences inserted into the hammerhead ribozyme are highlighted in *boldface italics*. The FRET-substrate carries 5′-fluorophore (F) and 3′-quencher (Q). The *arrow* indicates the FRET-substrate cleavage site. In the presence of HIV-1 RT (*gray*), the aptamer sequence adopts a pseudoknot, disrupting the formation of stem II (*left*). Addition of a potential HIV-1 RT inhibitor (*black*), which interferes with the interaction between HIV-1 RT and the aptamer by binding to HIV-1 RT, induces the active conformation of the hammerhead ribozyme (*middle*). Subsequent cleavage of the FRET-labeled substrate results in the generation of a fluorescent signal increment that can be detected in real time (*right*). (**B**) A_{rel} values of library compounds. Each *bar* represents the mean value from duplicate measurements for one compound from the library. The assay was performed in the presence of 100 μM of compound (*white bars*). The controls with the addition of DMSO were shown in *dark gray bars* and the average of DMSO controls was indicated in light *gray bar* in the *left*. The potential hit compounds with relative values (A_{rel}) higher than 0.4 were marked in *black*.

RNA-dependent primer elongation activity of HIV-1 RT and HIV-2 RT, while certain DNA polymerases or RTs from other organisms were not targeted or only marginally blocked *(12)*. One of these inhibitors was shown to interfere with the replication of wild-type HIV-1 and with that of a multidrug resistant strain. The utility of this methodology is clearly not limited to the screening of site-specific RT inhibitors. With the current availability of numerous aptamers that have therapeutic potential *(13, 14)* the principle of the strategy is broadly applicable to any protein/aptamer combination. Here we describe a detailed protocol for the screening assay and the in vitro analysis of the identified small molecules in a primer elongation assay.

2. Materials

2.1. Reporter Ribozyme-Based Screening Assay

1. Reporter ribozyme: FK-1 RNA(5'-GGG UCC UCU GAU GAG CUU CCG UUU UCA GUC GGG AAA AAC UGA AGC GAA ACU CGU-3') is transcribed from dsDNA synthesized by PCR using following appropriate DNA template (5'-*TCT AAT ACG ACT CAC TAT A*GG GTC CTC TGA TGA GCT TCC GTT TTC AGT CGG GAA AAA CTG AAG CGA AAC TCG T-3') and primers (forward primer; 5'-*TCT AAT ACG ACT CAC TAT A*-3' and reverse primer; 5'-ACG AGT TTC GCT TCA GTT TTT CC-3'). T7 promoter sequences are italicized in primer and template sequences. The synthetic DNA oligonucleotides are available from MWG.
2. FRET-labeled RNA substrate: Fluorescently labeled RNA (5'-FAM-ACG AGU CAG GAU U–TAMRA-3') is available from IBA, Göttingen.
3. HH buffer (5x): 250 mM Tris–HCl, pH 7.9, 125 mM NaCl. Store at 4°C. (*see* **Note 1**)
4. HIV-1 RT storage buffer (1x): 20 mM Tris–HCl, pH 7.8, 1 mM EDTA, 1 mM DTT, 15 % glycerol. Store at 4°C.
5. HIV-1 RT: Prepare small aliquots in HIV-1 RT storage buffer. Store at −80°C and avoid repeated freeze–thaw cycles.
6. Small molecule library (Comgenex, Hungary): Prepare 1 mM each stock solution of each compound in analytically pure DMSO. Store at −80°C. Stock solutions in 96-well plate are suitable for robotic manipulations. (*see* **Note 2**).
7. 40 mM $MgCl_2$: Store at 4°C.

2.2. Radioactive Labeling 5' End of DNA Primer

1. DNA primer to be labeled: synthetic oligonucleotide with dephosphorylated 5'end.
2. T4 polynucleotide kinase supplied with 10x buffer from New England Biolabs.
3. [γ-^{32}P]-ATP (10 mCi/mL) from Perkin Elmer.
4. MicroSpin G-25 columns from GE Healthcare.

2.3. Reverse Transcriptase (RT) Assay

1. Compounds that are identified as hit in the screening assay: prepare diluted solutions in DMSO.
2. RT reaction buffer (10x): 500 mM Tris–HCl, pH 8.0, 500 mM KCl. Store at 4°C.
3. 100 mM $MgCl_2$: Store at 4°C.
4. HIV-1 RT: Prepare small aliquots in HIV-1 RT storage buffer. Store at −80°C and avoid repeated freeze–thaw cycles.
5. 2.5 mM dNTP: 2.5 mM of dATP, dCTP, dGTP, and TTP (Roche). Store at −20°C.

6. 5'-[^{32}P]-end labeled complementary DNA primer: labeled primer (5'-[^{32}P]-GTG GTG CGA AAT TTC TGA C-3') prepared in **Section 2.2**.
7. DNA/RNA template: DNA template (5'-GTG CGT CTG TCA TGT CTG TCA GAA ATT TCG CAC CAC-3') is commercially available (MWG). Corresponding RNA template is transcribed from dsDNA synthesized by PCR using following DNA template (5'-*TCT AAT ACG ACT CAC TAT A*GT GCG TCT GTC ATG TCT GTC AGA AAT TTC GCA CCA C-3') and primers (forward primer; 5'-*TCT AAT ACG ACT CAC TAT A*-3' and reverse primer; 5'-GTG GTG CGA AAT TTC TGA C-3').
8. RNasin ribonuclease inhibitor from Promega.
9. STOP solution: 80 % formamide, 20 mM EDTA. Store at room temperature.

2.4. Denaturing Gel Electrophoresis

1. 70% Ethanol in squirt bottle.
2. 5% Dimethylchlorsilane (Fluka): Dilute in dichloromethane. Store at room temperature.
3. Denaturing acrylamide gel solution: Mix appropriate amount of gel solution B, C, and D.

   Gel solution B: 10x TBE in 8.3 M urea.
   Gel solution C: 25 % acrylamide/bisacrylamide (19:1) in 8.3 M urea.
   Ready-to-use solution is available from Roth (Rotiphorese® Sequencing gel concentrate).
   Gel solution D: 8.3 M urea.
4. 10% (w/v) Ammonium persulfate: Prepare fresh in water weekly and store at 4°C.
5. *N,N,N',N'*-Tetramethylethylenediamine (TEMED): Store at 4°C.
6. Sequencing gel apparatus (Sequi-Gen GT system) from Bio-Rad.
7. Running buffer: 1x TBE. Store at room temperature.
8. Dye markers: Add 0.25% of bromophenol blue and xylene cyanol to STOP solution.
9. Chromatography paper 3 mm from Whatman.

3. Methods

The reporter ribozyme FK-1 described in **Fig. 11.1** is a rationally engineered allosteric ribozyme, which contains an aptamer sequence that binds to an enzyme namely HIV-1 RT selectively in domain-specific manner (**1**). The active conformation of the ribozyme, in which the cleavage of the substrate RNA labeled with a fluorophore/quencher pair generates a measurable signal, is disrupted in the presence of HIV-1 RT and the FRET-labeled substrate cleavage reaction is blocked. Therefore, FK-1 has proven to be powerful tool which can report molecular interactions between HIV-1 RT and aptamer by fluorescent signal in real-time (**1**). The assay format is applicable to HTS to search for drug-like small molecules which displace the aptamer binding site of the HIV-1 RT (*see* **Fig.** 11.1).

The protocol described below to perform the screening assay is in principle optimized for the robot manipulation bearing sufficient reliability and sensitivity. However, it is still competent for manual operation. The automation by the liquid handling system comfortably allowed screening 2,500 compounds per 1 week. Subsequent studies of the identified active hit compounds involved the characterization of their inhibitory properties. The reverse transcriptase assay (RT assay), represented in 3.4., can evaluate effect of the compounds on polymerase activity derived from HIV-1 RT. The strategy is based on the primer-elongation assay with radio-labeled primer, which is extended depending on either DNA template or RNA template by enzyme. HIV-1 RT carries both DNA-dependent DNA polymerase activity (DDDP) and RNA-dependent DNA polymerase activity (RDDP). It is possible to determine IC_{50} values from multiple-dose measurement. Furthermore, comparing the IC_{50} values obtained with various reverse transcriptase and DNA polymerase from prokaryotic and eukaryotic, the selectivity of the potential inhibitor will be investigated.

3.1. Screening Procedure Using the Reporter Ribozyme

1. Set reaction condition for every compound to be screened as follows (for 25 μL volume). Perform a negative control reaction in the absence of HIV-1 RT per compound as well (*see* **Note 3**). Duplicate every reaction condition;

100 μM	compound from the library
10 nM	Reporter ribozyme FK-1
200 nM	FRET-labeled RNA substrate
200 nM	HIV-1 RT
1x	HH buffer
8 mM	$MgCl_2$

2. Prepare a mixture including the reporter ribozyme FK-1, FRET-labeled RNA substrate and HIV-1 RT (add HIV-1 RT storage buffer in negative control) in HH buffer. Transfer the mixture into glass vials (Pyrex) to equip on the robotic system (*see* **Note 4**).

3. Pre-mix the library compounds in DMSO. This step can be performed by robotic manipulation.
4. Combine the library compounds and the mixture from step 2 into 384-well plates by using an automated liquid handling system (Freedom EVO, Tecan) (*see* **Note 5**). As a positive control, perform standard ribozyme reaction without compound (add same volume of DMSO instead of the compound) per plate.
5. Agitate a plate shortly, and then pre-incubate for 15 min at 37°C in microplate fluorometer apparatus (Fluoroskan Ascent® FL, Termo Scientific). Initiate the ribozyme cleavage reaction by adding $MgCl_2$ at a final concentration of 8 mM to each reaction mixture (*see* **Note 6**) through dispenser in the Fluoroskan Ascent FL.
6. Monitor each reaction in real-time by measuring FAM-dye fluorescent emission at 520 nm for 15 min at 60 s intervals at a fixed excitation wavelength at 492 nm (Fluoroskan Ascent® FL, Termo Scientific).

3.2. Identification of Hit Compounds

1. Determine an initial velocity (V: fluorescence/min) for every reaction by plotting the increment of fluorescence intensity, multiplied by factor 10,000, versus time. Take a mean value from duplication.
2. Calculate a relative ribozyme activity (A_{rel}) for each compound by following equation;

$$A_{rel} = V_{RT+}/V_{RT-}$$

where A_{rel} = relative activity
V_{RT+} = mean value of initial velocity of ribozyme in the presence of HIV-1 RT
V_{RT-} = mean value of initial velocity of ribozyme in the absence of HIV-1 RT

3. Define the average of A_{rel} value of standard reaction without compound (DMSO). The compounds which show the value of A_{rel} > average of A_{rel} (DMSO standard) can be identified as hit compounds (*see* **Note 7**). An example of the A_{rel} values is shown in **Fig. 11.1B**.

3.3. Radioactive Labeling the 5′ End of the DNA Primer

1. Prepare a reaction mixture as follows (for 50 μL volume);

0.4 μM	DNA primer
1x	T4 PNK buffer
1 μCi/μL	[γ-^{32}P]-ATP
1 U/μL	T4 polynucleotide kinase

2. Incubate at 37°C for 60 min.
3. After heat denaturation at 100°C for 2 min, cool down on ice immediately.
4. Separate the labeled DNA from the unincorporated radioactive nucleotides by centrifugation through a MicroSpin G-25 column.

3.4. Reverse Transcriptase (RT) Assay

1. Set reaction conditions as follows (for 20 μL volume);

Compound at varying concentrations
(e.g., 0.5, 0.75, 1, 2, 3.5,5, 7.5, 10, 20, 35, 50 μM) (*see* **Note 8**)

1x	RT reaction buffer
14 nM	HIV-1 RT (*see* **Note 9**)
10 mM	$MgCl_2$
50 μM	dNTP
1U/ μL	RNasin (add this when RNA template is used)
50 nM	5'-[^{32}P]-end labeled complementary DNA primer (*see* **Note 10**)
100 nM	DNA template or RNA template

2. Mix 5'-[^{32}P]-end labeled complementary DNA primer and DNA template (or RNA template). Denature at 95°C (at 70°C for RNA template) for 5 min, then anneal by slow cooling to room temperature over 1 h.
3. Pre-incubate the compound at varying concentrations in the enzyme mixture containing HIV-1 RT, $MgCl_2$ and dNTP (and RNasin if RNA template is used) in RT reaction buffer at 37°C for 5 min.
4. Initiate the polymerase reaction by the addition of DNA template (or RNA template)/5'-[^{32}P]-DNA primer complex to the enzyme–compound mixture and incubate at 37°C for 15 min.
5. Deactivate the reaction by addition of 40 μL of STOP solution. Prior to gel analysis, denature the sample by heating at 95°C for 5 min, then cool down on ice immediately.

3.5. Analysis by Denaturing Gel Electrophoresis

1. The following instructions use the sequencing gel apparatus, including clamps, caster base, comb, and spacers from Bio-Rad (Sequi-Gen GT System).
2. Wash meticulously the glass plates with soap and water. Rinse well with deionized water and dry. Then, wet plates with 70% ethanol in a squirt bottle and wipe dry with a Kimwipe.

3. With the gloves, pore small amount of 5% dimethylchlorosilane solution on one side of each plate, then spread the solution over the plate carefully by using Kimwipe (*see* **Note 11**). Perform this step in a draft chamber. After the plates get dried, check finally for dust and other particulates.
4. Assemble gel plates with 0.4 mm thickness spacers and fix with clamps. Insert sandwich assembly into caster base.
5. Prepare 150 mL of 15% gel solution by mixing 15 mL of gel solution B, 90 mL of gel solution C, and 50 mL of gel solution D. Initiate the polymerization with 1,200 μL of 10% APS and 60 μL of TEMED. Pour the gel solution immediately between glass plates through the syringe attached to the caster base (*see* **Note 12**).
6. When the gel solution reaches the top of the short plate, insert the comb into the gel solution.
7. Polymerization completes 60–90 min. After polymerization, the gel can be used immediately or store at room temperature up to 1 day. To store the gel, place a wet paper towel over the comb and wrap the top of the gel with plastic wrap not to dry.
8. Fill the bottom reservoir of the gel apparatus with 1x TBE. Remove the bottom caster base and set the gel sandwich in the apparatus. Pour 1x TBE into the top reservoir. Remove the comb gently and rinse the wells by syringe.
9. Pre-run the gel for 30 min at 2,000 V. Then, rinse wells just prior to sample loading.
10. Load heat-denatured samples. Run the gel at 2,000 V for appropriate time. The course of electrophoresis can be monitored with a control lane containing marker dyes.
11. After electrophoresis is complete, drain the buffer from the top and bottom reservoirs and discard the liquid as radioactive waste.
12. Remove the gel sandwich from the apparatus and lay the gel plates flat. Remove the clamps. Slowly lift the top plate from the one corner, gradually increasing the angle until the top plate is completely separated from the gel. The gel should stick to the bottom plate.
13. Cut out the gel area required for the analysis by razor. The area can be estimated by the position of marker dyes (*see* **Note 13**).
14. Place the chromatography paper on the gel carefully. Peel the paper up off the plate from one edge and gradually curl the paper. The gel will be transferred to the paper by sticking. Place the paper and the gel on the gel dryer (Bio-Rad). Cover with plastic wrap. Dry the gel for 40 min at 80°C.

15. Place the dried gel in a cassette with a PhosphorImager screen in direct contact. After sufficient exposure time (at least 1 h for a clear image, over night exposure is recommended), scan the screen by PhosphorImager (Fujifilm, LFA-3000). An example of the results is shown in **Fig.** 11.2.
16. Analyze the data and apply OriginPro 7.5 software to determine the inhibitory concentrations that give half-maximal activity (IC_{50}) from the dose-dependent inhibition curves presented in **Fig.** 11.2

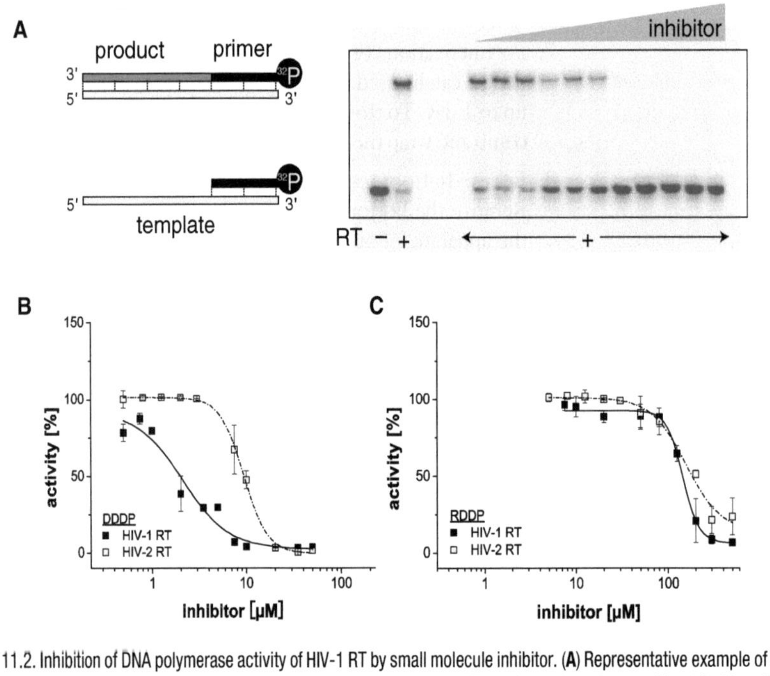

Fig. 11.2. Inhibition of DNA polymerase activity of HIV-1 RT by small molecule inhibitor. (**A**) Representative example of gel image resulting from RT assay. The control lane in the absence of HIV-1 RT (–) shows the primer position (*first lane from the left*). In the presence of HIV-1 RT (+), 5'-end labeled primer is extended employing either DNA or RNA template (*second lane from the left*). Polymerization reactions, which were able to be detected as elongated products, were inhibited by the addition of increasing concentrations of inhibitors (*right lanes*). The inhibitor concentrations ranged from 0.5 to 50 μM. (**B**) Dose–response curves for DNA-dependent DNA polymerization (DDDP) inhibition by a potential inhibitor (SY-3E4) are shown. The DDDP activity of HIV-1 RT and HIV-2 RT was assayed by using [^{32}P]-labeled DNA primer/DNA template. The symbols used in the graphs are as follows; HIV-1 RT (filled squares), HIV-2 RT (open squares). The mean values and the error bars are the results from two independent experiments. IC_{50} determined are 2.1 μM ± 0.58 for HIV-1 RT and 9.41 μM ± 0.21 for HIV-2 RT. (**C**) Dose–response curves for RNA-dependent DNA polymerization (RDDP) inhibition by the compound (SY-3E4) are shown. The RDDP activity of HIV-1 RT and HIV-2 RT was assayed by using [^{32}P]-labeled DNA primer/RNA template. IC_{50} values determined are 145 μM ± 5.06 for HIV-1 RT and 160 μM ± 19.1 for HIV-2 RT, respectively.

4. Notes

1. The buffer required for RNA manipulation such as HH buffer should be examined for RNase contamination. For this purpose, radioactive-labeled RNA in 25 μL of each buffer is incubated at least for 4 h at room temperature. After addition of PAGE loading buffer, the RNA degradation of each sample is analyzed on PAGE. The radioactive bands are visualized by exposure of the gel on a phosphor imager screen.
2. In order to avoid contamination among the compounds, it is important to open and close the lid of the plate very carefully.
3. For negative control, the same volume of HIV-1 RT storage buffer is added.
4. RNA and/or proteins tend to adsorb several plastic materials (e.g., reservoir, tube) required for robot manipulations. The primary test operations using plastic vials resulted in considerably varying data output even under the same assay conditions. Preparation of the reaction mixture in Pyrex grass vials which are adaptable to the robotic system is strongly recommended to obtain reproducible data output.
5. The samples are pipetted through the ceramic tips in the automated liquid handling system. This lowered the adsorption of the materials like RNA and/or proteins. A pipetting volume of less than 2.5 μL is not recommended for this system from Tecan. We programmed the procedure to pipet the enzyme mixture first (17.5 μL), and then pipet the library compound (2.5 μL). The liquid with higher viscosity such as the enzyme mixture is desirable to be pipetted first.
6. The ribozyme requires Mg^{2+} ions for its catalytic function. Therefore the cleavage reaction is initiated by the addition of $MgCl_2$.
7. The A_{rel} value resulting from the assay will depend on the design of the reporter ribozyme as well as the assay conditions. Standard reaction without compound (DMSO) usually gave the A_{rel} value of 0.2–0.3 depending on the plate. Therefore it is essential to include DMSO standard reaction in every plate. We usually performed at least eight standard reactions per plate to test if each pipet functions properly. In our assay, the compounds which show the value of $A_{rel} > 0.4$ can be identified as hit compounds.

8. We have synthesized the most potential hit compounds (identified in 3.1 and 3.2) for subsequent investigations. Re-synthesis of the authentic samples permits to confirm the activity and validity of the corresponding hits. To perform the kinetic assay with varying concentrations of the compounds synthesized, it is desirable to make several dilutions, from which the same volume can be taken for the assay.
9. The RT assay can also be performed in the presence of other RTs from HIV-2, avian myeloblastosis virus (AMV), and moloney murine leukemia virus (MMLV) as well as DNA polymerase such as Klenow fragment, and human DNA polymerase β. In those cases, use the buffer supplied from the manufacturer. The reaction time and the concentration of dNTP should be optimized.
10. Radioactively labeled DNA primer (from 3.3) can be optionally diluted into non-radioactive DNA primer solution to adjust adequate radioactivity for the PAGE analysis. We have generated a 3 μM stock solution of radioactive DNA primer by adding 50 μL of labeled DNA primer to 20 μL of 10 μM unlabeled primer, which was then used for the RT assay.
11. Silanization of the plates not only facilitated pouring a gel without bubbles but also prevented the gel from sticking to the plates during post-electrophoresis processing.
12. Air bubbles in the syringe should be avoided before pouring the gel solution and the formed air bubbles between the plates can be removed by tapping the glass plates.
13. The radioactivity of the samples in the gel is still detectable by a Geiger counter. This enables to define the gel area required for the further analysis.

Acknowledgment

The authors are grateful to Tobias Restle, Lübeck, for providing purified HIV-1 RT for screening.

References

1. Hartig, J.S. and Famulok, M. (2002) Reporter ribozymes for real-time analysis of domain-specific interactions in biomolecules: HIV-1 reverse transcriptase and the primer-template complex. *Angew. Chem. Int. Ed. Engl.* **41**, 4263–4266.
2. Tuerk, C., MacDougal, S. and Gold, L. (1992) RNA pseudoknots that inhibit

human immunodeficiency virus type 1 reverse transcriptase *Proc. Natl. Acad. Sci. U.S.A.* **89**, 6988–6992.

3. Jaeger, J., Restle, T. and Steitz, T.A. (1998) The structure of HIV-1 reverse transcriptase complexed with an RNA pseudoknot inhibitor. *EMBO J.* **17**, 4535–4542.
4. Chaloin, L., Lehmann, M.J., Sczakiel, G. and Restle, T. (2002) Endogenous expression of a high-affinity pseudoknot RNA aptamer suppresses replication of HIV-1. *Nucleic Acids Res.* **30**, 4001–4008.
5. Joshi, P. and Prasad, V.R. (2002) Potent inhibition of human immunodeficiency virus type 1 replication by template analog reverse transcriptase inhibitors derived by SELEX (systematic evolution of ligands by exponential enrichment). *J. Virol.* **76**, 6545–6557.
6. Jonckheere, H., Anne, J. and De Clercq, E. (2000) The HIV-1 reverse transcription (RT) process as target for RT inhibitors. *Med. Res. Rev.* **20**, 129–154.
7. Menendez-Arias, L. (2002) Targeting HIV: antiretroviral therapy and development of drug resistance *Trends Pharmacol. Sci.* **23**, 381–388.
8. Imamichi, T. (2004) Action of anti-HIV drugs and resistance: reverse transcriptase inhibitors and protease inhibitors. *Curr. Pharm. Des.* **10**, 4039–4053.
9. Joshi, P.J., Fisher, T.S. and Prasad, V.R. (2003) Anti-HIV inhibitors based on nucleic acids: emergence of aptamers as potent antivirals. *Curr. Drug Targets Infect. Disord.* **3**, 383–400.
10. Ng, E.W., Shima, D.T., Calias, P., Cunningham, E.T., Jr., Guyer, D.R. and Adamis, A.P. (2006) Pegaptanib, a targeted anti-VEGF aptamer for ocular vascular disease *Nat. Rev. Drug Discov.* **5**, 123–132.
11. Hartig, J.S., Najafi-Shoushtari, S.H., Grüne, I., Yan, A., Ellington, A.D. and Famulok, M. (2002) Protein-dependent ribozymes report molecular interactions in real time. *Nat. Biotechnol.* **20**, 717–722.
12. Yamazaki, S., Tan, L., Mayer, G., Hartig, J.S., Song, J.N., Reuter, S., Restle, T., Laufer, S.D., Grohmann, D., Kräusslich, H.G., Bajorath, J. and Famulok, M. (2007) Aptamer displacement identifies alternative small-molecule target sites that escape viral resistance. *Chem. Biol.* **14**, 804–812.
13. Nimjee, S.M., Rusconi, C.P. and Sullenger, B.A. (2005) Aptamers: an emerging class of therapeutics. *Annu. Rev. Med.* **56**, 555–583.
14. Que-Gewirth, N.S. and Sullenger, B.A. (2007) Gene therapy progress and prospects: RNA aptamers. *Gene Ther.* **14**, 283–291.

Chapter 12

Aptamers as Artificial Gene Regulation Elements

Beatrix Suess and Julia E. Weigand

Abstract

Conditional gene expression systems are important tools to identify the function of essential genes or in terms of gene therapy approaches. Small molecule-binding aptamers can be used for efficient control of gene expression by inserting them into the 5' untranslated region of an mRNA with the ligand-bound form inhibiting gene expression by interfering with translation initiation. However, only a small fraction of in vitro selected aptamers has the potential to act as regulator of gene expression which originates the necessity to develop screening systems for the identification of regulatory active aptamers. We describe here a simple and powerful yeast-based screening system which allows the rapid identification of small molecule-binding aptamers with the potential to act as artificial riboswitches for conditional control of gene expression.

Key words: Aptamer, engineered riboswitches, small molecule, screening, GFP, yeast.

1. Introduction

Gene regulation mediated by riboswitches is based on a direct RNA–ligand interaction *(1–3)*. Thereby, riboswitches offer several advantages. The interaction between the highly structured RNA element and the respective target ligand is often of remarkable affinity and specificity and the RNA accomplishes both sensory and regulatory functions and integrates the tasks formerly carried out by protein and RNA components together. Therefore, the principal of direct RNA–ligand interaction has been used to build up artificial riboswitches which can be used for conditional gene expression *(4, 5)*. One attempt bases on the insertion of aptamer sequences into the untranslated regions of an mRNA. These aptamers exploit the fact that small molecule-binding to the RNA alters their structures and subsequently inhibits the

Günter Mayer (ed.), *Nucleic Acid and Peptide Aptamers: Methods and Protocols, vol. 535*

DOI 10.1007/978-1-59745-557-2_12 Springerprotocols.com

translation of the downstream coding regions *(6, 7)*. Dependent on the insertion site the aptamer–ligand complex interferes either with binding of the 43S subunit to the cap-structure, acts as road block for the scanning ribosome *(8)* or inhibits 5' splice site recognition when located in an intron sequence *(9)*. The advantage of such artificial riboswitches is that they can be in principle designed for any ligand of choice. But, unfortunately, only a small subset of all in vitro selected aptamers display riboswitch activity in vivo *(6, 10)*.

In this chapter we describe a simple and powerful screening method to identify aptamers from in vitro selected RNA pools, which are regulatory active by inhibiting gene expression on the level of translation initiation in yeast.

2. Materials

2.1. Expression Vector

The vector pWHE601 is a yeast 2 μ plasmid (20–200 copies per cell) which carries the *ura3* gene as autotrophy marker for yeast, the gene for β-lactamase (*bla*) which confers resistance to the antibiotic ampicillin and an origin of replication for *Escherichia coli*. The vector constitutively expresses a *gfp* reporter gene from an *adh1* promoter. The 5' UTR of the *gfp* gene contains two unique restriction sites for directed insertion of aptamer sequences (6).

2.2. Yeast Strain, Medium

1. *Saccharomyces cerevisiae* strain RS453α: MATα *ade2-1 trp1-1 can1-100 leu2-3 leu2-112 his3-1 ura3-52 (11)* (*see* **Note 1**).
2. Minimal medium: 0.2% (w/v) yeast nitrogen base, 0.55% (w/v) ammonium sulfate, 2% (w/v) glucose, 12 μg/ml adenine, MEM amino acid (Gibco BRL). Autoclave for 10 min only and store at 4°C until use. For plates add 2% (w/v) agar–agar before autoclavation (*see* **Note 2**).
3. Stock solution (1,000-fold) of the aptamer ligand.

2.3. Instrumentation

1. Fluorescence stereomicroscope.
2. 96-well plate reader, e.g. SpectraFluor Plus fluorescence reader (Tecan, Crailsheim).
3. FastPrep 24 (MP Biomedicals, OH, USA).

2.4. Plasmid Preparation from Yeast

1. Lysis buffer: 2% (v/v) Triton X-100, 1% (w/v) SDS, 100 mM NaCl, 10 mM Tris–HCl pH 8.0, 1 mM EDTA pH 8.0. Autoclave and store at room temperature.
2. Glass beads: Ø 0.4 mm

3. PCI: Phenol/chloroform/isoamyl alcohol (25:24:1). Mix freshly. Chloroform/isoamyl alcohol mixture can be stored for up to 1 week at room temperature under a hood.
4. 100% and 70% Ethanol p.a. stored at –20°C. Mix freshly.

3. Methods

3.1. Construction of the Plasmid Pool

The 44 nt long 5' UTR of the *gfp* reporter gene contains singular restriction sites for *AflII* immediately upstream of the start codon and for *NheI* directly behind the start codon. These sites allow an insertion of the aptamers directly in front of the start codon which was determined as the most active position for regulation (*see* **Fig. 12.1**) (*6*).

In vitro selected aptamer pools are flanked by constant regions necessary for reverse transcription and PCR amplification. These constant regions are used for PCR amplification of cDNA from the last round of in vitro selection. Thereby, the *AflII* restriction site is introduced at the 5' end and the *NheI* restriction site at the 3' end of the aptamer pool via the used PCR primers. After digestion of the vector with *AflII* and *NheI*, the start codon is

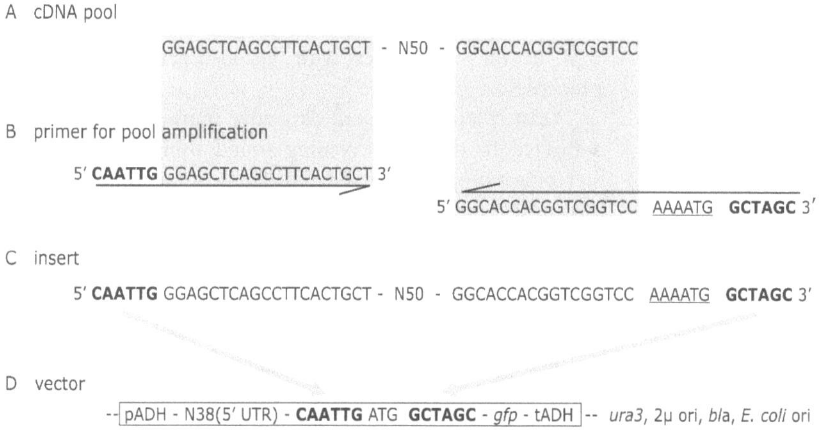

Fig. 12.1. Expression system. (**A**) DNA pool used for in vitro selection, the variable region (here N50) is flanked by a constant region necessary for reverse transcription and PCR amplification during in vitro selection. (**B**) Sequences of the respective primers used for pool amplification. The forward primer includes the sequence for the restriction site *AflII* (CAATTG, *bold*). The sequences of the start codon and the optimized Kozak sequence (AAAATG) together with the *NheI* restriction site (GCTAGC, *bold*) are introduced with the reverse primer. Note, for the sake of clarity, the reverse primer is also shown in the sense orientation. (**C**) PCR amplified insert. (**D**) Expression vector pWHE601. Genetic elements necessary for amplification in *E. coli* and yeast are given. The reporter gene is displayed as an *open bar* with the constitutive promoter and the terminator from the alcohol dehydrogenase gene (pADH and tADH), the 5' UTR and the coding region for *gfp*. Restriction sites used for pool insertion are indicated in *bold*.

cut out of the vector and has to be attached 3' to the aptamer pool by PCR mutagenesis. **Figure 12.1** displays schematically the 5' UTR of the vector, the PCR amplified aptamer pool and the cloning strategy (*see* **Note 3**).

Purify the PCR product, digest it with *AflII* and *NheI* and ligate it into the likewise digested vector pWHE601. Transform chemically or electrocompetent *E. coli* DH5α with the ligation product and prepare the plasmid pool by standard plasmid preparation methods. The starting diversity generated in *E. coli* should contain a minimum of 5 x 10^4 sequences (*see* **Note 4**).

3.2. Transformation of Yeast Cells

Transform yeast cells with the vector pool according to the protocol supplied with the Frozen-EZ Yeast Transformation II kit (Zymo Research, Orange, CA) and grow them at 28°C on solid minimal medium for 48 h. In order to ensure only one plasmid per cell, a range of DNA concentration should be used for transformation. In addition, the amount of yeast cells per plate should not exceed 500 (*see* **Note 5**).

3.3. Screening for Ligand-Dependent Changes in GFP Expression

The screening for ligand-dependent changes in GFP expression is a two-step process which is schematically shown in **Fig. 12.2**. First, candidates have to be identified which allow gene expression in the absence of the ligand. This step is necessary to eliminate candidates in which the ligand-free aptamer structure interferes with translation or premature start codons are introduced by the aptamer sequence. The second step results then in a subpopulation of thus candidates with ligand-dependent regulatory properties.

Yeast colonies, visible 2 days after transformation, can be subjected to the first screening round using a fluorescence stereomicroscope. Three different kinds of colonies are distinguishable. Non-fluorescent colonies: (i) Due to failed aptamer insertion no start codon is present and hence no *gfp* will be expressed; (ii) The aptamer folds into a stable structure already in the absence of the ligand which inhibits translational initiation; (iii) The introduction of premature start codons which are not in frame with the *gfp* open reading frame will lead to truncated, non-functional proteins. Slightly/medium fluorescent colonies: GFP is expressed, but at reduced levels compared to the original vector without an aptamer sequence. This is due to the insertion of an aptamer sequence which is partially folded. Bright fluorescent colonies: GFP is expressed at wild-type level. The vector was incompletely digested (only by one enzyme) and religated into the wild-type situation (*see* **Note 6**).

Only the second group of slightly/medium fluorescent colonies will be subjected to the further selection (**Fig. 12.2**). Transfer the colonies to 96-well plates containing 200 µl minimal medium

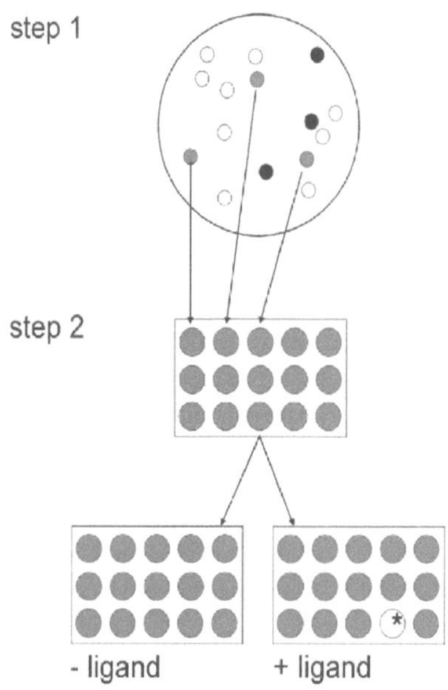

Fig. 12.2. Screening system. Three types of colonies are visible after transformation of yeast cells with the aptamer-containing plasmid pool: Non-fluorescent colonies (*white dots*), bright fluorescent colonies (*black dots*), slightly/medium fluorescent colonies (*gray dots*). Slightly/medium fluorescent colonies are transferred in 96-well plates and incubated for 24 h, then an aliquot is transferred to medium with and without the ligand and fluorescence is measured after 48 h of incubation at 28°C. The *asterisk* indicates a candidate with ligand-dependent decrease in fluorescence.

and incubate them for 24 h at 28°C. Transfer 20 μl aliquots of each sample into new plates with fresh medium with and without 100 μM ligand in a final volume of 200 μl (*see* **Note 7**). Perform fluorescence measurements 48 h after inoculation (*see* **Note 8**). Additionally, determine the optical density to correlate the fluorescence to the cell number (*see* **Note 9**). Calculate a regulation factor as a quotient of the fluorescence value without and with the ligand. Candidates with a regulatory factor of more than 1.2 will then be further processed (*see* **Note 10**). Streak out these candidates to single colonies and incubate for further 2 days at 28°C. Repeat the screening for three independent colonies from each positive candidate (following instructions beginning with the overnight culture).

3.4. Plasmid Preparation and Passage Through E. coli

S. cerevisiae is able to take up more than one plasmid during the transformation procedure. To verify that the observed regulation is due to only one single aptamer sequence, plasmids have to be isolated and passed through *E. coli*. Therefore, inoculate promising candidates in 4 ml of fresh minimal medium and grow them in test tubes at 28°C overnight under continuous shaking. Spin down 2 ml of the overnight culture in a 2 ml reaction tube with screw cap. Discard the supernatant and resuspend the pellet in 200 μl lysis buffer, 100 μl glass beads and 200 μl PCI. Lyse yeast cells either by using a FastPrep 24 (2x 30 s at 6 m/s) or by vortexing two times for 2 min. Cool down the cells on ice for 30 s between the two lysing steps. Spin down for 10 min with 13,000 rpm. Transfer the supernatant into a fresh 1.5 ml reaction tube and precipitate the DNA with Ethanol. Air dry the pellet and resolve it in 20 μl water. Use the complete 20 μl sample to transform *E. coli* DH5α, plate the complete transformation mixture on solid ampicillin-containing LB medium (*see* **Note 11**) and incubate overnight at 37°C. Restreak the obtained transformants and prepare plasmids from three independent *E. coli* colonies in case different plasmids had been existing in the original yeast cells. Transform yeast cells with the isolated plasmids and repeat the fluorescence measurements. Candidates which retain their regulation are subjected to sequence analysis.

3.5. Identification of the Regulatory Active RNA Element

In the most cases, only a small part of the inserted aptamer sequence confers regulation. Therefore, a truncation analysis has to be added to define the minimal active sequence. Perform a subcloning with a sequential truncation by ten nucleotides from both sides by PCR mutagenesis to narrow down the active sequence. This shortening normally enhances the regulatory activity due to higher expression levels of the downstream gene, more efficient folding of the aptamer and a closer location to the start codon. Perform secondary structure predictions (*see* **Note 12**) and validate the predicted secondary structure, tertiary interactions and interaction with the ligand by structural probing (*see* **Note 13**).

4. Notes

1. Do not keep yeast cells longer than 7 days on plate.
2. Prepare minimal medium without uracil since the *ura3* gene is used as auxotrophy marker on pWHE601.
3. We have used an optimized Kozak sequence for yeast (AAAATG) which allows efficient start codon recognition.
4. Exact data are missing about how many different sequences are left over after several rounds of in vitro selection. However, as far as our experiences goes, 5 x 10^4 candidates are sufficient to cover the remaining sequence space.
5. The number of colonies per plate should not exceed 500, so that single colonies are visible and easily to distinguish.

6. Use the plasmids pWHE601 and pVTU102 (parental vector of pWHE601 without the reporter gene) as control. The plasmid pWHE601 results in bright fluorescent colonies, whereas pVTU102 yield non-fluorescent colonies.
7. Shaking of the 96-well plates during incubation is not necessary. However, resuspend the yeast cells immediately before transferring them into fresh plates because they sink to the bottom of the well very rapidly.
8. For fluorescence measurements adjust the plate reader as follows:

Measurement mode:	Bottom measurement
Excitation wavelength:	485 nm
Emission wavelength:	510 nm
Gain (Manual):	60
Number of flashes:	3
Lag time:	0 μs
Integration time:	40 μs

9. Resuspend the cells before measurement of the optical density.
10. A factor of 1.2 appears pretty low, but the factor will increase when optimal growth conditions are used (shaking) or when measurements are performed in PBS-buffer which prevents the high self-fluorescence of minimal medium. Furthermore, only a part of the inserted sequence is responsible for the regulatory activity. Therefore, the factor will further increases with narrowing down the regulatory activity of the insert and positioning it close to the start codon *(6)*.
11. Spin down the transformation mixture before plating and resuspend the pellet in 200 μl LB medium. Plate the 200 μl on one agar plate.
12. Free available RNA folding programs:
http://www.tbi.univie.ac.at/~ivo/RNA/ *(12)*
http://frontend.bioinfo.rpi.edu/applications/mfold/ *(13)*
13. Analyses of the predicted secondary structures of the respective RNAs can be performed by chemical *(14)*, enzymatic *(15)* and in-line probing *(9)*.

Acknowledgment

This work was supported by the Volkswagenstiftung (I/79 950) and the Deutsche Forschungsgemeinschaft (SU 402/1-1).

References

1. Serganov, A. and Patel, D.J. (2007) Ribozymes, riboswitches and beyond: regulation of gene expression without proteins. *Nat. Rev. Genet.* **8**, 776–790
2. Tucker, B.J. and Breaker, R.R. (2005) Riboswitches as versatile gene control elements. *Curr. Opin. Struct. Biol.* **15**, 342–348
3. Winkler, W.C. and Breaker, R.R. (2005) Regulation of bacterial gene expression by riboswitches. *Annu. Rev. Microbiol.* **59**, 487–517
4. Bauer, G. and Suess, B. (2006) Engineered riboswitches as novel tools in molecular biology. *J Biotechnol.* **124**, 4–11
5. Buskirk, A.R. and Liu, D.R. (2005) Creating small-molecule-dependent switches to modulate biological functions. *Chem. Biol.* **12**, 151–161
6. Suess, B., Hanson, S., Berens, C., Fink, B., Schroeder, R., and Hillen, W. (2003) Conditional gene expression by controlling translation with tetracycline-binding aptamers. *Nucleic Acids Res.* **31**, 1853–18538
7. Werstuck, G. and Green, M.R. (1998) Controlling gene expression in living cells through small molecule-RNA interactions. *Science* **282**, 296–298
8. Hanson, S., Berthelot, K., Fink, B., McCarthy, J.E. and Suess, B. (2003) Tetracycline-aptamer-mediated translational regulation in yeast. *Mol. Microbiol.* **49**, 1627–1637
9. Weigand, J.E. and Suess, B. (2007) Tetracycline aptamer-controlled regulation of pre-mRNA splicing in yeast. *Nucleic Acids Res.* **35**, 4179–4185
10. Weigand, J.E., Sanchez, M., Gunnesch, E., Zeiher, S., Schroeder, R. and Suess, B. (2008) Screening for engineered neomycin riboswitches that control translation initiation. *RNA* **14**, 89–97.
11. Sauer, N. and Stadler, R. (1993) A sink-specific H+/monosaccharide co-transporter from Nicotiana tabacum: cloning and heterologous expression in baker's yeast. *Plant J.* **4**, 601–610
12. Hofacker, I.L. (2003) Vienna RNA secondary structure server. *Nucleic Acids Res.* **31**, 3429–3431
13. Mathews, D.H., Sabina, J., Zuker, M. and Turner, D.H. (1999) Expanded sequence dependence of thermodynamic parameters improves prediction of RNA secondary structure. *J. Mol. Biol.* **288**, 911–940
14. Ehresmann, C., Baudin, F., Mougel, M., Romby, P., Ebel, J.P. and Ehresmann, B. (1987) Probing the structure of RNAs in solution. *Nucleic Acids Res.* **15**, 9109–9128.
15. Hanson, S., Bauer, G., Fink, B. and Suess, B. (2005) Molecular analysis of a synthetic tetracycline-binding riboswitch. *RNA* **11**, 503–511.

Chapter 13

Aptamers and Biosensors

Thomas M. A. Gronewold

Abstract

The immobilization procedure to a biosensor surface has a major influence on the measurement results. To characterize the immobilization onto various biolayers, the interaction of DNA anti-thrombin aptamer with the protein thrombin was used as a model system. The aptamer was immobilized to a two-dimensional alkanethiol SAM via carboxylamide bonds and to a three-dimensional dextran matrix via streptavidin–biotin interaction. The calculated K_D values of about 260 and 267 nM, respectively, were comparable, while the amount of bound analyte varied by a factor of 2, depending on the accessibility of the immobilized aptamer. Differences in the specificity were shown by use of the similar protein elastase.

Key words: Surface acoustic wave sensor, DNA anti-thrombin aptamer, immobilization, alkanethiol SAM, dextran.

1. Introduction

Biosensors are devices for measuring the concentration or interaction of biological molecules. A versatile biosensor converts the molecular recognition processes of specific binders such as antibodies or aptamers, to mobile proteins or even whole cells into detectable, preferably electrical, signals. For on-line, real-time, label-free detection of binding effects and in situ measurements in the liquid phase, a whole range of biosensors is in use. Most widespread are optical sensors based on the Surface Plasmon Resonance (SPR) technology. As an alternative emerge Surface Acoustic Wave (SAW) sensors. The sensing technologies differ in their basic principles, but the produced curves resulting from the measurements are very similar, and the typical limit of detection is in the range of 1–20 pg/cm^2 for proteins binding to the sensor

Günter Mayer (ed.), *Nucleic Acid and Peptide Aptamers: Methods and Protocols, vol. 535*

DOI 10.1007/978-1-59745-557-2_13 Springerprotocols.com

surface. The biosensor principle is based on the detection of small mass changes that result from binding of a mobile molecule, the analyte, to an immobilized binding partner coupled to the active sensor surface, the ligand. Binding events are transduced into an electrical signal proportional to the additional mass loading on the sensor surface. Advanced sensors have several channels, enabling parallel measurements. Such sensor signals can be compared directly and reference sensors can be used to distinguish signals of varying fluid parameters as are variations in viscosity or ion concentrations, or unspecific binding events from sensor signals due to specific binding events. Depending on the binding events detected, reliable determination of kinetic constants as are the on-rate k_{on}, the off-rate k_{off}, and the equilibrium dissociation constant $K_D = k_{off}/k_{on}$ is enabled. The K_D value is a measure of how readily analytes sorb to the surface, how tightly they bind to the aptamer surface and how long the analytes remain to be bound.

In our approach, the S-sens® K5 system (Biosensor GmbH, Bonn, Germany) is used, which is based on a physical transducer of the Love-wave surface acoustic wave (SAW) sensor type, measuring at two fixed excitation frequencies *(1)* working in delay-line geometry. The propagation velocity of acoustic shear waves travelling through a guiding layer at the sensor surface is very sensitive to additional mass loading. The sensitivity for small mass depositions depends on changes in the phase velocity in the guiding layer *(1, 2)*. Advantage compared to other sensor types is the elimination of cross-sensitivities even under high damping conditions (fluids with high viscosity or with varying salt contents). Additionally can mass be discriminated from viscoelastic effects by use of both the phase shift and the amplitude shift of the surface acoustic wave *(1)*.

The surface of the SAW sensor chips can be modified by the customer. The sensitivity is determined by the sensor setup. The specificity of a sensor element for a certain analyte strongly depends on the modification of the sensor chip surface. Limiting are the specificity of the coupled ligands, but also the binding method used. Ligands can be coupled to the active sensor area using standard biochemical protocols, providing specific binding-sites for their analytes. Immobilization methods are based on (i) adsorption needing very little preparatory efforts; (ii) fixation to a membrane; (iii) covalent bonds via an interface layer of a spacer molecule with two functional ends; (iv) fixation in a matrix of linked molecules or in a gel coat. One end of such a spacer molecule (iii) binds covalently to the surface. Currently, almost all coupling methods onto the sensor surface are based on the extremely stable sulfur–gold bond *(3)*. Sulfur-containing alkanethiols form two-dimensional self-assembled monolayers (SAMs) on gold or other noble surfaces. The other end of the spacer molecule presents a functional group (*see* **Table 13.1**), which comes in contact with the sensor liquid, enabling chemical

Table 13.1
Bonds formed with commonly used alkanethiols. Alkanethiols with the formula $R-(CH_2)_n-SH$, $n \geq 10$ form stable SAMs with quasi-crystalline domains. Large head groups of alkanethiols as is the carboxyl terminus hinder the development of such quasi-crystalline structures

Alkanethiols with $n=10$	End group, R	Ligand	Bond
1-Decanthiol	–H	Hydrophobic groups, membranes	Unspecific interactions
1,9-Decane-dithiol	–SH	–SH	Disulfide
11-Mercapto-1-undecanol	–OH	$-PO_4$	Phosphate
11-Mercaptoundecanoic acid	–COOH	$-NH_2$	Carboxylamide

Source: Schreiber, *(17)*.

reactions with a ligand. The head groups determine the enabled binding chemistries and the characteristics of the surface formed. Hydroxy alkanethiols can be used as a basis for the formation of dextran layers (iv), forming a hydrophilic polymer hydrogel, which is highly flexible, non-crosslinked and extends 100–200 nm from the coupling surface under physiological conditions *(4)*. The three-dimensional matrix formed has a much higher immobilization capacity compared to two-dimensional surfaces. Diffusion of small molecules is fast enough not to be hindered. On 3-D matrices, the bound ligands have more degrees of freedom, enabling better binding of their analytes. Both the alkanethiols and the dextran surface are biocompatible, similar to separation media used in protein purification, are long-term durable, and stabilize even sensitive proteins. Dextran surfaces mostly do not require blocking agents and, by nature, show low unspecific binding, which significantly enhances the specificity. The embedding of ligands in a gel-matrix often enhances the biostability of the surface.

To demonstrate the differences in binding, two coupling chemistries are applied as an example. Used are two of the most common binding chemistries. Both are based on surfaces containing carboxyl groups, as alkanethiol head groups (as in iii), and as carboxymethyl-dextran (as in iv). The well-known DNA aptamer against thrombin *(5)* is used as the biological test system. Thrombin is multifunctional with hormone-like properties. As the last protease in the clotting cascade, thrombin plays a central role in thrombosis, haemostasis, blood-coagulation and in platelet activation *(6)*. Elastase is a related serine protease, thus utilized as a reference protein. It has been shown that linkage of the DNA aptamer to the sensor surface has no effect on the binding properties compared with filter binding experiments. The modified surface can be regenerated several times without loss of functionality *(7)*. Amino-functionalized ligands are coupled via carbodiimide chemistry to the sensor surface, forming carboxylamide bonds *(8)*.

2. Materials

2.1. Surface Cleaning

1. Precleaning solutions: 100% ddH_2O, 100% acetone and 100% isopropanol.
2. Piranha solution: 3:1 mixture of sulfuric acid and 30% hydrogen peroxide, either mixed before application, or the sulfuric acid is applied to the surface first, followed by the peroxide. Be careful, piranha solutions are harmful and have to be handled with extreme caution!
3. Ammonia–peroxide solution: A 5:5:1 mixture of ddH_2O, ammonia (32%) and hydrogen peroxide (30%).

2.2. Surface Preparation

1. Prepare a 2 mM stock solution of alkanethiols (**Table 13.1**). With each 200 μl of a 2 M alkanethiol stock solution, 10 ml of a 200 mM ethanolic solution can be prepared (dilution 1:50 v/v). Some alkanethiols are of low solubility.
2. Activation solution: 0.6 M epichlorhydrin in 0.4 N NaOH. Prepare 10 ml 0.4 N NaOH and add 1 ml epichlorohydrin. Add 10 ml diglyme (diethylene glycol dimethyl ether; *see* **Note 1**).
3. Dextran with a molecular weight of about 400,000–500,000 Da (from *Leuconostoc mesenteroides*, Sigma D1037) (Shorter dextran molecules might be applied, e.g. when the penetration depth of the sensor is reduced).
4. Dextran solution: 1 g dextran (**Step 3** above) in 3.3 ml 0.1 N NaOH.
5. Carboxylation solution: 5 g bromoacetic acid in 7.7 g 2 N NaOH.

2.3. Immobilization Procedures

1. Carboxyl activation solution: 400 mM 1-[3-(dimethylamino)propyl]-3-ethylcarbodiimide hydrochloride (EDC, 8.00907.0005, Merck) and 100 mM *N*-hydroxysuccinimide (NHS, H-7377, Sigma). Mix 1:1 immediately before use, resulting in a mixture containing 200 mM EDC and 50 mM NHS. Prepare solutions fresh as required.
2. 2 μM aqueous solution of anti-thrombin aptamer carrying a 5′-amino linker with the sequence 5′-NH_2-$(CH_2)_3$-GGT TGG TGT GGT TGG-3′, synthesized by Operon. The ligand might

be diluted in water, or in buffer at a pH close to the isoelectrical point.

3. 200 μg/ml Streptavidin (Molecular Probes, S888).
4. 2 μM aqueous solution of anti-thrombin aptamer with the sequence 5′-BioTEG- GGT TGG TGT GGT TGG-3′. A 5′-biotin moiety was attached via a 15-atom spacer through the 4-carboxy group of the aptamer. It was synthesized by Operon.
5. Phosphate buffered saline with $MgCl_2$ ($MgCl_2$-PBS): 140 mM NaCl, 2.7 mM KCl, 1 mM $MgCl_2$, 10 mM Na_2HPO_4, pH 7.4.
6. Capping solution: 1 mM ethanolamine-hydrochloride, pH 8.5. Important for a successful deactivation is the use of a redistilled ethanolamine (Aldrich, 41.100-0).

2.4. Protein Binding

1. Binding buffer: 20 mM Tris–HCl, pH 7.4, 140 mM NaCl, 5 mM KCl, 1 mM $MgCl_2$.
2. Human α-thrombin (Haematologic Technologies, Essex Junction, VT, USA), molecular weight 33,600 g/mol. Dilutions of 1–1,000 nM were performed in binding buffer **Section 2.4, Step 1**.
3. Elastase type I from porcine pancreas (E-1250, Sigma), molecular weight 25,900 g/mol. Dilutions of 1–1,000 nM were performed in binding buffer **Section 2.4, Step 1**.
4. Regeneration solution 0.1 NaOH in running buffer.

3. Methods

The functionalities of the aptamers and the specificity of the binding of analytes to the aptamer-modified sensor surface mostly depend on the stringency of the conditions applied to the SELEX procedure. When applied to sensors, aptamers bind directly to the gold surface, but without a proper orientation. An orientation is achieved by use of thiol-derivatized oligonucleotides coupled via carbodiimide chemistry to alkanethiol surfaces, or by binding to dextran surfaces. The amount of background noise from unspecific binding to dextran surfaces is significantly lower than to alkanethiol surfaces. Unspecific or unwanted binding can have a number of reasons. It can be attributed to increasing unspecificity of aptamers at excessive concentrations. Often bind very short aptamers, as is the 15-mer used, not only their cognate analyte, but also other molecules with similar patches. The modern techniques fabricate longer, but also highly specific aptamers.

The immobilization procedure might also affect the binding specificity to the surface itself. Basic principles for biosensors apply, e.g. the size ratio of ligand to analyte. To prefer is coupling of the smaller molecule. For very large ligands it becomes increasingly difficult to measure binding of very small molecules. Low amounts of small molecules result in very small signals, also influenced by the choice of the surface (–chemistry).

In general, the sensor chip surface is composed of gold. Its preparation with a carboxymethyl dextran layer is a process which allows forming hydrogels of different thicknesses, depending on the molecular weight of the dextran material used *(9)*. Since the S-sens K5 technology does not require gold as the sensitive surface, alternative coupling methods might be applied, e.g. binding of organosilanes to SiO_2 surfaces. The lifetime of a modified chip is limited. The first factor is the regeneration. For most oligonucleotides, the binding capacity only changes after the first few injections to remain stable afterwards. Antibody-modified surfaces are more challenging, since they are continuously decreasing in their binding capacity when regenerated. Second factor is the growth of bacteria and yeast. Dextran, for example and other organic materials bound to the surface are subject to decay (fouling) if stored in aqueous solutions outside the constant flow of the sensor. The sensor chips equipped with a dextran surface are to be stored against air at 4°C, even when oligonucleotides are bound. The three-dimensional structure of most proteins would be destroyed. After usage, the sensor chip surface can be stripped from all bound materials and especially organic residues by harsh methods as are chemical or plasma etching. This regenerates the surface completely down to the bare sensor surface without destroying the sensor chips.

3.1. Cleaning of the Sensor Chip Surface

1. The sensor chips are precleaned by three 3-min washing steps in an ultrasonic bath sonifier with precleaning solutions (*see* **Section 2.1**) removing salts and organic residues.
2. The washes are followed by etching in O_2 plasma formed in a TePla etcher at 300 W. Alternatively used are chemical etching procedures (**Steps 3** and **4** below).
3. Chemical stripping using piranha solution (*see* **Section 2.1**) for 1 min.
4. Chemical stripping by heating the chips in ammonia–peroxide solution (e.g. H_2O_2, 30%: H_2O: ammonia, 32% at 1:5:1) to a temperature of >65°C for 5 min.
5. After chemical stripping, the chips are carefully cleaned with ddH_2O. Dry with a stream of nitrogen or argon and store at 4°C.

3.2. Reuse of the Sensor Chips and Storage

1. Sensor Chips: The sensor chip surface is very resistant. The methods mentioned in **Section 2.1** are not harmful to the sensor chip. But frequent usage might mechanically wear off the layers on the chip and produce scratches. About 50 usages per quartz chip are feasible by careful usage.
2. Biological layers: When a biologically active surface is created in the S-sens K5 fluidics system, the sensor chip can be reused hundreds of times. This is only limited by the biosurface cre ated. Aptamer surfaces are simply regenerated by pH changes, e.g. by use of 0.1 N NaOH solution. In between, the chips are storable outside the sensor machine in plain water at 4°C for reuse. Aptamer surfaces might even be stored against air at 4°C for several years.
3. Aptamers: Stored at –20°C stable for many years without any loss in activity.

3.3. Preparation of Alkanethiols Monolayers

1. Freshly clean the sensor chip surface with solutions **Section 2.1, Step 1**. Each washing step is performed for 3 min in a sonifier.
2. Remove all remains off of the chip surface by chemical or plasma etching according to **Section 3.1**. Place the chip into an inert (glass) container.
3. Self-assembled monolayers are formed by adsorption of solution listed in **Section 2.2**. Cover the chip surface with about 2–5 ml of the solution (*see* **Note 2**). Store dark for at least 12 h and allow the formation of a SAM over night. Longer incubation times are required to ensure proper, densely packed SAMs.
4. Remove unbound alkanethiols: Sonicate for 3 min. Remove the thiol solution and wash the coated sensor chip three times with pure solvent ethanol. Dry the sensor chip with a stream of dry nitrogen or argon.
5. Store at 4°C. SAMs are very stable, but might oxidize over time at long storage.

3.4. Preparation of Dextran-Based Hydrogels

1. Prepare a SAM of 11-mercapto-1-undecanol (Sigma) (according to **Section 3.3**). Place each sensor chip into a small glass container. The total time to prepare the chips is 4 days depending on the long incubation times.
2. Activate the hydroxyl groups of the SAM with activation solution. Add 2.5 ml to each chip and incubate at room temperature for 4–5 h.
3. Wash three times with ddH_2O, then twice with 100% ethanol and three times with ddH_2O.
4. Prepare dextran solution and drip onto the sensitive surface. Incubate at room temperature over night. The thickness of the

resulting dextran layer depends on the molecular weight of the dextran. The dextran used forms a layer with a thickness of about 200 nm.

5. Wash thoroughly with 50°C ddH_2O (about 15 times).
6. Add 500 µl of carboxylation solution (*see* **Section 2.2, Step 5**) to each chip. Incubate at room temperature for about 4 h. The hydroxyl groups in the dextran layer form carboxymethyl groups.
7. Repeat **Section 3.4, Step 5**, and add 2.5 ml of carboxylation solution (*see* **Section 2.2, Step 5**) Incubate at room temperature over night. Repeat **Section 3.4, Step 5**.
8. Dry the sensor chips with dry N_2 or argon and store at 4°C.

3.5. Measurement Using the S-Sens® K5 Sensor

1. These instructions assume the use of an S-sens® K5 biosensor. Place the sensor chip with the interface into the sensor.
2. Start the continuous buffer flow at about 30–40 µl/min. The flow rate is limited to 10–300 µl/min (*see* **Note 3**). The higher the flow rate, the higher is the shear stress applied.
3. Take a spectrum between frequencies 145 and 155 MHz and choose two frequencies so that the phase difference is at about 180°. This results in maximal sensitivity.
4. Start measurement. Recorded as output signals are phase shift and amplitude of the propagating wave. The baseline stabilizes according to the surface within minutes. Dextran might be subject to swelling, increasing the time for stabilization.
5. At a constant baseline, inject samples into the buffer flow using the connected autosampler. A volume of up to 500 µl can be injected from an injection loop. The size of the flow chamber is about 2.4 µl per sensor element. The total contact time is the coefficient of amount of substance injected and flow rate, e.g. for 200 µl injected at 40 µl/min is the contact time about 300 s.
6. After injection, running buffer is pumped over the surface. Excess, non-hybridized analyte molecules are expelled from the cell. The next injection cycle can start.
7. From the recorded phase and amplitude signals, the amount of binding and viscosity changes, respectively, can be extracted. By use of a reference solution, the mass signal might be separated from the viscosity signal. An injection of 5% glycerol gives a viscosity signal in the range of most biological measurements.

3.6. Immobilization of Aptamers to a SAM

1. Insert a sensor chip with a 11-mercaptoundecanoic acid SAM into the S-sens® K5 and start measurement with ddH_2O as running buffer.

2. Inject carboxyl activation solution **Section 2.3, Step 1** (*see* **Note 4**).
3. Inject aptamer bearing a primary amine **Section 2.3, Step 2** (*see* **Fig. 13.1**). Carbodiimide chemistry is commonly used to couple molecules presenting a primary amine to sensor surfaces presenting carboxyl groups, forming carboxylamides (*see* **Note 5**).
4. Inject capping solution **Section 2.3, Step 6** (*see* **Note 6**).

3.7. Immobilization of Aptamers to a Dextran Surface Forming a Biotin–Streptavidin Complex

1. Insert a sensor chip with a carboxymethyl-dextran layer (**Section 3.4**) into the S-sens® K5 and start measurement with ddH_2O as running buffer (*see* **Note 7**).
2. Inject carboxyl activation solution.
3. Inject streptavidin solution.
4. Inject capping solution.
5. Exchange buffer to $MgCl_2$–PBS buffer.
6. Inject biotinylated aptamers (*see* **Fig. 13.1**).

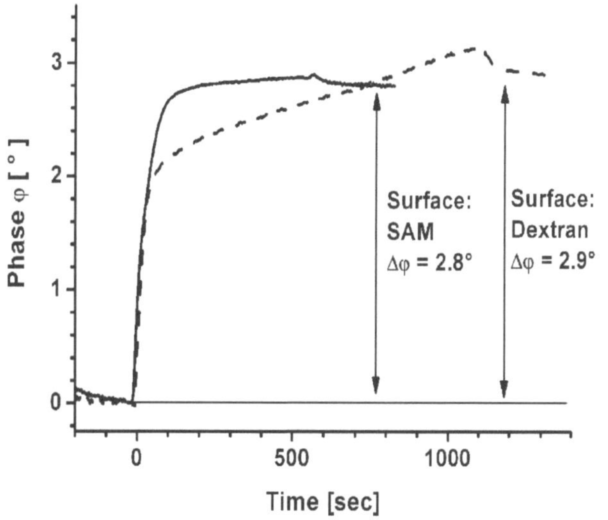

Fig. 13.1. Immobilization of ligand 2 μM DNA anti-thrombin aptamer at a flow rate of 40 μl/min in running buffer PBS-$MgCl_2$. The *dashed line* shows coupling of 200 μl of NH_2–$(CH_2)_3$–aptamer (**Section 2.3**, **Step 2**) to a COOH SAM via carbodiimide chemistry and the solid line 300 μl of biotinylated aptamer (**Section 2.3**, **Step 4**) to a streptavidin-modified dextran surface (**Section 3.7**). Both binding curves show an immediate increase. At 60 s after begin of injection, the SAM was saturated, while the amount of binding to the dextran surface was progressing until the injection stopped, indicating the larger binding capacity of the 3-D surface. The capacity was not saturated afterwards, but was limited by the amount presented. For the 2-D alkanethiol surface, it is necessary to adjust the amount of ligand. In general, higher total amounts can be immobilized to the 3-D surface.

3.8. Specific Binding of Proteins to a Biosensor Surface

1. Immobilize the aptamer according to **Section 3.6** or **3.7**.
2. Exchange running buffer to binding buffer, favouring specific binding of the cognate analyte in high amounts.
3. Inject sample fluid containing, or suspected of containing, analyte at desired concentrations. The mobile proteins elastase as a reference and analyte thrombin were injected.
4. Running buffer automatically was flowed over the surface to determine the off-rate of analytes detaching from the immobilized aptamers.
5. Regeneration solution is injected to preferably regenerate the sensor surface completely, detectable at the return of the signal down to baseline.
6. New cycles of binding and regeneration can follow starting with **Section 3.7. Step 5** (*see* **Fig. 13.2**).

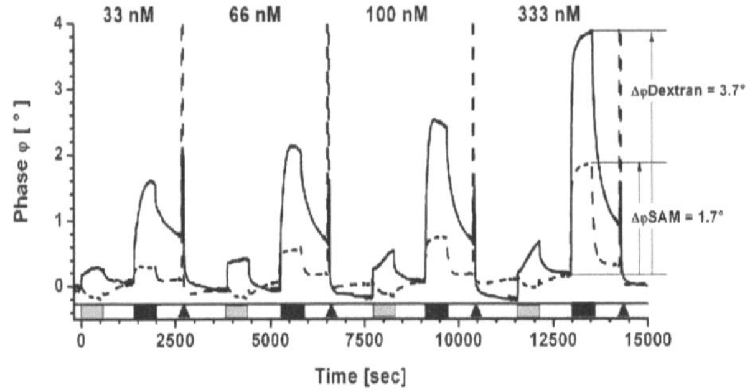

Fig. 13.2. Binding to SAM surfaces (*dashed line*) and dextran (*solid line*) modified by anti-thrombin aptamer (*see* **Fig. 13.1**). Buffer was exchanged to binding buffer. Elastase (*grey bars*) and thrombin (*black bars*) were injected at increasing concentrations from 1 to 1,000 nM. Subsequently, buffer was flowed over the surface to dissociate unbound protein and the surface was regenerated to baseline by an injection of 0.1 N NaOH (indicated by *triangles*). For clarity, injections at medium concentrations of 33, 66, 100 and 333 nM are displayed. The system is based on specific binding of the protein in the sample to surface-immobilized aptamers. In each injection, a specifically bound protein–aptamer complex is formed according to the available and reactive aptamers and to the K_D value of the ligand. Each analyte molecule binds according to the association constant, depending on the space spared by the previously bound analyte molecules. The surface was regenerated and the next round of binding was started. The binding experiments to the dextran surface resulted in a phase shift of 0.3° for 1 μM elastase. With a sensitivity of 515 [° cm^2 / μg], this equals to about 0.58 ng/cm^2 or 14 fmol/cm^2. The injection of 1 μM thrombin resulted in $\Delta\varphi = 3.7°$. This equals to about 7.2 ng/cm^2 and 196 fmol/cm^2. The same experiment on a SAM resulted in a phase shift of 1.7° for 1 μM thrombin, equalling to about 3.3 ng/cm^2 and 90 fmol/cm^2. The amount of thrombin bound to the SAM surface was only about 45% of the dextran surface. The total amount of immobilized aptamer differed by only about 4%. Thus, the additional amounts of bound thrombin to the 3-D dextran surface result from the approachability of immobilized aptamers from more sides compared to the flat, 2-D SAM. The fraction of elastase which equals to about 5–7% of the bound thrombin also is detectable in filter binding experiments and was already shown in previous sensor experiments *(7)*.

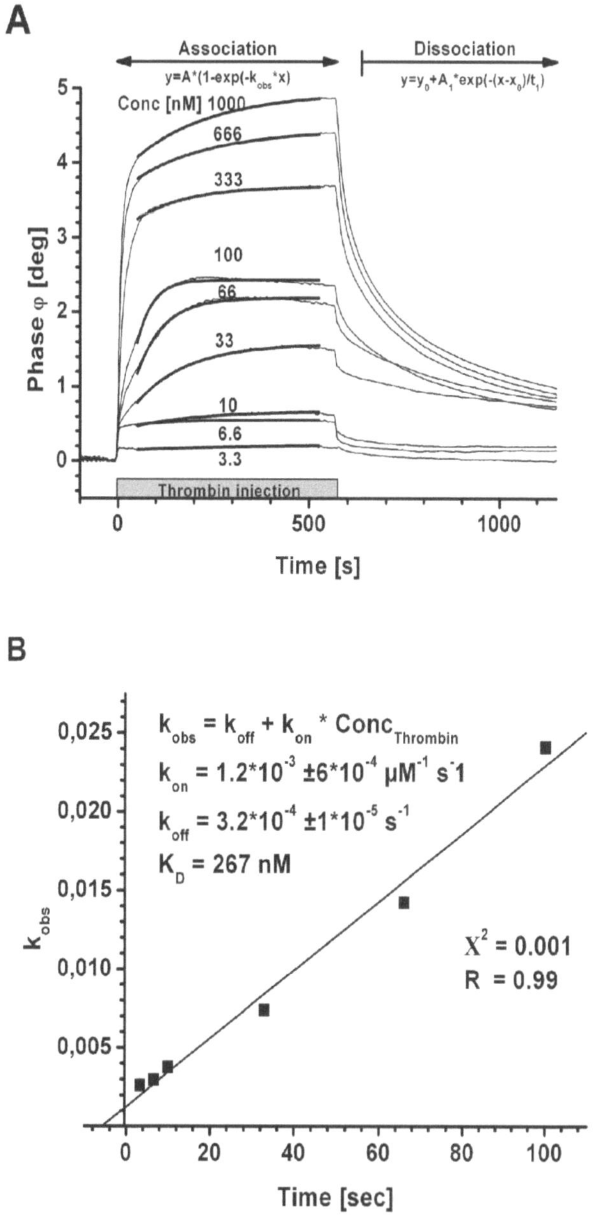

Fig. 13.3. Overlay plot and analysis of injections at different concentrations of thrombin taken from **Fig. 13.2**. (**A**). Overlay plot with combined signals of injections to a aptamer-modified dextran surface. Fits for association of thrombin were applied under the assumption of a 1:1 binding model with A = association signal, maximum binding extrapolated to infinite long injections, and k_{obs} = pseudo-first-order kinetic constant. Fits for dissociation of thrombin were applied, with y_0 = off-set, x_0 = end of injection and begin of buffer injection, A_1 = dissociation signal, and t_1 = decay constant (half-life

3.9. Evaluation of Data

1. Evaluation of bound masses (*see* **Fig. 13.2**) *(7)*: Differences in phase and amplitude are measured as an interval, e.g. at the starting and end point of an injection (*see* **Fig. 13.2**). The phase shift is assumed to be directly proportional to mass or fluid loading on the sensor surface and is identical for all five sensor elements. For proteins, after subtraction of the reference, bound masses can be calculated using a sensitivity of the S-sens® K5 515 ° $cm^2/\mu g$ assumed for proteins (determined analogous to *(10)*). With the molecular weight, the concentrations can be calculated.
2. Extraction of kinetic data: The program Fitmaster (Biosensor GmbH, Bonn, Germany) was used, based on the Origin Software 7.5G SR6 to overlay the binding events into a single plot and to extract kinetic data *(11)*. Binding of elastase and thrombin at concentrations from 1 to 1,000 nM were monitored (*see* **Fig. 13.3A**) and calculated k_{obs} values were plotted versus concentration of the injected fragments (*see* **Fig. 13.3B**) to extract kinetic data (*see* **Fig. 13.3**).

4. Notes

1. Diglyme is volatile and has a low surface tension.
2. In each step, enough material should be prepared for all sensor chips to ensure a constant concentration. Chip-to-chip variations make it more difficult to compare different measurements.
3. The speed at which the running buffer is pumped over the surface equals the product transport. It determines the contact time of analyte and aptamer. At increased speed, the length of contact time is decreased, and the shear force applied to surface-bound analytes and ligands is increased. Based on the shear force, strong, specific binding events are preferred, while weaker binding is reduced. This might be applied to small analytes, since large objects are affected stronger. Simultaneously, the amount of injected sample might increase, too.

Fig. 13.3. (continued) of complex). (**B**) The k_{obs} values extracted from the sensor signals in (**A**) plotted versus concentration of injected thrombin. A linear best fit was applied to the data using the equation shown with k_{on} = association rate constant (on-rate) and k_{off} = dissociation rate constant (off-rate). $K_D = k_{off}/k_{on}$ = dissociation constant. The fits applied resulted in a $k_{on} = 1.2*10^{-3} \pm 6*10^{-4}\ \mu M^{-1}\ s^{-1}$, a $k_{off} = 3.2*10^{-4} \pm 3*10^{-4}\ s^{-1}$, resulting in a $K_D = 267$ nM for the dextran surface. For the SAM, a $k_{on} = 2.8*10^{-3} \pm 1*10^{-3}\ \mu M^{-1}\ s^{-1}$, a $k_{off} = 7.3*10^{-4} \pm 4*10^{-5}\ s^{-1}$ were calculated, resulting in a $K_D = 260$ nM. The resulting K_D values of about 260 and 267 nM agree well with the previously obtained value in the range of 300–400 nM obtained with a preliminary model to the S-sens K5 sensor and with a number of alternative methods *(7)*.

Low amounts of an expensive, delicate or hard to obtain analyte are to be injected at reduced flow rates.

4. The fast reaction of EDC with carboxyl containing molecules forms an *O*-acylisourea as an amine-reactive intermediate. The intermediate is stabilized by the NHS, forming an amine reactive NHS–ester *(12, 13)*. The NHS increases the efficiency of EDC-mediated reactions.
5. The amount of bound ligand is directly related to the concentration of ligand applied and to the number of accessible activated esters on the sensor surface. Therefore, for 2-D surfaces as are SAMs, lower amounts are bound than to three-dimensional surfaces. The concentration of aptamers can be in the high nanomolar to low-micromolar range. Lower concentrations are used for larger ligands or analytes to avoid steric hindrance.
6. Small blocking agents can be added after conjugation to terminate the chemical reaction and to quench any unreacted primary amines. The compounds containing primary amines will result in modified carboxyl groups on the surface ("capping"). Examples for primary amines used are 1–50 mM of hydroxylamine or substances containing a primary amine such as lysine, glycine, ethanolamine, or Tris. Most commonly used is ethanolamine, *see* **Section 2.3, Step 6**.
7. Mostly, proteins are bound directly. But both the carboxymethyl-dextran layer and the oligonucleotide-based aptamers are negatively charged, resulting in an electrostatic repulsion. Salt will reduce the repulsion between them. But to bind high amounts of oligonucleotides, biotinylated aptamers are coupled to a CM-dextran chip surface with covalently immobilized streptavidin. The affinity of 244 Da vitamin H biotin for the bacterial streptavidin is very strong *(14, 15)*. Each tetrameric streptavidin molecule can bind up to four single biotin molecules with positive cooperativity between the subunits *(16)*. Advantage of this attachment system are (i) biotinylated compounds are bound with the correct orientation. (ii) Non-specific binding is reduced, since streptavidin has no carbohydrate group and an isoelectric point of 5. (iii) The complexes formed are extremely stable over a wide range of temperature and pH, unaffected by most organic solvents and even denaturing agents. (iv) Various biomolecules including proteins and antibodies can be biotinylated. Oligonucleotides can be synthesized with biotin moieties at almost every position and in any number. The position of linkage determines the efficiency of biological material interacting with an analyte. (v) The length of the tethering arms covalently attached to biotin determines the binding capacity, the accessibility and flexibility at solid–fluid interfaces and the binding kinetics for analytes.

Acknowledgments

I would like to thank Antje Baumgartner for technical assistance and Ulrich Schlecht for helpful discussions.

References

1. Perpeet, M., Glass, S., Gronewold, T., Kiwitz, A., Malavé, A., Stoyanov, I., Tewes, M. and Quandt, E. (2006) SAW sensor system for marker-free molecular interaction analysis. *Anal. Lett.* **39**, 1747–1757.
2. Du, J.K., Jin, X.Y., Wang, J. and Xian, K. Love-wave propagation in functionally graded piezoelectric material layer. *Ultrasonics* **46**(1) (2007):13–22.
3. Dubois, L.H. and Nuzzo, R.G. (1992) Synthesis, structure, and properties of model organic surfaces. *Annu. Rev. Phys. Chem.* **43**, 437–463.
4. Schäferling, M. and Kambhampati, D. (2004) Protein microarray surface chemistry and coupling schemes. In: Kambhampati, D. (ed.), Protein Microarray Technology. Wiley-VCH Verlag, Weinheim, pp. 11–37.
5. Bock, L.C., Griffin, L.C., Latham, J.A., Vermaas, E.H. and Toole, J.J. (1992) Selection of single-stranded DNA molecules that bind and inhibit human thrombin. *Nature* **355**, 564–566.
6. Stubbs, M.T. and Bode, W. (1993) A player of many parts: the spotlight falls on thrombin's structure. *Thrombosis Res.* **69**, 1–58.
7. Gronewold, T.M. A., Glass, S., Quandt, E. and Famulok, M. (2005) Monitoring complex formation in the blood-coagulation cascade using aptamer-coated SAW sensors. *Biosens. Bioelectron.* **20**, 2044–2052.
8. Duevel, R.V. and Corn, R.M. (1992) Amide and ester surface attachment reactions for alkanethiol monolayers at gold surfaces as studied by polariziation modulation FTIR. *Anal. Chem.* **64**, 337–342.
9. Löfås, S. and Johnnson, B. (1990) A novel hydrogel matrix on gold surfaces in surface plasmon resonance sensors for fast and efficient covalent immobilization of ligands. *J. Chem. Soc. Chem. Commun.* **21**, 1537–1544.
10. Schlensog, M., Gronewold, T., Tewes, M., Famulok, M. and Quandt, E. (2004) A love-wave biosensor using nucleic acids as ligands. *Biosens. Actuators.* **101**, 308–315.
11. Gronewold, T. Baumgartner, A., Quandt, E. and Famulok, M. (2006) Discrimination of single mutations in cancer-related gene fragments with a surface acoustic wave sensor. *Anal. Chem.* **78**, 4865–4871.
12. Grabarck, Z. and Gergely, J. (1990) Zero-length crosslinking procedure with the use of active esters. *Anal. Biochem.* **185**, 131–135.
13. Staros, J.W., Wright, R.W. and Swingle, D.M. (1986) Enhancement by N-hydroxysulfosuccinimide of water-soluble carbodiimide-mediated coupling reactions. *Anal. Biochem.* **156**, 220–222.
14. Kuntz, I.D., Chen, K., Sharp, K.A. and Kollmann, P.A. (1999) The maximal affinity of ligands, *Proc. Natl. Acad. Sci. USA*, **96**, 9997–10002.
15. Böhm, H.-J. (1994) The development of a simple scoring function to estimate the binding constant for a protein ligand complex of known 3-dimensional structure. *J. Comput. Aided Mol. Des.* **8**, 243–256.
16. Williams, D.H., Stephens, E. and Zhou, M. (2003) Ligand binding energy and catalytic efficiency from improved packing with receptors and enzymes. *J. Mol. Biol.* **329**, 389–399.
17. Schreiber, F. (2000) Structure and growth of self-assembling monolayers. *Prog. Surf. Sci.* 65, 151–256.

Chapter 14

Nanoparticles/Dip Stick

Yi Lu, Juewen Liu, and Debapriya Mazumdar

Abstract

Aptamers are single-stranded nucleic acids or peptides that can bind target molecules with high affinity and specificity. The conformation of an aptamer usually changes upon binding to its target analyte, and this property has been used in a wide variety of sensing applications, including detections based on fluorescence, electrochemistry, mass, or color change. Because native nucleic acids do not possess signaling moieties required for most detection methods, aptamer sensors usually involve labeling of external signaling groups. Among the many kinds of labels, inorganic nanoparticles are emerging as highly attractive candidates because some of their unique properties. Here, we describe protocols for the preparation of aptamer-linked gold nanoparticles (AuNPs) that undergo fast disassembly into red dispersed nanoparticles upon binding of target analytes. This method has been proven to be generally applicable for colorimetric sensing of a broad range of analytes. The sample protocols have also been successfully applied to quantum dots and magnetic nanoparticles. Finally, to increase the user friendliness of the method, the sensors have been converted into simple dipstick tests using lateral flow devices.

Key words: Aptamer, nanoparticle, sensor, colorimetric, lateral flow.

1. Introduction

1.1. Aptamers as Sensor Components

Aptamers are nucleic acids or peptides that can be selected to bind essentially any molecule of choice *(1, 2)*. With their versatile binding properties, aptamers have found important applications in many fields of research, including sensing *(3–10)*, drug screening *(11–13)*, therapeutics *(14–17)*, materials science, and nanotechnology *(18–20)*, some of which are detailed in various chapters of this book. In this chapter, we focus our discussion on analytical applications of nucleic acid aptamers. As components for sensors, aptamers possess many advantages. First, aptamers

Günter Mayer (ed.), *Nucleic Acid and Peptide Aptamers: Methods and Protocols, vol. 535*

DOI 10.1007/978-1-59745-557-2_14 Springerprotocols.com

targeting essentially any molecule of choice can be obtained through combinatorial selections *(1, 2, 4, 21)*, which provides a unique opportunity to construct a general sensing platform for a broad range of analytes. Second, aptamers, especially DNA aptamers are highly stable and can be denatured and renatured many times without losing their binding abilities, allowing a long shelf life. Third, nucleic acids have predictable base pairing interactions, which have been proven to be very useful for rational sensor design. On the other hand, such rational designs are difficult in making protein-based sensors. Finally, DNA with a broad range of chemical modifications can be chemically synthesized with relatively low cost.

Natural nucleic acids do not possess functional groups that can generate absorption in the visible region, fluorescence, magnetic or electrochemical signals. Therefore, to make aptamers into sensors, external signaling labels need to be applied. To achieve this goal, many organic fluorophores, chromophores, and electrochemically active labels have been employed. Although being effective in demonstrating the design of aptamer sensors, these organic molecule-based labels suffer from a number of limitations. For example, organic fluorophores photo bleach relatively quickly, while organic chromophores are not "bright" enough with the highest extinction coefficient being on the order of 10^5 $M^{-1}{\bullet}cm^{-1}$. As an alternative, recent advances in the preparation, characterization, and functionalization of inorganic nanoparticles allowed their applications in many fields of research, including replacing organic labels for biosensing applications. In this chapter we summarize recent progress towards using inorganic nanoparticles with varying properties for constructing a wide range of highly sensitive and selective aptamer sensors.

1.2. Physical Properties of Nanoparticles

Depending on the composition, size, and shape of inorganic nanoparticles, a wide range of properties can be obtained. We show here that inorganic metallic *(9, 22)*, semiconductor *(23)*, and magnetic nanoparticles *(24)* can all be assembled by aptamers to generate functional sensors with different detection modes. As an example to illustrate the sensor preparation process, metallic nanoparticles are used to narrate the protocol. Dispersed AuNPs (diameter from several nanometers to about 100 nm) display red colors resulting from their surface plasmons. In addition to such distance-dependent optical properties, AuNPs also possess very high extinction coefficients, which are usually 3–5 orders of magnitude higher than the brightest organic chromophores. Thiol-modified DNA can be attached to the surface of AuNPs and these functionalized nanoparticles can be crosslinked by complementary DNA to form blue-colored aggregates *(25)*. This process has been applied by Mirkin and co-workers to design highly sensitive and selective colorimetric sensors for DNA detection *(26)*. By using functional DNA (aptamers *(9, 27)*, catalytic DNA or DNAzymes *(28, 29)* and aptazymes *(30)*) that can recognize a diverse range of analytes, we demonstrate that the AuNP-based colorimetric detection method can be applied to detect many analytes beyond DNA. Fluorescent semiconductor nanocrystals are known as quantum dots (QDs). Compared to organic fluorophores, QDs are much less susceptible to photo bleaching. The emission wavelength of QDs can be tuned by varying their size, shape, and chemical composition, while keeping the excitation wavelength same, allowing multiplexed detection of many analytes in the same solution. Superparamagnetic nanoparticles (such as iron oxide) can affect the relaxation of water protons under a field, and such effects enable iron oxide nanoparticles as useful magnetic resonance imaging (MRI) contrast agents for biomedical imaging applications. We have demonstrated that all the above mentioned inorganic nanoparticles can be assembled by aptamers, and the properties of the nanoparticles are controlled by the target molecules of the aptamers *(23, 24)*. Because the method used for preparing these nanoparticle/aptamer assemblies is very similar among different nanoparticles, we use AuNPs as an example to describe the protocols.

1.3. Aptamer Assembled Nanoparticles for Sensing Applications

In addition to the generality among different nanoparticles, the method is also general to many aptamers. We have demonstrated AuNP-based aptamer sensors for various analyte, including adenosine, cocaine, potassium ions and their combinations *(9, 31)*. Here only adenosine sensors are described. The adenosine sensor consists of three components (*see* **Fig. 14.1**): two DNA-functionalized AuNPs (particles 1 and 2) and a linker DNA ($Linker_{Ade}$). The DNAs for AuNP 1 and 2 are attached to nanoparticles at its 3′ and 5′ end, respectively (*see* **Note 1**). The linker DNA is designed so that the 5′ end, which is complementary to the DNA attached to particle 1, is separated from the adenosine aptamer at its 3′ end by a pentanucleotide sequence (*see* **Note 2**). The DNA for particle 2 is complementary to the pentanucleotide and to the first seven nucleotides of the adenosine aptamer. There is a 12-adenine spacer (A_{12}) in DNA for AuNP 1 but not 2. The importance of this design is discussed in **Note 3**. In the absence of adenosine or in the presence of other molecules such as other nucleosides, the AuNPs are assembled at room temperature and appear purple as a result of the surface plasmon effect. In the presence of adenosine, however, the aptamer DNA switches its structure and binds two adenosine molecules *(6, 32, 33)*. As a result, only the pentanucleotide (in *gray*) in the linker DNA is left to bind particle 2. The five DNA base pairs

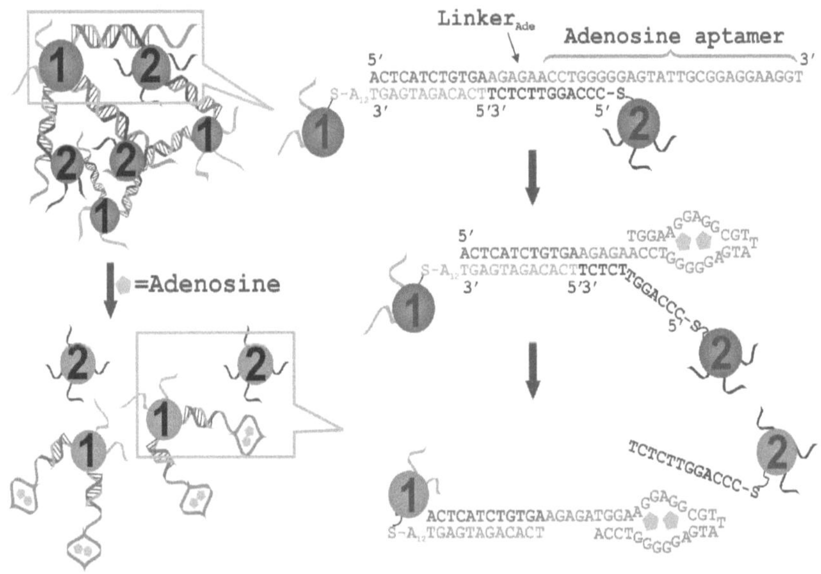

Fig. 14.1. Adenosine-induced disassembly of nanoparticle aggregates for colorimetric detection of adenosine. Nanoparticles 1 and 2 are functionalized with two different DNA molecules through thiol-gold chemistry. The two kinds of AuNPs are linked by $Linker_{Ade}$ to form aggregates. In the presence of adenosine, the AuNPs disassemble to give dispersed red nanoparticles. (Reproduced from ref. *9* with permission from Wiley).

are not strong enough to hold particle 2 at room temperature, leading to its dissociation and resulting in red individual $AuNP_s$ (**Fig. 14.2**).

An important feature of the design is that it is highly modular. Simple replacement of the adenosine aptamer DNA sequence with those of other aptamers allows one to obtain sensors for a diverse range of analytes. The AuNPs can also be replaced by other metallic nanoparticles, so that different color changes can be achieved in the presence of different analytes.

1.4. Dipsticks

The aptamer–nanoparticle-based colorimetric tests can be converted into user-friendly "dipstick" tests using lateral flow devices. This technology provides the reagents in a dry or nearly dry state immobilized on a pad, thus alleviating the need for precise transfer of solution-based reagents, which is often difficult for people without a scientific training. Several antibody-based dipstick tests are known, the home pregnancy tests being one of the most common use of this technology. The detection of DNA using lateral flow device has also been demonstrated *(34)*. By utilizing the lateral flow devices for aptamer-based detection, we have expanding the range of analytes that can be detected using

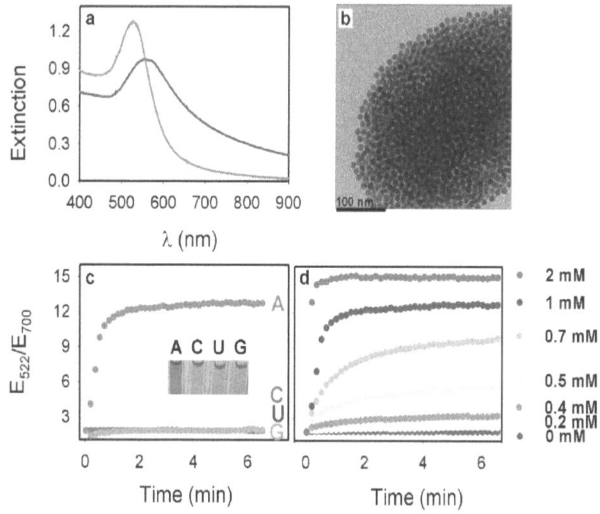

Fig. 14.2. Colorimetric detection of adenosine with aptamer-assembled nanoparticle aggregates. (**a**) UV–visible spectra of dispersed (*gray*) and aggregated (*black*) gold nanoparticles. (**b**) TEM of aptamer-linked gold nanoparticle aggregates. The scale bar is 100 nm. (**c**) Kinetics of color change of adenosine aptamer-assembled aggregates in the presence of 1 mM nucleosides. *Inset*: photograph of the four samples with designated nucleoside added. (**d**) Kinetics of color change of the aggregates with varying adenosine concentrations. (Reproduced from ref. *9* with permission from Wiley).

this simple platform *(35)*. We show that these devices are not only simpler to operate, but also more sensitive than solution-based tests owing to the integration of binding, separation, and detection on a simple test-paper-like platform with no background interference

The adenosine aptamer is used to build a model system to study aptamer-based lateral flow devices. The adenosine sensor consists of the same components as described in **Section 1.3**, except in this case approximately 50% of DNA on particles 1 contains a biotin moiety, which is denoted as a black star. The biotin modification allows the nanoparticles to be captured by streptavidin. The DNA functionalized AuNPs (particles 1 and 2) are assembled using a linker DNA, called $\text{Linker}_{\text{Ade}}$ (*see* **Fig. 14.3a**, top). Detailed DNA sequences, modifications and linkages are shown in **Fig. 14.3b**. A lateral flow device is constructed, consisting of four overlapping pads (wicking pad, conjugation pad, membrane and absorption pad) placed on a plastic backing. The aggregates are dried on the conjugation pad of the devise and streptavidin is applied on the membrane

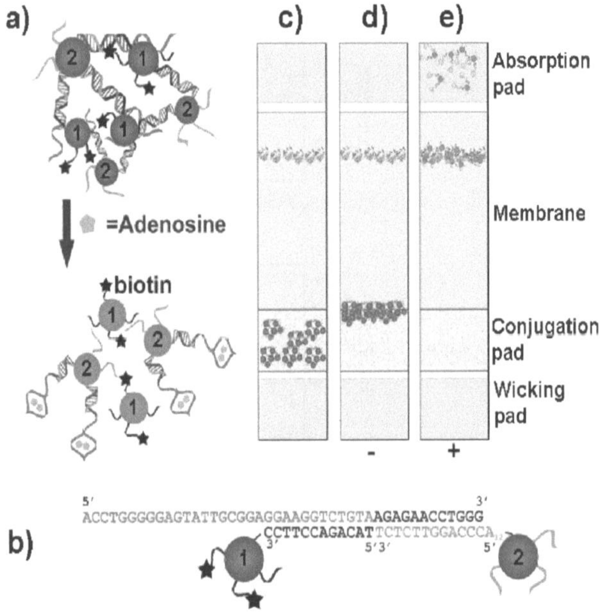

Fig. 14.3. Aptamer/nanoparticle-based lateral flow device. (**a**) Adenosine induced disassembly of nanoparticle aggregates into dispersed nanoparticles. Biotin is denoted as a *black star*. (**b**) DNA linkages in nanoparticle aggregates. Lateral flow devices loaded with the aggregates (on the conjugation pad) and streptavidin (on the membrane) before use (**c**), in a negative (**d**), or positive (**e**) test. (Reproduced from ref. *35* with permission from Wiley).

(*see* **Fig. 14.3c**). We hypothesize that nanoparticle aggregates are too large to migrate along the membrane, while dispersed nanoparticles can. If the device is dipped into a solution without adenosine, the aggregates would be re-hydrated and migrate to the bottom of the membrane, where they stop because of their large size (*see* **Fig. 14.3d**). In the presence of adenosine, the nanoparticles would be disassembled due to binding of adenosine by the aptamer (*see* **Fig. 14.3a**) *(6, 9)*. The smaller dispersed nanoparticles can then migrate along the membrane and be captured by streptavidin to form a red line (*see* **Fig. 14.3e**).

A novel aspect of the lateral flow device described here is that it takes advantage of the physical size difference of nanoparticles in various assembly states, and the fact that aggregated nanostructures do not move along the membrane, which provides a critical control for the performance of the device. This could be used as a new way of designing lateral flow devices where no covalent surface attachment is needed.

2. Materials

1. Oligonucleotides: All oligonucleotides are purchased from a commercial source (e.g., Integrated DNA Technologies Inc., (Coralville, IA)). The oligonucleotides are purified by HPLC or polyacrylamide gel electrophoresis (PAGE) to ensure high purity.
2. Other chemicals: Hydrogen tetrachloroaurate(III) ($HAuCl_4$), trisodium citrate dihydrate, tris-(2-carboxyethyl)phosphine hydrochloride (TCEP), tris-(hydroxymethyl) aminomethane (Tris), adenosine, cytidine, uridine, guanosine, and NaOH are purchased from Aldrich. Sucrose and concentrated HCl, HNO_3, and HOAc are purchased from Fisher. Streptavidin is purchased from Promega.
3. Buffers: Tris–acetate buffers are used in the experiments. 500 mM of Tris–acetate buffer stock of pH 8.2 is prepared by adding acetic acid (glacial) to 500 mM Tris solution until the desired pH value is achieved. The buffer stock solutions are incubated with metal chelating resin (iminodiacetic acid, sodium form, Aldrich) overnight to eliminate trace divalent metal ions. Finally, the buffer stock solutions are filtered through 0.2 μm syringe filters (Nalgene, Rochester, NY) and stored in a -20°C freezer.
4. Equipments: A two-neck flask (100 ml), a condenser and a stopper; hot plate with magnetic stirring and a stir bar; disposable scintillation vials (20 ml), polypropylene microcentrifuge tubes (1.7 ml; catalog no. MCT-175-C; Axygen Scientific), temperature-controlled UV–visible spectrophotometer (Hewlett-Packard 8453), quartz UV–visible cell (Hellma), 0.2-μm syringe filter (Nalgene), and Sep-Pak desalting column (Waters).
5. Lateral flow devices: Hi-Flow TM Plus Assembly Kit (Millipore Corporation, Belford, MA) was used to assemble the lateral flow devices. The kit contained: (a) Hi-Flow plus Cellulose Ester Membrane with a nominal capillary flow time of 90 s/4 cm and a nominal membrane thickness of 135 μm directly cast onto 2 mil polyester backing and placed on an adhesive card (60 * 300 mm), (b) Millipore cellulose fiber sample pads and (c) Millipore glass fiber conjugate pads.

3. Methods

3.1. Preparation of AuNPs

Preparation of high-quality AuNPs ensures the success of subsequent steps of the experiment. For current applications, we choose to synthesize ~13 nm diameter AuNPs for the following reasons.

First, the protocol for such synthesis is well-established and requires only a simple mixing step *(36)*. Second, the resulting AuNPs can be readily used for conjugation of thiol-modified DNA *(26)*, and the conjugates are usually highly stable against aggregation. Although we have demonstrated that larger nanoparticles of 40 nm diameter can also be used, it is more difficult to obtain DNA conjugates with comparable stability. Smaller nanoparticles (i.e., 5 nm diameter) are not recommended for this application because it is difficult to grow large AuNP aggregates linked by DNA.

1. Prepare 500 mL of aqua regia by mixing 3:1 concentrated HCl/HNO_3 in a large beaker in a fume hood. The color of the mixture changes to deep orange/red in several minutes. Be extremely careful when preparing and working with aqua regia. Wear goggles and gloves, and perform the experiment in a fume hood.
2. Soak a two-neck flask, magnetic stir bar, stopper, and condenser in the aqua regia solution for at least 15 min (*see* **Note 4**). The volume of the flask can vary depending on the scale of synthesis, and usually 100–500 mL is used. Rinse the glassware with copious amount of deionized water and then Millipore water.
3. Prepare 50 mM $HAuCl_4$ solution by dissolving the solid in Millipore water. Do not use metal spatula while weighing out the $HAuCl_4$. Filter the solution with a 0.2 μm pore size syringe filter. Prepare 38.8 mM trisodium citrate solution by dissolving the salt in Millipore water and filter the solution.
4. To prepare about 100 mL of AuNPs, add 98 ml of Millipore water into the two-neck flask. Add 2 ml of 50 mM $HAuCl_4$ solution so that the final $HAuCl_4$ concentration is 1 mM. Connect the water condenser to one neck of the flask, and place the stopper in the other neck. Put the flask on a hot plate to reflux while stirring.
5. When the solution begins to reflux, remove the stopper. Quickly add 10 ml of 38.8 mM sodium citrate, and replace the stopper. The color should change from pale yellow to grayish blue to deep red in 1 min. Allow the system to reflux for another 20 min.
6. Turn off the heating and allow the system to cool to room temperature (23–25°C) with stirring. The diameter of such prepared AuNPs is ~13 nm. The extinction value of the 520 nm plasmon peak is ~2.4, and the nanoparticle concentration is ~13 nM. The color of the solution should be burgundy red, and the AuNP shape should be spherical under transmission electron microscopy (TEM). The prepared nanoparticles are stable for months when stored in a clean container

(glass or plastic) at room temperature. Do not freeze the nanoparticles.

3.2. Functionalization of AuNPs with Thiol-Modified DNA

1. Soak two disposable scintillation vials (20 ml volume) in 12 M NaOH for 1 h at room temperature (*see* **Note 5**). Rinse the vials with copious amounts of deionized water and then Millipore water. Be extremely careful when preparing and working with concentrated NaOH. Wear goggles and gloves. When preparing 12 M NaOH solution, the temperature of the system increases significantly. Occasional stirring is needed to avoid the condensation of solid NaOH on the bottom of the container. The concentrated NaOH solution can be reused many times for soaking glass vials.
2. Prepare 10 mM fresh TCEP solution by dissolving a tiny crystal of TCEP in Millipore water.
3. Pipette 9 μl of 1 mM DNA1 into a microcentrifuge tube and 9 μl of 1 mM DNA2 into another one.
4. Add 1 μl of 500 mM acetate buffer (pH 5.2) and 1.5 μl of 10 mM TCEP to each tube to activate the thiol-modified DNA. Incubate the sample at room temperature for 1 h. This activation step is necessary because the thiol-modified DNA from IDT is shipped in the oxidized form with a disulfide bond.
5. Transfer 3 ml of the already prepared AuNPs to each of the two NaOH-treated glass vials, and then add the TCEP-treated thiol DNA with gentle shaking by hand.
6. Cap the two vials and store them in a drawer at room temperature for at least 16 h. Although all the operations described in this protocol can be carried out under light, it is advised to keep nanoparticles in the dark for long-term storage.
7. After the initial incubation, add 30 μl of 500 mM Tris–acetate (pH 8.2) buffer dropwise to each vial with gentle hand shaking. The final Tris–acetate concentration is 5 mM.
8. Add 300 μl of 1 M NaCl dropwise to each vial with gentle hand shaking. Cap the vials tightly and store them in a drawer for at least another day before use. These two types of functionalized AuNPs correspond to particles 1 and 2 in **Fig.** 14.1. When sealed tightly, the functionalized AuNPs can be stored at room temperature for a very long time (several months). However, slow degradation of the DNA on AuNPs may happen to change the properties of the AuNPs. *See* **Note 6** for discussion on the storage of DNA-functionalized AuNPs.

3.3. Preparation of Aptamer-Linked AuNP Aggregates

1. Transfer 500 μl of functionalized particles 1 and 2 into two 1.7-ml microcentrifuge tubes, respectively.
2. Centrifuge the two tubes at 16,110 *g* at room temperature (23–25°C) on a benchtop centrifuge for 15 min.
3. Gently remove the two tubes from the centrifuge. The supernatant should be clear, and the AuNPs should be at the bottom of the tubes. If a red color can still be observed in the supernatant, centrifuge for another 5 min.
4. Gently pipette off as much supernatant as possible to remove free DNA. Again, disperse the AuNPs in 200 μl of buffer containing 100 mM NaCl, 25 mM Tris–acetate, pH 8.2.
5. Centrifuge again for 10 min at 16,110 *g* at room temperature.
6. Remove the supernatant. Again, disperse the nanoparticles in 500 μl of buffer containing 300 mM NaCl, 25 mM Tris–acetate, pH 8.2. We found that most of the free DNA can be removed by two centrifugations. If desired, repeat **Steps 2–5** to remove more of the free DNA.
7. Mix the two nanoparticle solutions.
8. Mix 10 μl of 10 μM $Linker_{Ade}$ DNA with the nanoparticles so that its final concentration is 100 nM.
9. Incubate the nanoparticles at 4°C for at least 1 h. The solution color should change from red to purple. To obtain nanoparticle aggregates in high yield, incubate the solution longer. We found that after overnight incubation, almost all of the nanoparticles went into aggregates because no red color was observable in the supernatant. After long-term incubation, the aggregates may grow large enough to precipitate out of solution. These large aggregates, however, can still be used for sensing applications. With brief agitation by a pipette, the aggregates can be resuspended. The aggregates can be stored at 4 °C for weeks and still maintain their sensing activity.

3.4. Detection of Adenosine with Aptamer-Linked AuNPs

Before performing the detection reaction, it is important to optimize the experimental conditions such as temperature and ionic strength of the system so that a quick color change can be observed (*see* **Note 7**). The following protocol is focused on such an optimization process.

1. Centrifuge the AuNP aggregates at 800 *g* for 1 min at room temperature.
2. Remove the supernatant. Redisperse the aggregates in 500 μl buffer containing 300 mM NaCl, 25 mM Tris–acetate, pH 8.2.
3. Take 50 μl of the just prepared AuNP aggregates and dilute to 200 μl with buffer. The dilution buffer contains 25 mM Tris–acetate, pH 8.2 and 100 mM NaCl. After dilution, the

final NaCl concentration is 150 mM. The drop in NaCl concentration from 300 to 150 mM does not cause significant changes to the optical properties of the nanoparticles (only a slight increase in the extinction ratio was observed), because the melting temperature of the aggregates in 150 mM NaCl is still much higher than room temperature.

4. Transfer the diluted aggregates into a UV–visible cell; seal the cell with Parafilm.
5. Place the sealed cell in the temperature-controlled UV–visible sampling chamber, and measure the extinction spectra for a range of temperatures (e.g., from 15 to 60°C at intervals of 2°C). Allow at least 1 min for equilibration after reaching each designated temperature. Agitate the cell before taking each measurement to make sure that the aggregates are suspended homogeneously.
6. Record and plot the extinction at 260 nm versus temperature. Initially, the extinction may be constant or decrease slightly with increasing temperature. After reaching a certain temperature, the extinction begins to increase sharply.
7. Record the temperature at which the extinction starts to increase. The optimal temperature for detection is ~2–3°C below this.
8. Repeat **Steps 5–7** at different NaCl concentrations. Usually the optimal temperature increases with increasing NaCl concentration. Because experiments are the most convenient to conduct at room temperature, adjust the NaCl concentration to maintain the optimal temperature around room temperature. A good starting point for freshly functionalized nanoparticles is 150 mM NaCl.
9. Add 1 μl of 50 mM adenosine or any other nucleoside solution to 49 μl of nanoparticle aggregates with optimized NaCl concentration to observe color change. To completely dissolve 50 mM adenosine or guanosine, heat the solutions in a boiling water bath.
10. Upon addition of adenosine, the color of the sensor solution should change from purple to red. Addition of other nucleosides should not change the color of the solution. Under optimized conditions, (such as NaCl concentration and temperature) the color change should be instantaneous. Shown in **Fig. 14.2a** are the typical UV–visible spectra of nanoparticles in the dispersed (*gray curve*) and aggregated states (*black curve*). Upon disassembly, the extinction at the 522-nm plasmon peak increases while the extinction in the 700-nm region decreases. Therefore, the ratio of extinction at 522 nm to that at 700 nm can be used to quantify the assembly state and color of AuNPs, with a high ratio

indicating dispersed nanoparticles of red color. A typical TEM image of an aggregate is shown in **Fig. 14.2b**. Thousands of nanoparticles are linked by DNA to form a rigid network structure. The kinetics of sensor color change in the presence of 1 mM adenosine, cytidine, uridine or guanosine is presented in **Fig. 14.2c**, and a photograph of the samples is shown as an inset. Only the sample with added adenosine showed a rapid color change from purple to red, whereas the other three samples remained purple. The kinetics of color change also depends on the concentration of adenosine added, which can be used for quantitative analysis (*see* **Fig. 14.2d**).

3.5. Dipsticks

3.5.1. Preparation of Lateral Flow Devices

The lateral flow device was assembled using components of a Millipore Hi-Flow kit. The membrane used in this case had a nominal capillary flow time of 90 s/4 cm. If greater sensitivity is required in an assay, a membrane with a higher nominal flow time (up to 240 s/ 4 cm is available from Millipore) can be chosen; however, this will increase the time required for the test.

1. Cut out the absorption pad (15 * 300 mm) and the wicking pad (15 * 300 mm) from Millipore cellulose fiber sample pads, and the conjugation pad (13 * 300 mm) from Millipore glass fiber conjugate pad.
2. Attach the absorption pad, wicking pad, and conjugation pad to the adhesive card containing the membrane in a way as shown in **Fig. 14.4a.** (Note that this figure shows a single device of 8 mm width, where as the actual width of the assembled components on the adhesive card is 300 mm). The overlap for each pad should be ~2 mm. Cut the assembled components using a paper cutter into individual lateral flow devices with a width ~8 mm.

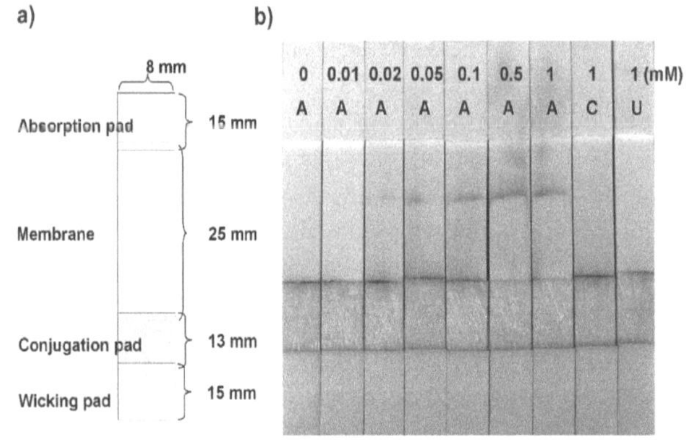

Fig. 14.4. (**a**) Assembly of a lateral flow device. (**b**) Test of the lateral flow device with varying concentrations of nucleosides. (Reproduced from ref. *35* with permission from Wiley).

3.5.2. Preparation of Nanoparticle Aggregates

3. In order to functionalize AuNPs with thiol modified DNA follow **Steps 1–8** of **Section 3.2** with a small change during **Step 4**. Instead of using 9 μL of DNA 1, use a mixture of 4.5 μL DNA 1 and 4.5 μL of DNA 1⋆. (DNA 1⋆ is the same as DNA 1, except for a biotin moiety on the 5′ end). We chose to use 50% biotinylated DNA because 100% led to low yield in nanoparticle aggregates (<20%), while 10% led to inadequate streptavidin capture (data not shown).
4. Follow **Steps 1–8** of **Section 3.3**.
5. Incubate the nanoparticles at 4°C for 6 h. Dark purple precipitants form at the bottom of the tube (*see* **Note 8**).
6. Centrifuge the nanoparticle aggregates at 800 *g* for 1 min at room temperature.
7. Remove the supernatant and redisperse the aggregates in a solution containing 8% sucrose, 200 mM NaCl, 25 mM NaCl and 25 mM Tris–acetate, pH 8.2 (*see* **Note 9**).

3.5.3. Application of Reagents on Lateral Flow Device and Detection

8. Apply 6 μL of the nanoparticle aggregates on the conjugation pad. Apply 2 μL of 10 mg/mL streptavidin on the membrane by a 2 μL pipette to form a thin line.
9. Dry at room temperature for at least 3 h.
10. In order to test the dipstick, prepare solutions with varying concentrations of adenosine (0–2 mM) in a buffer containing 100 mM NaCl, 25 mM Tris–acetate, pH 8.2. To completely dissolve adenosine, heat in a boiling water bath.
11. Dip the wicking pad into the buffer solution containing adenosine (or any other nucleosides, as negative controls) for ~20 s or until the conjugation pad is fully hydrated and the liquid starts to migrate on the membrane.
12. Remove the device from the buffer and lay it down on a flat plastic surface for flow to continue.
13. If adenosine is present at a concentration above the detection limit, a red line will be seen on the membrane. If adenosine is not present, no red line is seen. **Figure 14.4b** shows the typical test results as the concentration of adenosine is increased (note that the red line is seen as a *dark grey line* in grayscale, *see* **Fig. 14.4b**). A line can be seen starting from 0.05 mM adenosine and an increase in the intensity of the color is seen as the concentration of adenosine is increased. Other nucleosides (cytidine, C and uridine, U) do not show a red line. For comparison, the solution-based reaction was carried out using the same batch of aggregates. In order to observe a color change by naked eyes, ~0.5 mM adenosine was required (data not shown) and therefore the dipsticks have a tenfold greater sensitivity than the solution-based tests.

4. Notes

1. It is well documented in the literature that when functionalizing AuNPs with DNA, introducing a poly-adenine spacer between the thiol group and the DNA sequences for cross-linking allows the formation of more stable nanoparticle aggregates due to the increased distance among nanoparticles. Due to the presence of an aptamer overhang in the system (*see* **Fig. 14.1**), we found significant overhang/spacer interactions on AuNP surface when the overhang and poly-A spacer are aligned on the same side, which inhibits the adenosine-induced disassembly *(9, 37)*. To maintain the stability benefit from the poly-A spacer and to avoid the inactivation effect, only one poly-A spacer is used and the spacer is on the opposite side of the aptamer overhang.
2. The most important element in designing such stimuli-responsive aggregates is the number of base pairs between the linker DNA and the DNA on particle 2, and the number of nucleotides from the aptamer sequence that are involved in binding to particle 2. The binding should be strong enough to hold the nanoparticles together but still allows the aptamer to bind target molecules and undergo rapid structure switching. Systematic studies on this topic have been reported by Li and co-workers *(6, 38)*.
3. When designing the DNA sequences to attach to AuNPs, it is important to avoid self-complementarity. During centrifugation, even partially complementary base pairing interactions from the same kind of nanoparticles can induce significant aggregation of AuNPs that can be dispersed only in very low ionic strength buffer (i.e., pure water). Therefore, if insoluble precipitation (in 100 mM NaCl containing buffer) is observed after centrifugation, try using pure water and check the sequence of DNA.
4. Care should be taken to make sure that no contamination is introduced during AuNP synthesis. Low-quality AuNPs can result in poor DNA conjugation, which will induce nanoparticle precipitation during later processing steps such as centrifugation. Obtaining high-quality nanoparticles is the first important step towards the success of the experiment.
5. If the glass vials are not treated with concentrated NaOH, nanoparticles tend to stick to the surface of the vials, especially after addition of NaCl to the particles. If this problem occurs, the effective concentration of nanoparticles decreases.
6. DNA-functionalized nanoparticles can still be used to form aggregates even after storing at room temperature for months,

although their properties may change slightly with the passage of time as a result of events such as degradation of DNA. These changes could affect the properties of the prepared aggregates. To be sure that the results are consistent, use freshly functionalized AuNPs.

7. The nanoparticles are linked by DNA base-pairing interactions. A certain concentration of NaCl is required to stabilize the nanoparticle aggregates. The higher the salt concentration, the more stable the aggregates. The stability of nanoparticle aggregates can be evaluated by melting studies. Aptamer binding to its target analyte, on the other hand, is less dependent on NaCl. We find that doubling the NaCl concentration under optimized conditions can inhibit the color change. Therefore, it is important to optimize the salt concentration of the system to observe fast color change.
8. If the aggregates prepared for the lateral flow device are too large, then the disassembly on the device maybe difficult. Therefore the nanoparticles should not be left to aggregate for an extended period of time.
9. For lateral flow devices, it is important to optimize the sucrose and NaCl concentration for drying the aggregates on the conjugation pad. Sucrose is important for preserving the DNA-linked aggregates and aiding their rehydration. If sucrose is not added, then the aggregates do not disassemble. We tested sucrose concentration in the range of 0–30% and chose 8% sucrose as the optimum condition for drying our aggregates, as it produced a dark red line for adenosine containing samples and no red line for the blank samples. When 30% sucrose was used for drying, a faint red band could be seen even with the blank samples *(35)*. As discussed in **Note** 7, optimizing the NaCl concentration is important to obtain a balance between stabilizing the DNA base pairing interactions and achieving fast disassembly in the presence of analyte.

Acknowledgment

This material is based on work supported by the U.S. Army Research Laboratory and the U.S. Army Research Office under grant number DAAD19-03-1-0227, by the National Science Foundation through the Science and Technology Center of Advanced Materials for Purification of Water with Systems (WaterCAMPWS) and the Nanoscale Science and Engineering Center (NSEC) program.

References

1. Ellington, A.D. and Szostak, J.W. (1990) *In vitro* selection of RNA molecules that bind specific ligands. *Nature* **346**, 818–822.
2. Tuerk, C. and Gold, L. (1990) Systematic evolution of ligands by exponential enrichment: RNA ligands to bacteriophage t4 DNA polymerase. *Science* **249**, 505–510.
3. Breaker, R.R. (1997) DNA aptamers and DNA enzymes. *Curr. Opin. Chem. Biol.* **1**, 26–31.
4. Famulok, M. (1999) Oligonucleotide aptamers that recognize small molecules. *Curr. Opin. Struct. Biol.* **9**, 324–329.
5. Stojanovic, M.N., de Prada, P. and Landry, D.W. (2000) Fluorescent sensors based on aptamer self-assembly. *J. Am. Chem. Soc.* **122**, 11547–11548.
6. Nutiu, R. and Li, Y. (2003) Structure-switching signaling aptamers. *J. Am. Chem. Soc.* **125**, 4771–4778.
7. Huang, C.-C., Huang, Y.-F., Cao, Z., Tan, W. and Chang, H.-T. (2005) Aptamer-modified gold nanoparticles for colorimetric determination of platelet-derived growth factors and their receptors. *Anal. Chem.* **77**, 5735–5741.
8. Navani, N.K. and Li, Y. (2006) Nucleic acid aptamers and enzymes as sensors. *Curr. Opin. Chem. Biol.* **10**, 272–281.
9. Liu, J. and Lu, Y. (2006) Fast colorimetric sensing of adenosine and cocaine based on a general sensor design involving aptamers and nanoparticles. *Angew. Chem. Int. Ed.* **45**, 90–94.
10. Willner, I. and Zayats, M. (2007) Electronic aptamer-based sensors. *Angew. Chem. Int. Ed.* **46**, 6408–6418.
11. Blank, M. and Blind, M. (2005) Aptamers as tools for target validation. *Curr. Opin. Chem. Biol.* **9**, 336–342.
12. Famulok, M. and Mayer, G. (2005) Intramers and aptamers: applications in protein-function analyses and potential for drug screening. *ChemBioChem* **6**, 19–26.
13. Elowe, N.H., Nutiu, R., Allali-Hassani, A., Cechetto, J.D., Hughes, D.W., Li, Y. and Brown, E.D. (2006) Small-molecule screening made simple for a difficult target with a signaling nucleic acid aptamer that reports on deaminase activity. *Angew. Chem. Int. Ed.* **45**, 5648–5652.
14. Osborne, S. E., Matsumura, I. and Ellington, A.D. (1997) Aptamers as therapeutic and diagnostic reagents: problems and prospects. *Curr. Opin. Chem. Biol.* **1**, 5–9.
15. Nimjee, S.M., Rusconi, C.P. and Sullenger, B.A. (2005) Aptamers: an emerging class of therapeutics. *Annu. Rev. Med.* **56**, 555–583.
16. Lee, J.F., Stovall, G.M. and Ellington, A.D. (2006) Aptamer therapeutics advance. *Curr. Opin. Chem. Biol.* **10**, 282–289.
17. Famulok, M., Hartig, J.S. and Mayer, G. (2007) Functional aptamers and aptazymes in biotechnology, diagnostics, and therapy. *Chem. Rev.* **107**, 3715–3743.
18. Liu, Y., Lin, C., Li, H. and Yan, H. (2005) Aptamer-directed self-assembly of protein arrays on a DNA nanostructure. *Angew. Chem. Int. Ed.* **44**, 4333–4338.
19. Lu, Y. and Liu, J. (2006) Functional DNA nanotechnology: emerging applications of dnazymes and aptamers. *Curr. Opin. Biotechnol.* **17**, 580–588.
20. Famulok, M. and Mayer, G. (2006) Chemical biology: aptamers in nanoland. *Nature* **439**, 666–669.
21. Wilson, D.S. and Szostak, J.W. (1999) *In vitro* selection of functional nucleic acids. *Annu. Rev. Biochem.* **68**, 611–647.
22. Liu, J. and Lu, Y. (2006) Preparation of aptamer-linked gold nanoparticle purple aggregates for colorimetric sensing of analytes. *Nature Protocols* **1**, 246–252.
23. Liu, J., Lee, J.H. and Lu, Y. (2007) Quantum dot encoding of aptamer-linked nanostructures for one pot simultaneous detection of multiple analytes. *Anal. Chem.* **79**, 4120–4125.
24. Yigit, M.V., Mazumdar, D., Kim, H.-K., Lee, J.H., Odintsov, B. and Lu, Y. (2007) Smart "turn-on" magnetic resonance contrast agents based on aptamer-functionalized superparamagnetic iron oxide nanoparticles. *ChemBioChem* **8**, 1675–1678.
25. Mirkin, C.A., Letsinger, R.L., Mucic, R.C. and Storhoff, J.J. (1996) A DNA-based method for rationally assembling nanoparticles into macroscopic materials. *Nature* **382**, 607–609.
26. Storhoff, J.J., Elghanian, R., Mucic, R.C., Mirkin, C.A. and Letsinger, R.L. (1998) One-pot colorimetric differentiation of

polynucleotides with single base imperfections using gold nanoparticle probes. *J. Am. Chem. Soc.* **120**, 1959–1964.

27. Liu, J. and Lu, Y. (2006) Smart nanomaterials responsive to multiple chemical stimuli with controllable cooperativity. *Adv. Mater.* **18**, 1667–1671.
28. Liu, J. and Lu, Y. (2003) A colorimetric lead biosensor using DNAzyme-directed assembly of gold nanoparticles. *J. Am. Chem. Soc.* **125**, 6642–6643.
29. Liu, J. and Lu, Y. (2004) Accelerated color change of gold nanoparticles assembled by dnazymes for simple and fast colorimetric pb2+ detection. *J. Am. Chem. Soc.* **126**, 12298–12305.
30. Liu, J. and Lu, Y. (2004) Adenosine-dependent assembly of aptazyme-functionalized gold nanoparticles and its application as a colorimetric biosensor. *Anal. Chem.* **76**, 1627–1632.
31. Liu, J. and Lu, Y. (2005) Stimuli-responsive disassembly of nanoparticle aggregates for light-up colorimetric sensing. *J. Am. Chem. Soc.* **127**, 12677–12683.
32. Hermann, T. and Patel, D. J. (2000) Adaptive recognition by nucleic acid aptamers. *Science* **287**, 820–825.
33. Hartig, J.S., Najafi-Shoushtari, S.H., Gruene, I., Yan, A., Ellington, A.D. and Famulok, M. (2002) Protein-dependent ribozymes report molecular interactions in real time. *Nat. Biotechnol.* **20**, 717–722.
34. Glynou, K., Ioannou, P.C., Christopoulos, T.K. and Syriopoulou, V. (2003) Oligonucleotide-functionalized gold nanoparticles as probes in a dry-reagent strip biosensor for DNA analysis by hybridization. *Anal. Chem.* **75**, 4155–4160.
35. Liu, J., Mazumdar, D. and Lu, Y. (2006) A simple and sensitive "dipstick" test in serum based on lateral flow separation of aptamer-linked nanostructures. *Angew. Chem. Int. Ed.* **45**, 7955–7959.
36. Handley, D.A. (1989) Methods for synthesis of colloidal gold. In: *Colloidal Gold Principles, Methods, and Applications.* Hayat, M.A. (ed.), Academic Press, San Diego, ca, pp. 13–32.
37. Liu, J. and Lu, Y. (2007) Non-base pairing DNA provides a new dimension for controlling aptamer-linked nanoparticles and sensors. *J. Am. Chem. Soc.* **129**, 8634–8643.
38. Nutiu, R. and Li, Y. (2004) Structure-switching signaling aptamers: Transducing molecular recognition into fluorescence signaling. *Chem. Eur. J* **10**, 1868–1876.

Chapter 15

In Vivo Imaging of Oligonucleotidic Aptamers

Bertrand Tavitian, Frédéric Ducongé, Raphaël Boisgard, and Frédéric Dollé

Abstract

In this chapter we present the methods developed in our laboratory for in vivo imaging of oligonucleotidic aptamers. These methods relate to (i) the labelling of aptamers with fluorine-18, a positron emitter, (ii) Positron Emission Tomography imaging of laboratory animals with [^{18}F]aptamers and (iii) labelling with fluorescent dyes and optical imaging of aptamers in mice.

Key words: Aptamer, oligonucleotide, in vivo imaging, positron emission tomography, PET, fluorescence, confocal fibred fluorescence imaging, fluorine-18.

1. Introduction

Oligonucleotidic aptamers are produced by a molecular evolution process based on iterative selection–amplification steps known as the SELEX technology *(1, 2)*. Some of these aptamers can bind cellular targets in vivo *(3–6)* and therefore have potential applications as diagnostic or therapeutic agents *(7, 8)*. The translation of aptamers from in vitro ligands into in vivo applications requires a precise exploration of their biodistribution, pharmacokinetics and targeting capacities that can be explored by in vivo molecular imaging methods *(9)*. Molecular imaging can serve as a tool to explore the biostability and bioavailability of aptamers with therapeutic potential *(10–13)*. Another objective is the application of aptamers as molecular contrast agents for non-invasive imaging techniques such as optical imaging or radioisotopic methods *(14, 15)*.

The first in vivo imaging study with an oligonucleotidic aptamer was reported in 1997 by the group of NeXstar Pharmaceutical

Günter Mayer (ed.), *Nucleic Acid and Peptide Aptamers: Methods and Protocols, vol. 535*

DOI 10.1007/978-1-59745-557-2_15 Springerprotocols.com

(16). An aptamer raised against human neutrophil elastase was used to image inflammation in a reverse passive Arthus reaction model in rats. The aptamer labelled with technetium-99m (^{99m}Tc), a gamma-emitting isotope, was injected intravenously and the animals were imaged on a gamma camera. The aptamer achieved a target-to-background ratio significantly higher and more rapidly than an antibody directed against the same target. The conclusion of that report was that aptamer ligands could become useful as diagnostic imaging agents and may offer significant advantages over monoclonal antibodies.

Nine years later, a second example of in vivo aptamer imaging appeared in the literature *(17)*. In this study, an aptamer raised against human tenascin-C, an abundant extracellular matrix protein, was used to visualize U251 human glioma xenograft tumours in nude mice. This aptamer was also labelled by ^{99m}Tc, intravenously injected and animals were imaged on a gamma camera. Quantification of radioactivity in mouse tissues 1 h after injection showed that 2.7% of the injected dose was recovered per gram of tumour.

Image contrast and the rapid pharmacokinetics of aptamers observed in mice support the interest of Positron Emission Tomography (PET) imaging of this type of oligonucleotide. PET is a sensitive, functional nuclear imaging technique that permits repeated non-invasive assessment and quantification of specific biological and pharmacological processes at the molecular level in humans and animals. It is the most advanced technology currently available for studying in vivo molecular interactions in terms of distribution, pharmacokinetics and pharmacodynamics. Recently, new small animal-dedicated PET tomographs with improved spatial resolution have appeared for the imaging of small laboratory rodents. The last generation of small animal PET cameras (e.g. eXplore VISTA from GE or Inveon from Siemens) achieve better than 1.4 mm axial resolution at the centre of the field of view. Clinical cameras with lower spatial resolutions (typically, 4.5 mm) can also be used for PET studies, whenever resolution is not an issue. Their larger aperture with respect to that of dedicated small animal scanners allows imaging several rats simultaneously in the same field of view, thereby optimizing the use of the radiolabelled aptamer batch. Clinical cameras can also accommodate large experimental animals such as swine or non-human primates.

Molecular PET imaging requires the preparation of a positron-emitting radiolabelled molecular probe. For this purpose, the short-lived positron-emitter fluorine-18 ($T_{1/2}$: 109.8 min) is becoming increasingly the radionuclide of choice due to its adequate physical and nuclear characteristics as well as to the successful use in clinical oncology of 2-[^{18}F]fluoro-2-deoxy-D-glucose

([^{18}F]FDG). Fluorine-18 is available at most medical cyclotron facilities.

Isotopic labelling with fluorine-18 implies the replacement of a native fluorine-19 atom with the corresponding positron-emitting isotope (fluorine-18), a strategy being by definition limited to chemical structures already containing fluorine. However, no more than a few dozen of molecules in living nature contain fluorine. Fluorine-18-labelled probes are therefore often designed by hydroxyl-for-fluorine or hydrogen-for-fluorine replacement in the parent structure, a strategy called *foreign labelling* which has permitted the preparation of hundreds of chemical structures labelled with this radioisotope, although almost exclusively of low-molecular weight (<500 Da). Indeed, the direct labelling of macromolecules with fluorine-18 is, leaving aside some exceptions, not feasible due to the sensitivity of these structures towards generally severe radiofluorination conditions. Labelling is therefore usually performed by conjugation of a prosthetic group, carrying the radioisotope, with a reactive function of the macromolecule. This strategy has the advantage of allowing chemical routes requiring drastic chemical conditions for the preparation of the fluorine-18-labelled prosthetic group entity (in fact, a fluorine-18-labelled reagent), while the conjugation with a macromolecule can then be performed in a second step using mild conditions that preserve the macromolecule's structure.

Several fluorine-18-labelled reagents have been described for the coupling to amino-acid-based ligands (peptides and proteins). Most of them have been designed for coupling with an amino function of an amino acid residue (*N*-terminal α-NH_2 or internal lysine ε-NH_2) or an alkylamine linker. The activated ester *N*-succinimidyl 4-[^{18}F]fluorobenzoate ([^{18}F]SFB) is the most popular prosthetic group, acting through an acylation reaction. A smaller number of fluorine-18-labelled reagents for the coupling to the carboxylic acid function (C-terminal α-CO_2H or internal glutamic/aspartic acid β/γ-CO_2H) have also been designed. Mainly due to the poor regioselectivity observed for the coupling of these reagents with macromolecules (which often leads to partial loss of the latter's biological properties), considerable attention has been paid in the last decade to the design and development of sulphur-selective fluorine-18-labelled reagents. Two classes of sulphur-selective fluorine-18-labelled reagents have been described, the haloacetamides (dedicated to the alkylation of phosphorothioate monoester functions (–O–P(O)(–OH)–SH)) and the maleimides (dedicated to the alkylation of pure thiol (sulphhydryl) functions (–SH)).

[^{18}F]FPyBrA (*N*-[3-(2-[^{18}F]fluoropyridin-3-yloxy)-propyl]-2-bromoacetamide) *(18)* and [^{18}F]FPyME (1-[3-(2-[^{18}F]fluoropyridin-3-yloxy)propyl]pyrrole-2,5-dione) *(19, 20)* are two [^{18}F]fluoropyridinyl-based *(21–24)* selected reagents (the first one being a bromoacetamide and the second one a maleimide), which have proven to be efficient in thioconjugation reactions for the fluorine-18-labelling of oligonucleotide-based ligands, including aptamers *(25)* (*see* **Fig. 15.1**).

Both reagents can be prepared via a three-step pathway in less than 2 h from cyclotron-produced [^{18}F]fluoride and [3-(3-*tert*-butoxycarbonylamino-propoxy)-pyridin-2-yl]-trimethyl-ammonium trifluoromethanesulphonate.

On the other hand, in vivo imaging technologies based on photonic technologies that have been developed for small animal research during the last decade could significantly accelerate biomedical discovery by enabling expeditious tests to identify potent aptamers for diagnostic or therapy. Here, the animal is the test bed for the pre-clinical in vivo assessment of treatment efficiency, targeting sensitivity and specificity, bio-distribution, and long term effects. We routinely use two complementary in vivo fluorescence imaging systems to evaluate the biodistribution of fluorescent aptamers: (i) whole body fluorescence imaging and (ii) in vivo confocal fibred microscopy.

Whole body imaging allows rapid and semi-quantitative comparisons of the tissue biodistribution of aptamers in superficial tissues such as xenografts. Fluorescence imaging is undergoing rapid progress and quantitative tomographic fluorescence imaging systems for mice have recently been described *(26)*. In contrast to PET, fluorescence whole body imaging can image aptamer distributions over very long time periods (up to weeks) in mice, and perform the analysis of hundreds of compounds at costs and throughputs that are not conceivable with PET. However, the images obtained with this technique are most of the time planar (2D) and quantification is not as precise as with PET.

Although it has a field of view smaller than a square mm, in vivo confocal fibred microscopy is precious to complement the images obtained with PET or whole body fluorescence imaging which are of low spatial resolution (ca. 1–2 mm at best). Fibred confocal microscopes can reach spatial resolutions of a few μm, allowing to quantitatively document the uptake, distribution and penetration of aptamers tagged with a fluorescent dye at the cellular level.

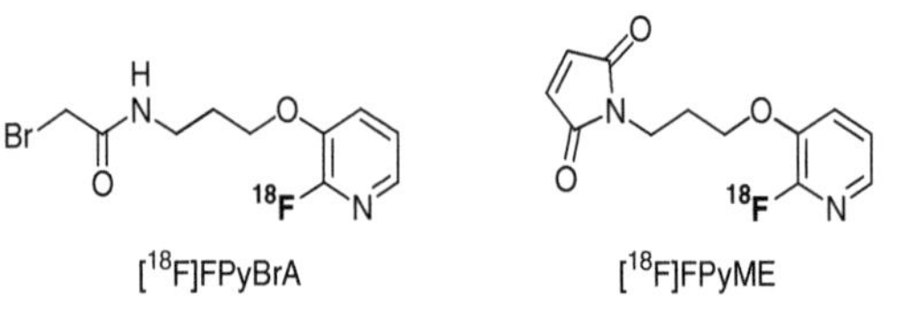

Fig. 15.1. [^{18}F]FPyBrA (*N*-[3-(2-[^{18}F]fluoropyridin-3-yloxy)-propyl]-2-bromoacetamide) *(left)* and [^{18}F]FPyME (1-[3-(2-[^{18}F]fluoropyridin-3-yloxy)propyl]pyrrole-2,5-dione) *(right)* for thioconjugation reactions of oligonucleotide-based ligands.

2. Materials

2.1. Chemicals

Aptamers (chemically synthesized and gel purified) containing a 3'-phosphorothiate monoester group or alternatively, a thiol function (via an alkyl linker chain) can be purchased from oligonucleotide suppliers. In our laboratory we order oligonucleotides from Eurogentec (Belgium). AlexaFluor C5 maleimide dyes are purchased from Invitrogen. Additional chemicals are purchased from standard commercial sources (Aldrich-, Fluka- or Sigma (France), for example) and are used without further purification, unless stated otherwise.

2.2. Radiochemical Labelling of Aptamers with Fluorine-18

1. [3-(3-*tert*-Butoxycarbonylamino-propoxy)-pyridin-2-yl]-trimethyl-ammonium trifluoromethanesulphonate is the common reagent for the preparation of [^{18}F]FPyBrA or [^{18}F]FPyME described below. It is synthesized in two steps from commercially available (3-hydroxypropyl) carbamic acid *tert*-butyl ester and 2-dimethylamino-3-hydroxypyridine as previously described *(18, 19)*. ^{1}H NMR (CD_2Cl_2, 298 K): δ: 8.10 (bd, J=3.3 Hz, 1H); 7.66 (d, J=8.1 Hz, 1H); 7.60 (dd, J=6.1 and 4.2 Hz, 1H); 4.31 (t, J=6.3 Hz, 2H); 3.71 (s, 9H); 3.31 (q, J=6.3 Hz, 2H); 2.12 (q^5, J=6.3 Hz, 2H); 1.38 (s, 9H).^{13}C NMR (CD_2Cl_2, 298 K): δ: 156.6 [C]; 147.7 [C]; 142.6 [C]; 139.0 [CH]; 129.0 [CH]; 124.6 [CH]; 121.2 [q, J=319 Hz, CF_3]; 79.3 [C]; 68.1 [CH_2]; 54.8 [$3 \times CH_3$]; 37.5 [CH_2]; 30.0 [CH_2]; 28.4 [$3 \times CH_3$].
2. FPyBrA (*N*-[3-(2-fluoropyridin-3-yloxy)-propyl]-2-bromoacetamide) is the chromatography standard for the preparation of [^{18}F]FPyBrA. It is synthesized in three steps from (3-hydroxypropyl)carbamic acid *tert*-butyl ester and 2-fluoro-3-hydroxypyridine as previously described *(18)*. ^{1}H NMR (CD_2Cl_2, 298 K): δ: 7.73 (dd, J=3.3 and 1.8 Hz, 1H); 7.32 (td, J=7.8 Hz, J< 1.5 Hz, 1H); 7.14 (dd, J=7.8 and 4.8 Hz, 1H); 4.12 (t, J=6.0 Hz, 2H); 3.85 (s, 2H); 3.49 (q, J=6.3 Hz, 2H); 2.05 (q^5, J=6.3 Hz, 2H). ^{13}C NMR (CD_2Cl_2, 298 K): δ: 166.2 [C]; 153.8 [d, J^1_{F-C}= 235 Hz, C]; 142.3 [d, J^2_{F-C}=25 Hz, C]; 137.5 [d, J^3_{F-C}=13 Hz, CH]; 123.0 [CH]; 122.3 [CH]; 67.7 [CH_2]; 37.8 [CH_2]; 29.5 [CH_2]; 28.9 [CH_2]. MS (DCI/NH_4^+) $C_{10}H_{12}Br_1F_1N_2O_2$: 292 [M + H$^+$]; 290 [M + H$^+$].
3. FPyME (1-[3-(2-fluoropyridin-3-yloxy)propyl]pyrrole-2,5-dione) is the chromatography standard for the preparation of [^{18}F]FPyME. It is synthesized in three steps from (3-hydroxypropyl)carbamic acid *tert*-butyl ester and 2-fluoro-3-hydroxypyridine as previously described *(19)*. ^{1}H NMR (CD_2Cl_2,

298.0 K): δ: 7.69 (bd, J=3.0 Hz, 1H); 7.27 (t, J=9.0 Hz, 1H); 7.11 (dd, J=9.0 and 3.0 Hz, 1H); 6.69 (s, 2H); 4.04 (t, J=6.0 Hz, 2H); 3.72 (t, J=6.0 Hz, 2H); 2.11 (q^5, J=6.0 Hz, 2H). ^{13}C NMR (CD_2Cl_2, 298.0 K): δ: 171.2 (2 × C); 154.1 (d, $J^1_{F\text{–}C}$= 235 Hz, C); 142.4 (d, $J^2_{F\text{–}C}$=25 Hz, C); 137.7 (d, $J^3_{F\text{–}C}$=13 Hz, CH); 134.5 (2 × CH); 123.2 (CH); 122.2 (CH); 67.5 (CH_2); 35.4 (CH_2); 28.4 (CH_2). MS (DCI/NH_4^+) $C_{12}H_{11}F_1N_2O_3$: 251 [M + H$^+$].

4. [^{18}F]fluoride is produced by a nuclear reaction, often the proton irradiation (18 MeV) of a > 97% oxygen-18-enriched water target (cyclotron: [^{18}O(p,n)^{18}F]).
5. [^{18}F]FPyBrA (*N*-[3-(2-[^{18}F]fluoropyridin-3-yloxy)-propyl]-2-bromoacetamide) is synthesized in three steps from [3-(3-*tert*-butoxycarbonylamino-propoxy)-pyridin-2-yl]-trimethylammonium trifluoromethanesulphonate and [^{18}F]fluoride as previously described (18). Purified by HPLC (λ=254 nm) on a semi-preparative SiO_2 Zorbax® Rx-SIL column (Hewlett Packard, 250 × 9.4 mm, porosity: 5 μm) using CH_2Cl_2/EtOAc (50:50 (v/v)) at a flow rate of 5 mL/min (R_t: 7.0–8.0 min).
6. [^{18}F]FPyME (1-[3-(2-[^{18}F]fluoropyridin-3-yloxy)propyl]pyrrole-2,5-dione) is synthesized in three steps from [3-(3-*tert*-butoxycarbonylamino-propoxy)-pyridin-2-yl]-trimethylammonium trifluoromethanesulphonate and [^{18}F]fluoride as previously described (19). Purified by HPLC (λ = 254 nm) on a semi-preparative SiO_2 Zorbax® Rx-SIL column (Hewlett Packard, 250 × 9.4 mm, porosity: 5 μm) using heptane / EtOAc (60:40 (v/v)) at a flow rate of 5 mL/min (R_t: 10.0–10.5 min).

2.3. PET Cameras for the Imaging of [^{18}F] aptamers

In our laboratory we use a clinical research camera, the HRRT from Siemens, for whole body pharmacokinetics imaging of aptamers simultaneously in several rats, and a small animal-dedicated PET (Focus from Siemens) for aptamer targeting studies in one mouse at a time.

1. Imaging of multiple rats on a clinical camera. The HRRT camera (High Resolution Research Tomograph, Siemens) presents a spatial resolution of 2.3–2.8 mm depending on the reconstruction algorithm. This resolution provides sufficiently precise information on the pharmacokinetics and biodistribution of aptamers at the organ level in rats, with the advantage that the field of view of the HRRT allows imaging of four rats simultaneously. Therefore, one can acquire data with statistical value (n=4) to evaluate pharmacokinetics in just "one shot" (one radiosynthesis, and one imaging acquisition). This is precious given that a major limiting factor in experimental

PET studies, apart from access to the cameras, is radiolabelling which is always extemporaneous.

2. High-resolution PET imaging of mice. Whenever spatial resolution is favoured, for example in studies of aptamer targeting against a tumour epitope, a small animal-dedicated PET camera is preferred to a clinical camera. For studies of aptamer targeting in mice, the use of a high-resolution PET camera is recommended. We use the Siemens FOCUS which has a spatial resolution of 1.35 mm at the centre of the field of view.

2.4. Fluorescence Imaging of Aptamers

1. Whole body fluorescence imaging of mice. Live whole body fluorescence recording is performed with the PhotonImager® from Biospace Lab, France. This imaging system uses a photon counting technique coupled with a third generation cooled intensified CCD camera (1080 × 1440 pixels) and a F 1.4 objective lens covering a field of view of either 16 × 12 cm or 8 × 6 cm, depending on the platform position used. The camera is mounted on top of a light tight chamber. Grey scale video images can be taken before and after recording the fluorescence images for superimposition of the animal body with the fluorophore emission. The instrument operates at up to 25 frames per second (40 ms exposure time) but longer integration times can be selected at the end of the experiment for data analysis and replay. Depending on the magnification (distance to object), the resulting "intrinsic" resolution is either 100 μm (two mice stand) or 200 μm (five mice stand), but the final working resolution depends on the tissue absorption and depth of the organs and is in the order of 1–2 mm. As a general rule, optical wavelengths in the near infrared range are less absorbed by living tissues and fluorescent dyes emitting over 650 nm are preferred for detection through biological tissues.
2. Fluorescence imaging at the cellular level. This is performed with the fibred confocal microscope CellVizio488® from Mauna Kea Technologies, France. The apparatus consists in a flexible sub-millimetric probe containing optical fibres that carry light from a continuous laser source at 488 nm to the living tissue. Fluorescence emitted after excitation of the fluorophores present in the tissue is sent back to the apparatus, where a dedicated set of algorithms reconstructs images in real time at a rate of acquisition of 12 frames per second. We usually use a MiniO/30 probe with 12,218 optical fibres, a field of view: 175 × 125 μm, lateral resolution of 1.8 μm, and axial resolution of 5 μm at 37 μm of depth.

3. Methods

3.1. Fluorine-18-Labelling of Aptamers with [^{18}F]FPyBrA or [^{18}F]FPyME

Conjugation of [^{18}F]FPyBrA (*N*-[3-(2-[^{18}F]fluoropyridin-3-yloxy)-propyl]-2-bromoacetamide) or [^{18}F]FPyME (1-[3-(2-[^{18}F]fluoropyridin-3-yloxy)propyl]pyrrole-2,5-dione) with the aptamer uses efficient, fast and relatively mild reaction conditions (such as a mixture of MeOH (or DMSO) and an aqueous buffer, the temperature being dependent on the reagent used), both compatible with the chemical stability of the oligonucleotide and the half-life of fluorine-18. Conjugated fluorine-18-labelled oligonucleotides ([^{18}F]aptamer) can be purified by reverse phase high-performance liquid chromatography (RP-HPLC) or by gel filtration (*see* **Note 1**).

3.1.1. Conjugation of the [^{18}F]reagent with the Aptamer

3.1.1.1. Option A: Using [^{18}F]FPyBrA

1. Redissolve [^{18}F]FPyBrA (freed from HPLC-solvents by concentration to dryness at 65–75°C under a gentle nitrogen stream) in 500 μL of MeOH (Solution A).
2. Dissolve 0.5–1.0 mg of the aptamer to be labelled in 500 μL of PBS (100 mM, ~pH 7.5) (Solution B).
3. Add the *Solution B* to the reaction vessel containing [^{18}F]FPy-BrA (Solution A).
4. Tightly close the reaction vessel.
5. Heat the reaction mixture at 120°C for 10–15 min.
6. Cool down the reaction mixture to about 30°C.

3.1.1.2. Option B: Using [^{18}F]FPyME

1. Redissolve [^{18}F]FPyME (freed from HPLC-solvents by concentration to dryness at 65–75°C under a gentle nitrogen stream) in 100 μL of DMSO (Solution A).
2. Dissolve 0.5–1.0 mg of the aptamer to be labelled in 900 μL of PBS (100 mM, ~pH 7.5) (Solution B).
3. Add the Solution B to the reaction vessel containing [^{18}F]FPyME (Solution A).
4. Vortex gently the reaction mixture at room temperature for 10–15 min.

3.1.2. Examination of the Yield of Conjugation of the [^{18}F]reagent with the Aptamer

Technique: thin-layer chromatography (TLC)

1. Dissolve 1 mg of FPyBrA or FPyME (standard) in 10 mL of EtOAc (Solution C).
2. Apply 2–10 μL of Solution C to a pre-coated plate of silica gel $60F_{254}$ (Merck).
3. Apply an aliquot (2–10 μL) of the crude reaction mixture to the plate.
4. Place the plate in an appropriate chromatographic tank.

5. Elute the plate with pure EtOAc over a path of 8 cm.
6. Take the plate out of the tank and allow it to dry in air for 1 min.
7. Determine the distribution of radioactivity using a suitable radiodetector (for example, a Berthold TraceMaster 20 automatic TLC linear analyser). There should be only two radioactive spots on the TLC: The first one at $R_f = 0.0$ corresponds to [^{18}F]aptamer and should represent more than 75% of the total radioactivity; the second one at $R_f > 0.7$ (comparison with standard by UV) corresponds to the [^{18}F]reagent.

3.1.3. Purification of the [^{18}F]aptamer (Removing Uncoupled [^{18}F]reagent)

Technique: RP-HPLC purification (option A) / gel filtration (option B)

3.1.3.1. Option A: RP-HPLC Purification on a C-18 Column (For Example, a μBondapak® Waters Column) Followed by Desalting on Sephadex® G25 Medium (For Example, a NAP-10™ Pharmacia Column)

1. Prepare and equilibrate the column with the appropriate initial mobile phase until a stable baseline is obtained according to current practice.
2. Inject the above-mentioned total crude reaction mixture onto the column.
3. Elute the column using a dedicated mixture of solvents (for example, a time-variable combination of aq. 50 mM triethylammonium acetate, pH 7.4 (TEAA-buffer) and acetonitrile (ACN)) at an optimal flow rate (for example, 6.0 mL/min). Typical gradient conditions: linear 5 min from 95:5 to 90:10 (TEAA-buffer /ACN, v/v) then linear 15 min from 90:10 to 75:25 (v/v) and wash-out 5 min at 50:50 (v/v).
4. Collect the fraction containing [^{18}F]aptamer.
5. Concentrate (if needed) the HPLC-fraction containing [^{18}F]aptamer to a volume of 1 mL.
6. Prepare and equilibrate the column according to manufacturer's instructions.
7. Add the above-mentioned solution of [^{18}F]aptamer to the column and allow it to enter the gel bed completely.
8. Place an appropriate vessel for sample collection underneath the column outlet.
9. Elute the desired [^{18}F]aptamer with 1.5 mL of, for example, water.

3.1.3.2. Option B: Gel Filtration Purification on Sephadex® G25 Medium (For Example, a NAP-10™ Pharmacia Column)

1. Prepare and equilibrate the column according to manufacturer's instructions.
2. Add the above-mentioned total crude reaction mixture onto the column and allow it to enter the gel bed completely.

3. Place an appropriate vessel for sample collection underneath the column outlet.
4. Elute the desired [^{18}F]aptamer with 1.5 mL of, for example, water or saline (0.9% aq. NaCl).

3.1.4. Formulation of the [^{18}F]aptamer for i.v. Injection

Dilute with saline (0.9% aq. NaCl) the solution containing the purified [^{18}F]aptamer to the concentration desired for i.v. injection.

3.2. PET Imaging of Aptamers

3.2.1. Animal Ethics

Animal experiments are subject to approval by an animal ethics committee prior to experimentation. Depending on the regulations applied at your Institution, you should seek approval from your local veterinarian or animal committee. Explain the reasons for undertaking the study and justify the number of animals that will be used. Anticipate all procedures in order to limit unnecessary pain to the animals. Animal experimentation should only be conducted by trained personnel under the authority of a researcher with registered certification for animal laboratory science.

3.2.2. Pharmacokinetic Studies in Rats

1. Select four rats of the appropriate strain (Wistar, Sprague-Dawley or other), gender and weight. The body of rats weighing 250–300 g is fully visible in the field of view of the camera.
2. Anesthetize the animals by placing them gently in a chamber filled with 3% isoflurane in a flow of medical-grade oxygen, or of a mixture of 2/3 nitrous oxide (medical grade) and 1/3 oxygen. Alternatively, one may use other anesthetics such as a mixture of ketamine and xylazine (9:1) delivered intraperitoneously. In our experience, gaseous anesthesia with isoflurane is best, both for monitoring and animal comfort. Be aware, however, that anesthetized animals are not in their normal physiological state: changes in heart rate, respiration, blood flow, blood pressure and temperature are always observed under any type of anesthesia.
3. From now on, maintain animals under 1.5–2.5% isoflurane during the complete transmission and acquisition scanning periods. Permanent precise monitoring of isoflurane anesthesia is mandatory for comfort and safety of the animals. In addition, exposure of the experimenters to isoflurane is not recommended and should be avoided. In our laboratory, this is achieved through the use of coaxial breathing nozzles into which the nose of the animal is introduced (Minerve, France). The anesthetic mixture expired by the animals is re-aspirated in the nozzle and carried back to an activated carbon cartridge

which retains isoflurane. Cartridges (Minerve, France) are discarded after 1,000 h of utilization.

4. Carefully clean the tail of the rat with your favorite medical disinfectant such as medical grade 70% alcohol.
5. Introduce a sterile 24 gauge Teflon catheter equipped with a sharp edge needle and filled with sterile physiological serum containing 0.01% heparin (SPSH) into the tail vain of each rat. Remove the needle carefully and check that the catheter is correctly inserted into the vein: blood should visibly reflux in the catheter.
6. Inject ca. 0.1 mL of SPSH and secure the catheter with its plug. Tape the catheter to the tail of the animal with adhesive plaster: gently slide a 3 cm-long piece of adhesive, sticky face towards you, between the catheter and the tail, up to the entry point of the catheter in the tail; flip the left side of the tape over the catheter and make it adhere to the tail; flip the right side of the adhesive over needle and tail so as to realize a half knot and secure it to the tail.
7. At the end of this step, control the venous access by injecting no more than 100 μL of SPSH with a 1 mL sterile syringe.
8. Place the animals inside the field of view of the HRRT camera. Because spatial resolution depends on the distance from the centre of the field of view of the tomograph, we use a home-made animal holder with four equidistant positions from the centre of the field of view in order to ensure similar spatial resolutions for all four animals. Animal holders should ideally be made of a low-attenuation material, such as carbon-embedded beehive materials.
9. Once all four rats are placed in the field of view of the camera, start the acquisition of the attenuation map. This will typically take 8 min using the rotating external source of germanium of the HRRT.
10. Measure the radioactivity A_{full} of the syringe containing the [^{18}F]aptamer in a dose calibrator (Capintec) and note precisely the time of the measurement t_{full}.
11. Inject ca. 15 MBq of the [^{18}F]aptamer using a lead-shielded 1 mL syringe and note precisely the time of injection t_{inj}. Rinse immediately with 100 μL SPSH. The total volume of liquids injected i.v. to the rat should represent less than 500 μL in order to avoid blood dilution artifacts.
12. Start the PET acquisition immediately after the first injection of the [^{18}F]aptamer into the fist rat. Repeat Section 3.2.2, **Steps 10–12**. for each animal i, noting carefully the injection times $t_{\text{inj}(i)}$.

13. Count the radioactivity of each empty syringe $A_{empty\,(i)}$ for each rat i in the same dose calibrator as before (Capintec) and again note precisely the time of the measurements $t_{empty(i)}$. *See* **Note 4.2** for measuring the actual dose delivered to each rat.
14. After injection of all four rats, acquire PET scans in 3D mode during 180 min or longer if necessary.

3.2.3. Targeting Studies in Mice

Basically, the protocol for mice is the same as that for rats, except that:

- Only one mouse is placed inside the field of view of the focus camera.
- No catheter is used. Intravenous injection of the [^{18}F]aptamer is done directly with a 29G (insulin-type) syringe.
- Animals are injected on the bench outside the field of view of the camera and then placed in the field of view. Acquisition starts immediately after the mouse is positioned in the field of view.
- The radioactivity dose is in the order of 7.4 MBq of [^{18}F]aptamer.
- The total volume injected is limited to 100 μL.
- Images are corrected for attenuation by calculation based on an emission map.

3.2.4. Image Reconstruction and Analysis

1. Images are reconstructed by Ordered Subset Expectation Maximization (OSEM) using four iterations of 16 subsets.
2. The sequence of time frames for image reconstruction is defined by the user. Choice of time frames is based on the in vivo pharmacokinetics of the aptamer, i.e. the kinetics of tumour uptake and the clearance from the surrounding tissues, which depends on the aptamer and the target. The objective in defining time frames is to obtain the highest possible contrast of the target tissue versus surrounding tissues. In our experience, this is generally achieved in mice at time points later than 1 h after intravenous administration of the [^{18}F]aptamer. Acquisition times as long as 6–8 h or more after injection may be required for some aptamers. Keep in mind, however, that the signal in the images decreases exponentially, so that 6 h after injection, only roughly 10% of the radioactivity remains in the animal, and less than 5% at 8 h. All images are corrected automatically for radioactive decay.
3. In our laboratory, image analysis is realized using the Anatomist software (http://brainvisa.info/index.html). Alternatively, machine-based software or other freewares (such as VINCI: http://www.nf.mpg.de/vinci/) can be used.
4. Regions of interest are drawn in three dimensions in order to delineate major organs and the radioactivity concentrations are

plotted along Time versus Activity Curves (TACs). This procedure known as *segmentation* is time consuming and requires basic anatomical knowledge and a good deal of patience when a series of regions need to be drawn from different planes and different animals. Several algorithms for accelerating segmentation and improving its accuracy are presently under development: see for instance, the LMA method *(27)*, a semi-automated method for the segmentation of PET images based on Brainvisa/Anatomist software.

5. Once the PET image has been segmented, values of radioactivity concentrations are normalized by the injected dose and expressed as *percentage of the injected dose per volume of tissue* (%ID/mL). This allows the comparison between animals which have received different doses of [^{18}F]aptamer.
6. Finally, pharmacokinetic parameters can be calculated from the time–activity curves using classical pharmacokinetics softwares or home-made Excel spreadsheets.

3.3. Analysis of [^{18}F]aptamer Degradation in Rat Plasma

TACs derived from the PET images express the radioactivity found in the animals organs, irrespective of the nature of the molecule that carries the fluorine-18 atom. Because aptamers can be rapidly transformed into other chemical entities after they have been introduced inside the body, any interpretation of the biodistribution depicted by a PET study should separate the signal linked to the intact [^{18}F]aptamer from its fluorine-18-labelled metabolite fractions if present at any time point during the course of the study. Therefore, it is a rule of thumb to perform separately radiometabolic dosages in blood samples taken at various time points during the PET acquisition. Blood sampling from a mouse placed in a PET camera is very delicate since the mouse should not be moved during the whole acquisition. Therefore, we usually perform additional metabolic studies on supplementary animals injected with a more important dose of tracer.

1. Collect 200–500 μL of blood through the tail vein catheter of a 350 g Wistar rat injected with 37 MBq (1.0 mCi) of [^{18}F]aptamer in a previously weighted glass tube containing dry heparin lithium salt (for example, Vacutainer from BD) and place immediately on ice.
2. Centrifuge rapidly at 1,500 *g* for 5 min at 4°C. Separate plasma from blood cells.
3. Measure the radioactivity concentration in the plasma sample in a gamma counter such as the Cobra II Auto gamma, Packard Bioscience Company.
4. Weigh again the tube after counting. The difference between the weights of the empty tube and the tube containing the plasma indicates the weight (ca. volume) of plasma in the

sample. Once the measurements have been repeated for several time points and decay-corrected, their plot will give you the plasmatic TAC of the [^{18}F]aptamer.

5. Add one volume of phenol/$CHCl_3$ (v/v) and four volumes of Tris–EDTA (pH 7.4) to 100–200 μL of rat plasma.
6. Centrifuge at 13,000 rpm (12,500 *g*) for 10 min and keep the supernatant.
7. Extract pellets again with an additional 200 μL of Tris–EDTA and centrifuge as above.
8. Collect supernatants, vacuum-dry and dissolve in a known volume of triethylammonium acetate buffer (TEAA) before injection onto an analytical RP-HPLC column similar to the one used for preparation of the [^{18}F]aptamer (*see* **Section 3.1.3.1**).

3.4. Fluorescence Imaging of Aptamers

When using a thiol-protected aptamer (–S–S–$(CH_2)_6OH$), generate first a free thiol function as described in **Section 3.4.1, Step 1.** If a free thiol function is already present, move directly to **Step 3**.

3.4.1. Fluorescent Labelling of Aptamers with Alexa Fluor C5 Maleimides

1. Incubate 2 nmol of the aptamer during 15 min in 30 μL of 15 mM sodium citrate buffer pH 7, 150 mM NaCl, 10 mM DTT (in order to reduce the disulphide bridge).
2. Remove excess of DTT in a Bio-Spin 6 SSC column (Biorad).
3. Add 20 nmol of Alexa Fluor C5 maleimide to the 2 nmol of the aptamer (10 molar equivalent) and incubate 1 h at 60°C. We use Alexa Fluor 680 for whole body fluorescence imaging and Alexa Fluor 488 for fibred confocal microscopy.
4. Remove any unincorporated dye using a Bio-Spin 6 SSC (Biorad).
5. Quantify the conjugation yield by measuring the absorbance at 260 nm and at the dye's maximum absorption wavelength (γ_{max}). Calculation software can be found at: http://probes.invitrogen.com/resources/calc/dyebaseratio.html or performed manually: in that case the absorbance of the dye at 260 nm should be taken into account (*see* **Note 3**).
6. The labelled aptamer can be used directly or stored at –20°C.

3.4.2. Whole Body Fluorescence Imaging of Aptamers

1. Hair is incompatible with whole body fluorescence imaging. Use mice without fur, such as nude mice, or shave the animals.
2. Allow the CCD camera of the PhotonImager to cool down to –22°C by starting the cooler 30 min before the experiments.
3. Induce anaesthesia of the mice by 5% isoflurane under a flow of medical oxygen and maintain thereafter at an isoflurane concentration of 2–2.5%. During anaesthesia, animals are maintained normothermic (body temperature: 36.7 ± 0.5°C,

mean±SD) through the use of the heating stage of the PhotonImager.

4. Inject the fluorescent aptamer in the tail vein of the animal with a 29 G (insulin-type) syringe. Up to five animals can be imaged at the same time, but it is highly advisable to include one non-injected animal for the control of autofluorescence contribution.
5. Acquire fluorescence images using the appropriate excitation/emission filters included in the PhotoImager. For AlexaFluor 680 labelled aptamers, we recommend 660 nm for fluorescence excitation, 550 nm for background excitation and a longpass filter higher than 714 nm for emission. It is also advisable to acquire a black and white photography of the mice in order to better localize fluorescence in the animals (*see* **Note 4**).
6. Fluorescence recordings are performed at different times post-injection using the integration mode of the machine. Bleaching can become a problem if continuous excitation of the dyes is performed during long periods, which is generally not necessary. In our experience, sufficient contrast is obtained in acquisition times of a few seconds to a few minutes, and we have performed repeatedly daily acquisitions for up to 3 weeks.
7. Data are analysed using the software Photovision from Biospace Lab. For each fluorescence recording, data obtained at an excitation of 660 nm is corrected by the background data obtained from excitation at 550 nm. The software requires definition of an internal non-fluorescent zone by the user which we place on the non-injected control animal.
8. Regions of interest are drawn over the body of the mice using Photovision in order to measure the fluorescence intensity.
9. Using these values, draw fluorescence intensity time curves in several organs such as liver, heart, kidneys, lungs, bladder, etc. and tumors in the case of xenografted nude mice.

3.4.3. In Vivo Fluorescence Imaging of Aptamers at the Cellular Level

1. Switch the Cellvizio laser on 30 min before the start of the experiment.
2. Calibrate the Flexiprobe according to manufacturer's instructions.
3. Induce anesthesia of the mice by 5% isoflurane under a flow of medical oxygen and maintain thereafter at an isoflurane concentration of 2–2.5%. During anaesthesia, animals are maintained normothermic (body temperature: 36.7 ± 0.5°C, mean ± SD) through the use of an homeothermic blanket (Harvard apparatus, Hollington, MA, USA).

4. Using sterile scalpel or dissection scissors, perform a small incision of the skin in order to give access to the organs of interest.
5. Inject the fluorescent-labelled aptamer in the tail vein of the animal with a 29 G (insulin-type) syringe.
6. Place the endoscope in contact with the tissue of interest and record fluorescence imaging. Several recordings of the same organ can be made at different times after injection (*see* **Note 5**).
7. Quantify the imaging data using either the software ImageCell from Maunakea Technologies™ or the freeware ImageJ (http://rsb.info.nih.gov/ij).

4. Notes

1. Radiosyntheses using fluorine-18 ([^{18}F]FPyBrA, [^{18}F]FPyME and [^{18}F]aptamer), including the HPLC and gel purifications, are performed in a 7.5-cm-lead shielded cell using a computer-assisted Zymate robot system (Zymark corporation, USA).
2. The actual dose delivered to a rat is given by the difference between the radioactivity measured in the syringes before and after injection, taking into account the decay of radioactivity between the times of measurements and the actual time at which this rat has been injected. Radioactivity decays in time following an exponential law, in which the radioactive half life T is the time necessary for radioactivity to decay by a factor of two:

$$A_t = A_o.2^{-t/T}$$

T is a constant depending on the isotope and equals 109.8 min for fluorine-18.

The dose of radioactivity that has been actually injected into each animal at its time of injection t_{inj}.

A_{inj} is given by the following formula:

$$A_{inj} = A_{full} \cdot 2^{-(t_{full}-t_{inj})/T} - A_{empty} \cdot 2^{-(t_{inj}/t_{empty})/T}$$

It is convenient for simple calculation of $A_{inj}$to create an Excel sheet in which one introduces the measured radioactivity values and times of measurements. An example can be found in **Table 15.1.**

With T= 109.8 being the half life of fluorine-18, Excel will automatically calculate the value of $A_{inj}$when the following function is entered in B3:

$$= \text{B2*POWER}(2; -1^*((\text{C2} - \text{C3})/109.8)$$
$$- \text{B4*POWER}(2; -1^*((\text{C3} - \text{C4})/109.8)$$

Table 15.1
Excel file in which one introduces the measured radioactivity values and times of measurements

	A	B	C
1		Activity	Time
2	Syringe preinjection	A_{full}	t_{full}
3	Injection	(A_{inj})	T_{inj}
4	Syringe after injection	A_{empty}	t_{empty}

3. Always use fresh solution of Dye-maleimide. The yield of fluorescent labelling is generally in the order of 50–75%. Efficiency of the labelling may be reduced if the 3'-extremity of the aptamer is involved in its structure. In that case labelling conditions can be changed to 70% DMF. If the yield is less than 50%, the quality of the oligonucleotide provided by the manufacturer should be controlled using electrophoresis.
4. Using the PhotonImager in integration mode instead of counting mode allows to reach higher excitation intensities.
5. Whole body fluorescence imaging and in vivo confocal fibred microscopy can be used in a number of different combinations. For instance, it is often interesting to scan a given organ of interest at high resolution with the CellVizio after biodistribution of the aptamer has been documented with the PhotonImager. Postmortem imaging can also be performed with the CellVizio prior to sampling of tissues for histology.

Acknowledgements

Studies relating to in vivo imaging of aptamers in our laboratory are supported by the EMIL and DIMI European networks, the Agence Nationale de la Recherche, the Institut National du Cancer, the Cancéropôle Ile de France, the Fondation de France and the Association pour la Recherche sur le Cancer.

References

1. Ellington, A.D. and Szostak, J.W. (1990) In vitro selection of RNA molecules that bind specific ligands. *Nature* **346**, 818–822.
2. Tuerk, C. and Gold, L. (1990) Systematic evolution of ligands by exponential enrichment : RNA ligands to bacteriophage T4 DNA polymerase. *Science* **249**, 505–510.
3. Blank, M., Weinschenk, T., Priemer, M. and Schluesener, H. (2001) Systematic evolution of a DNA aptamer binding to rat brain tumor

microvessels. Selective targeting of endothelial regulatory protein pigpen. *J. Biol. Chem.* **276**, 16464–16468.

4. Hicke, B.J., Marion, C., Chang, Y.F., Gould, T., Lynott, C.K., Parma, D., Schmidt, P.G. and Warren, S. (2001) Tenascin-C aptamers are generated using tumor cells and purified protein. *J. Biol. Chem.* **276**, 48644–48654.
5. Daniels, D.A., Chen, H., Hicke, B.J., Swiderek, K.M. and Gold, L. (2003) A tenascin-C aptamer identified by tumor cell SELEX: systematic evolution of ligands by exponential enrichment. *Proc. Natl. Acad. Sci. U S A*, **100**, 15416–15421.
6. Cerchia, L., Duconge, F., Pestourie, C., Boulay, J., Aissouni, Y., Gombert, K., Tavitian, B., Franciscis, V.D. and Libri, D. (2005) Neutralizing aptamers from whole-Cell SELEX inhibit the RET receptor tyrosine kinase. *PLoS Biol.* **3**, e123.
7. Pestourie, C., Tavitian, B. and Duconge, F. (2005) Aptamers against extracellular targets for in vivo applications. *Biochimie* **87**, 921–930.
8. Cerchia, L. and de Franciscis, V. (2007) Nucleic acid-based aptamers as promising therapeutics in neoplastic diseases. *Methods Mol. Biol.* **361**:187–200.
9. Ntziachristos, V., Leroy-Willig, A. and Tavitian, B. (Eds.) (2007) *Textbook of*In Vivo *Imaging in Vertebrates.* John Wiley & Sons, Chichester.
10. Boisgard, R., Kuhnast, B., Vonhoff, S., Younes, C., Hinnen, F., Verbavatz, J.M., Rousseau, B., Fürste, J.P., Wlotzka, B., Dollé, F., Klussmann, S. and Tavitian, B. (2004) In vivo biodistribution and pharmacokinetics of [^{18}F]-labelled Spiegelmers: a new class of oligonucleotidic radiopharmaceuticals. *Eur. J. Nucl. Med. Mol. Imaging* **32**, 470–477.
11. Younes, C.K., Boisgard, R. and Tavitian, B. (2002) Labelled oligonucleotides as radiopharmaceuticals: pitfalls, problems and perspectives. *Curr. Pharm. Des.* **8**, 1451–1466.
12. Tavitian, B., Terrazzino, S., Kuhnast, B., Marzabal, S., Stettler, O., Dolle, F., Deverre, J.R., Jobert, A., Hinnen, F., Bendriem, B., Crouzel, C. and Di Giamberardino, L. (1998) In vivo imaging of oligonucleotides with positron emission tomography. *Nat Med.* **4**, 467–71.
13. Tavitian, B. (2003) In vivo imaging with oligonucleotides for diagnosis and drug development. *Gut* **52**, 1–7.
14. Lucignani, G. (2006) Aptamers and in-beam PET for advanced diagnosis and therapy optimisation. *Eur. J. Nucl. Med. Mol. Imaging* **33**, 1095–1097.
15. Perkins, A.C. and Missailidis, S. (2007) Radiolabelled aptamers for tumour imaging and therapy. *Q. J. Nucl. Med. Mol. Imaging* **51**, 292–296.
16. Charlton, J., Sennello, J. and Smith, D. (1997) In vivo imaging of inflammation using an aptamer inhibitor of human neutrophil elastase. *Chem. Biol.* **4**, 809–16.
17. Hicke, B.J., Stephens, A.W., Gould, T., Chang, Y.F., Lynott, C.K., Heil, J., Borkowski, S., Hilger, C.S., Cook, G., Warren, S. and Schmidt, P.G. (2006) Tumor targeting by an aptamer. *J. Nucl. Med.***47**, 668–678.
18. Kuhnast, B., de Bruin, B., Hinnen, F., Tavitian, B. and Dollé, F. (2004) Design and synthesis of a new [^{18}F]fluoropyridine-based haloacetamide reagent for the labeling of oligonucleotides: 2-bromo-*N*-[3-(2-[^{18}F]fluoro-pyridin-3-yloxy)-propyl]-acetamide. *Bioconj. Chem.* **15**, 617–627.
19. de Bruin, B., Kuhnast, B., Hinnen, F., Yaouancq, L., Amessou, M., Johannes, L., Samson, A., Boisgard, R., Tavitian, B. and Dollé, F. (2005) 1-[3-(2-[^{18}F]Fluoropyridin-3-yloxy)propyl]pyrrole-2,5-dione: Design, synthesis and radiosynthesis of a new [^{18}F]fluoropyridine-based maleimide reagent for the labeling of peptides and proteins. *Bioconj. Chem.* **16**, 406–420.
20. Dollé, F. Preparation of[18F]-labeled maleimides, their use for marking macromolecules for medical imaging. WO2004002984.
21. Dolci, L., Dollé, F., Jubeau, S., Vaufrey, F. and Crouzel, C. (1999) 2-[^{18}F]fluoropyridines by no-carrier-added nucleophilic aromatic substitution with [^{18}F]FK-K_{222} – a comparative study. *J. Labelled Comp. Radiopharm.* **42**, 975–985
22. Karramkam, M., Hinnen, F., Vaufrey, F. and Dollé. F. (2003) 2-, 3- and 4-[^{18}F]fluoropyridine by no-carrier-added nucleophilic aromatic substitution with K[^{18}F]F-K_{222} – a comparative study. *J. Labelled Comp. Radiopharm.* **46**, 979–992.
23. Dollé, F. (2005) Fluorine-18-labelled fluoropyridines: advances in radiopharmaceutical

design. *Curr. Pharm. Design*, **11**, 3221–3235.

24. Dollé. F. (2007) [^{18}F]fluoropyridines: from conventional radiotracers to the labelling of macromolecules such as proteins and oligonucleotides. In: *PET Chemistry: The Driving Force in Molecular Imaging*, Ernst Schering Research Foundation Ed., Springer Verlag, Berlin: Heidelberg, **62**, 113–157.
25. Friebe, M., Dollé, F., Kuhnast, B., Hinnen, F., Boisgard, R., Tavitian, B., Dinkelborg, L. and Hecht, M. 18F-labelled aptamers. EP07075801.
26. Graves, E.E., Yessayan, D., Turner, G., Weissleder, R., and Ntziachristos, V. (2005) Validation of in vivo fluorochrome concentrations measured using fluorescence molecular tomography. *J. Biomed. Opt.* **10**, 44019 .
27. Maroy, R., Boisgard,R., Comtat, C., Frouin, V., Cathier, P., Duchesnay, E., Dollé, F., Nielsen, P., Trébossen, R., and Tavitian, B. (2008) An unsupervised method for segmentation of rodent whole-body dynamic PET images based on voxel dynamics. *IEEE Trans. Med. Imaging*, in press.

Part II

Peptide Aptamers

Chapter 16

Selection of Peptides Interfering with Protein–Protein Interaction

Annette Gaida, Urs B. Hagemann, Dinah Mattay, Christina Räuber, Kristian M. Müller, and Katja M. Arndt

Abstract

Cell physiology depends on a fine-tuned network of protein–protein interactions, and misguided interactions are often associated with various diseases. Consequently, peptides, which are able to specifically interfere with such adventitious interactions, are of high interest for analytical as well as medical purposes. One of the most abundant protein interaction domains is the coiled-coil motif, and thus provides a premier target. Coiled coils, which consist of two or more α-helices wrapped around each other, have one of the simplest interaction interfaces, yet they are able to confer highly specific homo- and heterotypic interactions involved in virtually any cellular process. While there are several ways to generate interfering peptides, the combination of library design with a powerful selection system seems to be one of the most effective and promising approaches. This chapter guides through all steps of such a process, starting with library options and cloning, detailing suitable selection techniques and ending with purification for further down-stream characterization. Such generated peptides will function as versatile tools to interfere with the natural function of their targets thereby illuminating their down-stream signaling and, in general, promoting understanding of factors leading to specificity and stability in protein–protein interactions. Furthermore, peptides interfering with medically relevant proteins might become important diagnostics and therapeutics.

Key words: Activator protein-1, affinity chromatography, coiled coil, leucine zipper, library design, phage display, protein design, protein engineering, protein-fragment complementation assay, selection technology, semi-rational design, bacterial surface display.

1. Introduction

Peptide aptamers are short peptides generated to interfere with protein interactions. Our lab focuses on the generation of peptides that are able to interfere with protein–protein interactions primarily mediated by coiled coil sequences. Another application

Annette Gaida, Urs B. Hagemann, Dinah Mattay, and Christina Räuber **contributed equally and are listed alphabetically.**

Günter Mayer (ed.), *Nucleic Acid and Peptide Aptamers: Methods and Protocols, vol. 535*

DOI 10.1007/978-1-59745-557-2_16 Springerprotocols.com

is the targeting of ligand receptor interaction. In contrast to the classic view of peptide aptamers, which are usually inserted into a protein scaffold, these peptides are "stand-alone", and we call them interfering peptides (iPEP).

Approximately 3–5% of amino acids in proteins are estimated to adopt coiled-coil folds *(1)*. Coiled coils have one of the simplest dimerization interfaces, yet they can mediate highly selective protein associations *(2, 3)* and play key roles in virtually every physiological process. Furthermore, they are widely used in the protein engineering field *(4, 5)*. Coiled coils consist of two or more α-helices wrapped around each other in a left-handed supercoil *(2, 3, 6)*. Their sequences are characterized by a conserved seven-residue repeat, $(abcdefg)_n$. The first (a) and fourth (d) position, which form the dimerization interface between the helices, are generally occupied by hydrophobic amino acids and only a very restricted repertoire of polar and charged residues *(6, 7)*. The other amino acids in the repeat are mostly polar or charged and can form inter- and intrahelical interactions that contribute to the stability and specificity of complex formation. Proline is largely excluded to preserve the helical architecture.

We specifically target coiled-coil domains of proteins relevant to biology or medicine with small helical peptide probes. Important targets include the oncoproteins c-Jun, c-Fos, and c-Myc. We use a semi-rational design approach in combination with different selection systems to generate such peptides *(6)*. Libraries are designed using mixed codons at protein interface positions, mainly heptad positions a, d, e and g *(8–10)*. Selection is carried out using phage display *(11)* or a protein-fragment complementation assay (PCA) based on the enzymatic activity of murine dihydrofolate reductase *(8, 12)*. The latter assay was also modified to include competitive and negative design aspects *(13)*. Other intracellular selection assays such as the Ras-recruitment system have also been tested *(14)*. Another approach is the display on cell surfaces, which is also discussed in this chapter *(15, 16)*. Such a semi-rational approach helps define the fundamental principles guiding protein–protein interaction while at the same time generating tight-binding peptides targeted to scientifically or medically relevant proteins. Such peptides are ideal for signaling studies and are used as effective tools to explore and manipulate function in vivo using proteomics approaches.

This chapter briefly summarizes different libraries that can be used for the generation of peptide aptamers (**Section 3.1**) and focuses on the subsequent selection of suitable peptides. We provide detailed protocols accompanied with a short theoretical introduction for selection methods in vitro (**Section 3.2**), inside cells (**Section 3.3**) and on the surface of cells (**Section 3.4**). Furthermore, we briefly touch the selection of D-peptide aptamers

(**Section 3.2**). The chapter ends with one general example for purifying selected aptamers (**Section 3.5**).

2. Materials

2.1. Library Construction

2.1.1. Fill-in Reaction

1. Overlapping oligonucleotides (e.g., from Microsynth, Operon).
2. Klenow buffer (10x): NEB buffer 2 (10x, New England Biolabs).
3. Klenow fragment (10 U/μl; New England Biolabs).
4. dNTPs (deoxynucleosidetriphosphates; 10 mM; Genaxxon).

2.2. Phage Display

1. Shaking incubator (25–37°C).
2. Roller mixer.
3. 4 ml Immunotubes (e.g., Nunc).
4. *E. coli* ER 2738 (*see* **Note 1**).
5. Phage library (e.g., NEB C7C from New England Biolabs or *(11)*) or individually designed phagemid library (examples in the lab are coiled coil libraries or a mutated EGF library in phagemid vector pAK100 *(17)*).
6. Helper phage (e.g., M13KO7, VCSM13, Stratagene; see manufacturers information for amplification of helper phage).
7. Glycerol, autoclaved.
8. LB-medium: 10 g Bacto-Tryptone, 5 g Yeast–Extract, 5 g NaCl, add ddH_2O to 1 l.
9. LB-agar (1.5%): 15 g Bacto-Agar per liter LB-medium.
10. Tetracycline (Tet) 1,000x stock solution: 20 mg/ml in ddH_2O, sterile filtered through 0.2 μm.
11. Kanamycin (Kan) 1,000x stock solution: 70 mg/ml in ddH_2O, sterile filtered through 0.2 μm.
12. PEG6000-NaCl: 2.5 M NaCl, 20% (w/v) poly ethylene glycol (PEG)-6000.
13. TBS: 50 mM Tris–HCl (pH 7.5), 150 mM NaCl.
14. Top agar (0.7%): 0.35 g Agar per 50 ml LB medium.
15. IPTG/X-Gal stock solution: 1.25 g IPTG (isopropyl-β-D-thiogalactoside), 1 g X-Gal (5-bromo-4-chloro-3-indolyl-β-D-galactoside), dissolved in 25 ml dimethylformamide (wrap in aluminum foil) (*see* **Note 2**).
16. IPTG/X-Gal plates: 15 g agar per liter LB-medium, use IPTG/X-Gal stock solution 1:1,000 (wrap in aluminum foil).

17. Coating buffer: 0.1 M $NaHCO_3$.
18. Blocking buffer: TBS containing 4% (w/v) BSA (*see* **Note 3**).
19. Washing buffer: TBS, 0.05% (v/v) Tween (*see* **Note 3**).
20. Elution buffer: 0.1 M glycine–HCl (pH 2.2), 100 mM NaCl.
21. Neutralization buffer: 1 M Tris–HCl.

2.3. DHFR Protein-Fragment Complementation Assay

1. Shaking incubator (25–37°C).
2. Double yeast-trypton (DYT) medium: 16 g trypton, 10 g yeast extract, 5 g NaCl, add ddH_2O to 1 l.
3. Enriched DYT medium: DYT medium, 0.5% (w/v) glucose, 2.5 mM KCl, 10 mM $MgCl_2$.
4. M9 minimal medium: 0.1 mM $CaCl_2$, 0.048 mM Na_2HPO_4, 0.022 mM KH_2PO_4, 8.6 μM NaCl, 0.019 mM NH_4Cl, 2 mM $MgSO_4$, 0.4% glucose (*see* **Note 4**).
5. Ampicillin (Amp) 1,000x stock solution: 100 mg/ml in EtOH, sterile filtered through 0.2 μm.
6. Chloramphenicol (Cm) 1,000x stock solution: 25 mg/ml in EtOH, sterile filtered through 0.2 μm.
7. Kanamycin (Kan) 1,000x stock solution: 50 mg/ml in ddH_2O, sterile filtered through 0.2 μm.
8. Trimethoprim (TMP) 500x stock solution: 500 μg/ml in MeOH, sterile filtered through 0.2 μm.
9. IPTG 1,000x stock solution: 1 M in ddH_2O, sterile filtered through 0.2 μm.

2.4. Bacterial Surface Display

1. Shaking incubator (25–37°C).
2. 96-well microtiter plate.
3. 137-mm diameter Nitrocellulose membrane filters.
4. *E. coli* GI826 (Invitrogen).
5. FliTrx™ Peptide display library with randomized dodecamer peptide (or pFliTrx™ peptide display vector with multiple cloning site; Invitrogen).
6. Target molecule (conjugated to carrier protein).
7. Casamino acids (with low tryptophan content).
8. Non-fat, dry milk.
9. DNase I.
10. Lysozyme.
11. 10X M9 salts: dissolve 60 g Na_2HPO_4, 30 g KH_2PO_4, 5 g NaCl and 10 g NH_4Cl in 900 ml deionized water, adjust pH to 7.4 with 10 M NaOH and add water to 1 l (final volume). Autoclave solution.

12. 1 M $MgCl_2$: dissolve 20.33 g in 100 ml deionized water and autoclave the solution.
13. Ampicillin (Amp) 1,000x stock solution: dissolve 1 g ampicillin in 10 ml deionized water and pass solution through sterile filter.
14. 20% (w/v) Glucose: dissolve 20 g glucose in 100 ml deionized water and pass solution through sterile filter.
15. 10 mg/ml L-Tryptophan: dissolve 100 mg L-tryptophan in 10 ml deionized water, pass the solution through sterile filter and store it in the dark at 4°C.
16. 5 M NaCl: dissolve 29.22 g in 100 ml deionized water. Autoclave solution.
17. 1 M Tris: dissolve 121.1 g Tris in 900 ml deionized water, adjust pH to 7.4 with 37% HCl and add water to 1 l (final volume).
18. IMC Medium: mix 2 g casamino acids (low amounts of tryptophan!) with 875 ml water and autoclave mixture. Add 100 ml 10x M9 salts, 1 ml $MgCl_2$, 25 ml 20% glucose and 1 ml 100 mg/ml ampicillin aseptically (store at 4°C).
19. IMC medium for induction: add aseptically 10 mg/ml L-tryptophan to the stock solution of IMC-medium (final concentration: 100 μg/ml L-tryptophan).
20. RM medium: Mix 20 g casamino acids, 10 ml 100% glycerol and 890 ml deionized water and autoclave solution. When solution is cooled down add aseptically 100 ml 10x M9 salts, 0.5 ml 1 M $MgCl_2$ and 100 mg/ml ampicillin.
21. RMG-Amp plates: dissolve 20 g casamino acids, 15 g agar in 875 ml deionized water. Autoclave solution. Add 100 ml 10x M9 salts, 25 ml 20% glucose, 1 ml $MgCl_2$ and 1 ml 100 mg/ml ampicillin when solution is cooled to ca. 55°C (before preparing plates).
22. RMG-Amp-Trp plates: dissolve 20 g casamino acids, 15 g agar in 855 ml deionized water. Autoclave solution. Add 100 ml 10x M9 salts, 25 ml 20% glucose, 1 ml $MgCl_2$, 1 ml 100 mg/ml ampicillin and 20 ml 10 mg/ml L-tryptophan when solution is cooled to ca. 55°C (before preparing plates).
23. TS-buffer: mix 30 ml 5 M NaCl with 50 ml 1 M Tris–HCl (pH 7.5) and 920 ml deionized water.
24. Lysis-buffer: mix 99.5 ml TS-buffer with 1 mg non-fat dry milk, 0.5 ml 1 M $MgCl_2$ and add 2 mg lysozyme and 100 U DNase I.

2.5. Purification of Peptide Aptamers

1. GSH matrix: 1 ml-GSTrap (GE Healthcare) or GSH coupled to sepharose beads (GE Healthcare).

2. Reversed phase column: Jupiter Proteo column, 4 μm particle size, 90 Å pore size, 250 × 10 mm (Phenomenex).
3. Syringe filters: 0.45 μm PVDF filters (Roth).
4. Centrifuge: e.g., Sorvall RC5B with SS34 rotor.
5. Peristaltic pump or HPLC system coupled to UV detector.
6. GST binding buffer: 10 mM Na_2HPO_4, 1.8 mM KH_2PO_4, 140 mM NaCl, 2.7 mM KCl (pH 7.3).
7. Elution buffer: 50 mM Tris–HCl, pH 8.0, 10 mM reduced glutathione (GSH).
8. FXa buffer: 20 mM Tris–HCl, pH 8, 50 mM NaCl, 1 mM $CaCl_2$.
9. RP-HPLC: Acetonitrile (ACN, HPLC-grade, Roth), 0.1% trifluoric acid (TFA, Roth) and ddH20, 0.1% TFA.
10. FXa enzyme: 1 U/ μg protein (Qiagen).

3. Methods

3.1. Library Construction

Libraries allow for simultaneous testing of a high number of variants in one screen or selection. There are several possibilities to find the best adapted library for selection of peptide aptamers regardless of the selection method chosen. Definitely the easiest is to purchase a complete library from a company, but in some instances, a more target specific designed library is required. In this case, residue positions in the peptide sequence are randomized via degenerated codon usage and the oligonucleotides have to be synthesized with mixed bases at the randomized positions *(6, 9)*. Randomization can be tailored by the mixture of nucleotides as well as their ratio. Shorter libraries (up to about 80 bases) can be generated with one library oligonucleotide. The complementary strand is generated in a Klenow fill-in reaction from a short oligonucleotide pairing with the 3' end of the library oligonucleotide. Longer libraries can be generated by two overlapping library oligonucleotides but as a consequence require a non-randomized stretch as hybridization region for the two oligonucleotides. The double strand is similarly generated in a Klenow fill-in reaction.

A different strategy to generate libraries is by error-prone PCR and/or DNA shuffling, which has been detailed in another publication of this series *(18)* and elsewhere *(19)*.

After generation of the double-stranded library fragment, it can be ligated into the adequate vector and transformed into bacteria to collect clones. Either chemical transformation or electroporation can be chosen, the latter harboring the benefit of higher transformation efficiencies, but in this case the DNA needs to be desalted. A two- to tenfold overrepresentation of the library is desirable to ensure a complete representation of the library. For assessing the library quality after transformation, a simple calculation can be performed to determine the number of variants not represented in the collected pool (E), which should be as small as possible:

$$E = n \cdot (1 - \frac{1}{n})^m,$$

where n is the theoretical library size and m is the number of collected clones *(20)*.

In order to verify the expected distribution of the randomized positions, the DNA of individual clones should be sequenced.

3.1.1. Overlapping DNA Fragments

To get double-stranded fragments for long genetic libraries it may be necessary to align two overlapping single-stranded fragments and fill in the overhangs by reaction with Klenow fragment *(8, 21)*. The overlapping part should ideally harbor an annealing temperature between 50 and 60°C. It should be considered to introduce appropriate restriction sites at both termini of the library to ensure further cloning steps into the adequate vector.

1. Mix the following ingredients for the fill-in reaction in a total volume of 20 µl:

   25 pmol of each oligonucleotide
   2 µl Klenow buffer
   2 µl dNTPs
   add ddH_2O to 20 µl
2. Incubate the mixture at 94°C for 3 min.
3. Slowly decrease the temperature from 94 to 37°C at a rate of 0.5°C/s (*see* **Note 5**).
4. At 37°C, add 0.5 µl Klenow fragment and incubate for 1 h.
5. The fill-in product should be purified via an agarose gel. It then can be digested by restriction endonucleases and cloned into an adequate vector.

3.2. Phage Display

Phage Display is a technique to display peptides or proteins on the surface of filamentous phages by directly coupling the genotype with its phenotype (*see* **Fig. 16.1A**). Phage display was invented by George Smith in 1985 *(22)* and has evolved to one of the most used techniques for in vitro screening ranging from small peptides to proteins, like antibodies Fab fragments, binding to virtually any desired target *(23, 24)*. Filamentous phages f1, M13 and fd, which all infect *E. coli* are usually used for phage display. They are referred to as Ff phages, and their genomes are to 98% identical to each other *(25)*. As there is a vast amount of information

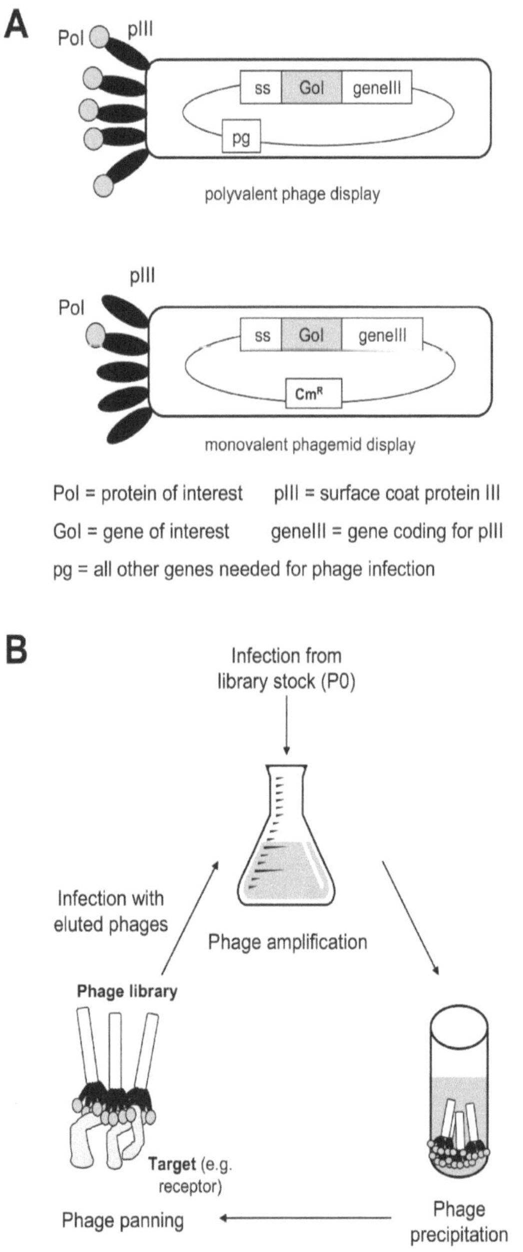

Fig. 16.1. (**A**) Simplified illustration of polyvalent phage display, showing the phage plasmid, containing a signal sequence (ss), all phage genes and the gene of interest fused to a coat protein (here gene**III**); and monovalent phagemid display, showing the phagemid vector with a signal sequence (ss), the gene of interest fused to a coat protein (here gene**III**) and an additional selection marker (here chloramphenicol, Cm). (**B**) First round of phage amplification, precipitation and panning, starting with P0, the unselected library pool.

available in the literature *(25–27)*, the principles of phage display will be explained only briefly. Variations of phage display such as Proside *(28)* and selectively infective phages (SIP) *(29, 30)* have also been reported but are less frequently used and are thus not detailed in this chapter.

In phage display, peptide or protein libraries (*see* **Section 3.1**) can be fused to the coat proteins of phages (mostly geneIII protein), which are displayed on the surface of the phage particle. All the proteins needed for cell infection and assembly of phage particles are encoded in the single-stranded phage genome. By fusing the gene of interest to a gene that encodes for one of the coat proteins (mostly geneIII), the protein of interest will be displayed on the phage particle upon expression with the coat protein (*see* **Fig.16.1A**).

In phage display with geneIII-fusion, the number of displayed peptides or proteins should be equivalent to the number of geneIII proteins, resulting in a so-called polyvalent display (**Fig.16.1A**), which causes an avidity effect. If this is not desired, a variant of the classical phage display, the so-called phagemid display can be performed. In this technique a phagemid vector is used that only encodes the fusion gene, e.g., geneIII, and no other phage genes. In addition to the Ff origin, phagemid vectors have an *E. coli* origin for replication and an antibiotic cassette to allow propagation in *E. coli*. To generate phages, superinfection of phagemid-carrying *E. coli*, with a helper phage, an Ff phage encoding for all phage genes, but with a compromised phage origin is needed. As a consequence, the helper phage itself is packed inefficiently, and generated phages almost exclusively have the phagemid vector packaged. After infection, the resulting phages display wild type geneIII proteins, assembled from structural proteins encoded by the helper phage and a reduced amount of the geneIII-fusion protein encoded by the phagemid vector. Depending on expression level, this statistically results in a monovalent display (*see* **Fig. 16.1A**) of the peptide or protein of interest and allows screening for high affinity.

While phage display is usually used for selection of small peptides from five to thirty amino acids, phagemid display can be used to select whole antibodies or proteins. Although phage display is faster, the advantage of the phagemid system is the monovalent display, allowing one to select for peptides with a high affinity. However, due to the high level of displayed wild-type proteins, the display level of the fused library is often rather low *(25)*.

One round of phage display always consists of the panning on the target, including washing and elution steps, and the subsequent amplification and purification of the phages (**Fig. 16.1B**). The first round is referred to as panning round 1. To select for specific binders, several panning rounds (always repeating the steps mentioned above), are accomplished subsequently. Usually, four panning rounds are sufficient to enrich for the best binders in the pool.

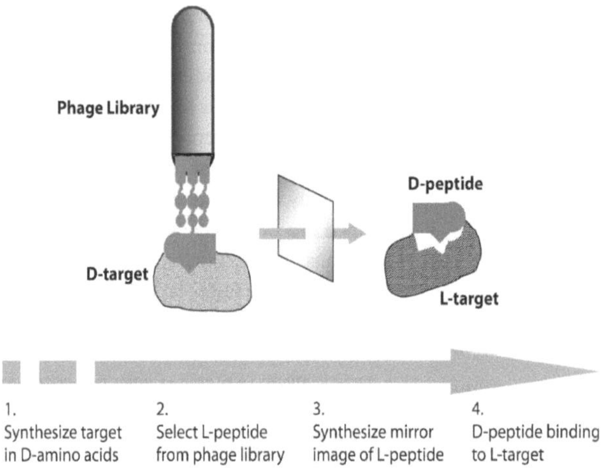

Fig. 16.2. Principle of mirror image phage display. Peptides composed of D-amino acids (D-peptides) binding their natural L-targets can be generated through phage display of L-libraries binding to targets synthesized in D-amino acids. Because of the mirror-image relationship between D- and L-enantiomeric peptides, the D-counterparts of the selected L-peptides will bind the natural L-target sequence.

For simplicity, phages derived from the panning are referred to as **E** (for Elution) and the number of the panning round, that is, phages eluted in panning round 1 are referred to as **E1** and so on. Phages derived from the amplification are referred to as **P** (for Panning), with **P0** being the unselected pool of the initial amplification round.

A deviation from the common used strategy in phage display is the so-called mirror image phage display (*see* **Fig. 16.2**) *(31, 32)*. This technique enables the generation of D-peptides binding to targets of interest and it relies on the mirror symmetry properties of naturally occurring L-amino acids in comparison to their enantiomeric D-counterparts. Peptides, which only contain D-amino acids are less susceptible to proteolytic degradation *(33)* and might therefore be advantageous in the design of therapeutical peptides. The difference to the commonly used phage display lies within the target of interest, which is synthesized in D-amino acids. After selection from an L-peptide library, the identified binder, an L-peptide, is converted into D-amino acids, resulting in a D-peptide binding to the native L-target.

3.2.1. Library Preparation

3.2.1.1. Commercially Available Libraries, e.g., NEB C7C PhD Library

1. Store library stock according to manufacturer's information.
2. Before starting the first panning round, amplify the phages from the original stock and keep them frozen at –20°C.
3. Perform the amplification protocol as described in **Section 3.2.2.**

3.2.1.2. Self-Made Libraries

1. For phage display, ligate the generated library (*see* **Section 2.1**) into, e.g., MKE13 vector, available from NEB, by using appropriate restriction sites (EagI and Acc65I or KpnI for this vector) (*see* **Note 6**).
2. For phagemid display, ligate the generated library (*see* **Section 2.1**) into, e.g., pAK100 *(17)* by using appropriate restriction sites.
3. Transform the library into *E. coli* ER 2738 cells (NEB) and plate them on LB-agar plates containing 20 µg/ml tetracycline (*see* **Note 1**). For phagemid display the plates also need to contain the selection marker of the phagemid vector (*see* **Note** 7).
4. At the next morning, carefully pool the cells by adding some LB medium and scratching them from the plate using, e.g., a Drigalski spatula, dilute them in LB-medium and gently shake them for about 5 min at 37°C to ensure homogeneous cell suspension.
5. Optionally, glycerol stocks can be made from the pool by mixing cell culture with autoclaved glycerol in a ratio of about 2:1. Glycerol stocks are stored at –80°C.
6. Dilute the culture with LB-medium to an OD_{600} below 0.3.
7. Perform the amplification protocol as described in **Section 3.2.2**, starting at **Step 5** for phage display or **Step 3** for phagemid display.

3.2.2. Phage Amplification

WARNING: Phages are highly infectious for bacterial cells. Work responsible and accurate to avoid cross-contamination (*see* **Note 8**)!

3.2.2.1. Phage Display

1. Inoculate 40 ml of LB-medium containing 20 µg/ml tetracycline with an overnight culture of *E. coli* ER 2738 cells to an OD_{600} of 0.2 (*see* **Note 1**).
2. Use a 500 ml Erlenmeyer flask without baffles (*see* **Note 9**).
3. Grow the culture at 37°C on an orbital shaker to an OD_{600} of 0.5.
4. Infect the cells with 1×10^{11} phages of the library. Directly after infection leave the culture for 15 min without shaking at 37°C (*see* **Note 10**).
5. Let the cultures grow on the shaker for about 4–6 h (*see* **Note 11**) and continue with **Section 3.2.3**.

3.2.2.2. Phagemid Display

1. Inoculate an overnight culture of ER 2738 cells, which harbor the phagemid library to an OD_{600} of 0.2 (*see* **Note 1**). Make sure to add the respective antibiotics.
2. Use a 500 ml Erlenmeyer flask without baffles (*see* **Note 9**).

3. Grow the culture at 37°C on an orbital shaker to an OD_{600} of 0.5.
4. Add helper phages at a ratio of 1:20 (bacteria to phages) and leave the culture without shaking for 15 min at 37°C (*see* **Note 10**).
5. Put the culture onto a shaker, let it grow for about 60–70 min and add kanamycin at a concentration of 70 μg/ml.
6. Let the culture grow for 4–6 h (*see* **Note 11**) and continue with **Section 3.2.3**.

3.2.3. Phage Precipitation

1. Transfer the culture into 50 ml polypropylene (PP)-tubes and spin down at 5,000 × *g*, 4°C for 10–30 min. The longer the centrifugation step, the cleaner the phage pellet will be in the end.
2. After centrifugation, transfer the supernatant, which now contains the phages, into a fresh 50 ml PP-tube and add 1/6 of the volume of PEG6000-NaCl to precipitate the phages. Invert the tube carefully to allow mixing. If the amplification was successful, the phages might already be seen as a cloudy suspension.
3. Leave the precipitation at 4°C over night. A white fluffy precipitation should be visible next morning.
4. Spin down the precipitation for 45 min at 4°C and 5,000 × *g*.
5. Discard the supernatant and resuspend the pellet in 1 ml ice-cold TBS. Transfer the phages into a 1.5 ml reaction tube and spin down for 10 min at 15,000 × *g*. Transfer the supernatant to a new reaction tube, add 1/6 of the volume of PEG6000-NaCl and incubate on ice for at least 1 h for a second precipitation step.
6. Spin down for 45 min, at 4°C and 15,000 × *g*. Discard the supernatant and resuspend the pellet in 1 ml ice-cold TBS. Spin down for additional 10 min to remove all traces of bacteria and transfer the supernatant to a new reaction tube.
7. Determine the phage concentration either at OD_{269} in a UV/VIS spectrophotometer (*see* **Note 12**) or by plating phage dilutions on IPTG/X-Gal plates (only for phage display, *see* **Section 3.2.4**).

3.2.4. Phage Titering

Phage titering with plaques is used for phage display only. When using phages encoding an inducible α-complementation of lacZ, e.g., M13 derivatives, a more convenient blue/white screening is possible, as described below. For phagemid display, *E. coli* ER 2738 cells can be infected with phages and plated on LB-agar plates containing the respective selection encoded on the phagemid vector, and the phage titer is determined by counting colony-forming units (cfu).

1. Inoculate 10 ml of LB-medium containing 20 μg/ml tetracycline with an overnight culture of *E. coli* ER 2738 cells to an OD_{600} of 0.2 and let them grow to an OD_{600} of 0.5 (*see* **Note 1**).
2. Preheat a water bath to 40°C and put as many test tubes as needed, that is one for each dilution, into the water bath.
3. Boil top agar until it liquefies and pipette 4 ml into each test tube.
4. Let IPTG/X-Gal plates adjust to a temperature of 37°C
5. Prepare phage dilutions in TBS to a final volume of 10 μl: For amplified phages start at a higher dilution, e.g., 1:1 × 10^{10}, 1:1 × 10^{12}, 1:1 × 10^{14}, for eluted phages start with lower dilution, e.g., 1:1 × 10^{4}, 1:1 × 10^{6}, 1:1 × 10^{8}, as the number of phages will be lower.
6. To each dilution add 90 μl of the *E. coli* ER 2738 culture, mix gently and incubate for 5–10 min at 37°C without shaking.
7. Pipette each dilution to one of the prepared reaction tubes containing the top agar and mix by vortexing very carefully.
8. Immediately pour the top agar onto the prewarmed IPTG/X-gal agar plates and spread the top agar evenly by tilting the plate carefully.
9. Proceed the same way with the other dilutions.
10. Wait until the top agar has hardened and incubate the plates at 37°C. Wrap the plates in aluminum foil as X-Gal is light sensitive.
11. At the next morning blue plaques can be seen in the bacterial lawn and the titer of phages can be determined (*see* **Note 13**).

3.2.5. Phage Panning

1. Immobilize the target of interest in 1.5 ml coating buffer in an immunotube or well of a microtiter plate. The protein or peptide concentration should be between 3 and 100 μg/ml. Incubate under gentle agitation at 4°C over night on a roller mixer.
2. Remove the supernatant of the immunotube next morning and wash 3 × with TBS. Use as much buffer as needed to fill up the immunotube completely.
3. Incubate the immunotube with blocking buffer by filling the immunotube completely. Incubate for 2 h at room temperature under gentle agitation.
4. Inoculate 40 ml of LB-medium containing 20 μg/ml tetracycline with an overnight culture of *E. coli* ER 2738 cells to an OD_{600} of 0.2 (*see* **Note 1**).
5. Wash the immunotube 3–5 x with washing buffer. Use as much buffer as needed to fill the tube completely.

6. Add 1×10^{11} phages in 1 ml washing buffer containing 1% (w/v) BSA (*see* **Note 3**) and incubate for 50 min at room temperature under gentle agitation.
7. Remove the phages carefully by pipetting to avoid contamination and wash the tube 10 × with washing buffer by discarding the supernatant and subsequently filling the entire tube with washing buffer (*see* **Note 14**).
8. Add 1 ml elution buffer into the tube and incubate for 10–30 min at room temperature under gentle agitation. Transfer the eluate into a fresh 1.5 ml reaction tube and neutralize with neutralization buffer immediately (*see* **Note 15**).
9. Infect the *E. coli* ER 2738 culture with the eluted phages at an OD_{600} of 0.5. Keep a small rest of the eluted phages (e.g., 10 µl to determine the phage titer or concentration; see **Section 3.2.3**, **Step** 7 or **Section 3.2.4**).
10. After infection, leave the culture for 15 min at 37°C without shaking. Proceed as described in **Section 3.2.2**.
11. Repeat the whole procedure until a dominating sequence has settled in the library pool (*see* **Note 16**).

3.2.6. Evaluation and Characterization

In the case of phage display single stranded or double stranded DNA can be prepared for sequencing. In the case of phagemid display, double stranded DNA is used for sequencing.

3.2.6.1. Sequencing

1. Infect an *E. coli* ER 2738 culture with phages from panning rounds to be investigated.
2. Let the culture grow for approximately 5 h (*see* **Note 17**).
3. Perform a standard DNA preparation and sequence the DNA. Note that either single phage clones or phage pools can be sequenced. In the pool sequences, the distribution of randomized codons and hence amino acids can be followed during the selection process. With increasing selection rounds, one favored sequence should dominate the pool. When using defined libraries with certain positions randomized to a set of amino acids (as opposed to libraries generated by error-prone PCR or DNA shuffling), it is even possible to quantify mixed codons and relate these to the amino acid distribution to assess the kinetics of selection *(8, 10)*.

3.2.6.2. Phage ELISA

1. Prepare phage pools or single phages as described in **Sections 3.2.2** and **3.2.3**. Single phages can be generated by inoculating an *E. coli* ER 2738 culture with a single plaque, picked from an IPTG/X-Gal plate. Grow culture over night and proceed as described in **Section 3.2.2.1**. For phagemid display, inoculate LB-medium containing the respective

concentrations with a single colony of an agarose plate, containing the phagemid pool. Grow culture over night and proceed as described in **Section 3.2.2.2**.

2. Immobilize a 96-well microtiter plate with the target that was used for selection.
3. Perform an ELISA with 1×10^{12} phages per well.
4. Detect with an anti-phage antibody (e.g., anti-M13 antibody for using a phage display library generated in an M13-vector) (*see* **Note 18**).

An increase of phages binding to the target should be seen from panning round 1 to the last round. As control, unspecific binding should be tested as well.

3.3. DHFR Protein-Fragment Complementation Assay

The Protein-fragment complementation assay (PCA) described in this chapter is carried out in bacterial cells. However, applications of PCA have also been reported for mammalian cells *(34)* and plant cells *(35)*. The assay relies on the principle that the survival of cells simultaneously expressing complementary fragments of the enzyme murine dihydrofolate reductase (mDHFR) is dependent on the correct folding and interaction of these fragments *(12)*. The mDHFR is genetically dissected into two rationally designed fragments (mDHFR1 and mDHFR2, *see* **Note 19**), each of which is fused to a gene of interest or target gene, and reconstitution of enzymatic activity is obtained through interaction of these fusion proteins (*see* **Fig. 16.3**). The endogeneous procaryotic DHFR is specifically inhibited by trimethoprim, preventing biosynthesis of purines, thymidylate, methionine, and pantothenate, and therefore cell division. In this manner, only cells in which interacting proteins provide proper folding of mDHFR can survive. The most efficient interacting proteins lead to a higher growth rate for this clone, which can be enriched in repeated turns of the growth competition assay in liquid culture.

Non-specific and unstable library members are removed during the assay, leaving those with strong interaction capacities *(8)*. The assay is very sensitive, since a few numbers of molecules of reassembled mDHFR are sufficient for bacterial survival *(34)*. Consequently, the assay is qualified to detect proteins that are expressed at extremely low levels or even for screening of weak interactions. Furthermore, it is also an excellent strategy for a large-scale library-*versus*-library selection *(9)*. Additionally, we have further developed the assay to select not only for interaction affinity but also for specificity by simultaneous expression of homologues peptides, which compete with protein libraries for an interaction with the target molecule *(13)*. Library members binding to their target, and promoting cell growth, must outcompete competitor interactions with the target (i.e., competition) and evade binding to

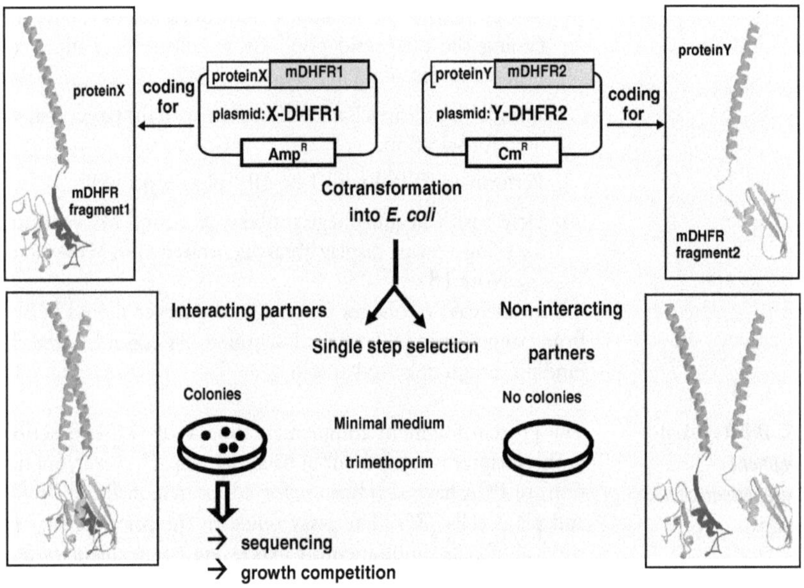

Fig. 16.3. Principle of the DHFR protein-fragment complementation assay (PCA). Protein X (in this case an α-helical peptide) is fused to mDHFR1 on a plasmid containing the gene for Ampicillin (Amp) resistance, protein Y is fused to mDHFR2 on a plasmid with the Chloramphenicol (Cm) resistance gene. Only cells which are cotransformed with both plasmids and have interacting partners can rebuild active mDHFR and survive on minimal medium plates containing trimethoprim. Further steps include sequencing and the growth competition assay.

the competitors (i.e., negative design). We term this a competitive and negative design initiative (CANDI).

1. Clone one vector encoding the fusion protein X-mDHFR1 and a second vector containing the fusion protein Y-mDHFR2, where X and Y are genes for specific target proteins, peptides or libraries (*see* **Note 20**). For the CANDI procedure, a third vector ending the competing peptide is used (*see* **Note 21**).
2. Transform the vectors (200 ng each) into BL21 gold cells (*see* **Note 22**). After electroporation of 100 μl cells, immediately add 1 ml enriched DYT medium, shake for 1 h at 37°C.
3. To reduce carryover of complex nutrients, wash the cells with M9 medium by gently pelleting the cells, removing the supernatant and resuspending the cells in M9 medium.
4. Plate cells in appropriate dilution series to facilitate counting and incubate at 37°C for 24–48 h:
   a. 1, 5, and 10 μl on LB-agar (+ Amp, Cm, Kan) (*see* **Notes 23** and **24**).
   b. 1, 5, and 10 μl on M9-agar (+ Amp, Cm, Kan, TMP, IPTG).

c. Remaining amount on M9 agar (+ Amp, Cm, Kan, TMP, IPTG).

For the CANDI procedure Tet needs to be added as selection marker (*see* **Note 24**).

5. Count the colonies from the dilution series on the 1, 5, and 10 μl plates to determine their theoretical number from the LB plates and their actual number from the M9 plates. When the required theoretical number of colonies (overrepresentation of each clone of the library tenfold) is reached, combine the colonies by pooling them into ~10 ml of M9 medium and shake at 37°C for 5 min to disrupt cell clumps.
6. Prepare glycerol stocks with 30% (v/v) glycerol and store at –80°C. For sequence analysis, DNA from library pools or single colonies can be prepared and sequenced (*see* **Note 25**).
7. Inoculate 100 ml M9 medium (+ Amp, Cm, Kan, TMP, IPTG, and Tet in the case of CANDI) to a start OD_{600} of 0.0001 and grow them at 37°C until an OD_{600} of 0.2–0.5 is reached (*see* **Notes 26** and **27**).
8. Repeat **Steps 6** and **7** until a unique sequence has settled.

3.4. Bacterial Surface Display

Parallel to phage display another display technology emerged during the last 20 years: Bacterial surface display enables the presentation of peptides and proteins on the cell surface of gram-negative and gram-positive bacteria. Like phage display, this technology provides physical linkage of geno- and phenotype. By selection of the target specific peptide displayed on the cell surface, also whole bacteria and their genetic information are obtained. A great variety of different display systems and their applications (from biocatalysts to vaccine delivery) has been published *(15)*. Although phage display is the more common technology for screening of peptide libraries, there was a successful approach by Lu et al. in 1995 *(16)*. Their so called "FLITRX peptide library" is commercially available from Invitrogen and will be described here.

Lu et al. *(16)* introduced a randomized oligonucleotide encoding for a dodecamer peptide [$(XNN)_{12}$; N (any nucleotide); X (any nucleotide with G:A:C:T = 7:7:7:3] into the *trxA* (thioredoxin) gene. Expression of this construct results in modified thioredoxin molecules harboring a conformationally constrained peptide library in their active site loop. Thioredoxin is a cytoplasmatic protein. For cell surface display the fusion construct was inserted into a non-essential region of the flagella *fliC* gene. The obtained *FLITRX* gene (*fliC* + *trxA*+DNA-sequence for peptide library) is under control of the P_L promoter at the pFliTrx vector. The bacterial chromosome of *E. coli* GI826 contains a bacteriophage λ *c*I repressor gene controlled by a *trp* (tryptophan) promoter.

In the absence of tryptophan, the λ *c*I repressor gene is expressed and the repressor can bind to the operator region upstream of the P_L promoter and inhibits expression of *FLITRX* gene. By adding tryptophan to the culture medium, *c*I repressor expression is blocked and *FLITRX* expression is induced. As a result, *E. coli* lacking wild type *fliC* carried thousands of FLITRX proteins at their flagella. In this way the peptide library is presented to the extracellular environment.

The following procedure (*see* **Fig. 16.4**) describes the selection of target-specific peptides with the "FliTrx^{TM} Random Peptide Display Library" (*Invitrogen*). It is also possible to clone an individually designed peptide library via provided multiple cloning sites into the pFliTrx^{TM} vector (*Invitrogen*) and to perform a selection procedure.

Characterization of the selection can be performed by pool sequencing (*see* **Section 3.2.6**). Lu et al. described a screening on nitrocellulose membranes *(16, 36)* for a preselection of target specific clones. They used the FliTrx Peptide Library for epitope mapping of antibodies. In this case antibodies for detection of positive clones already exist. If the peptide library is applied to a target without specific detection reagents provided, it is recommended to fuse or conjugate the target molecule to a peptide-tag or protein which has

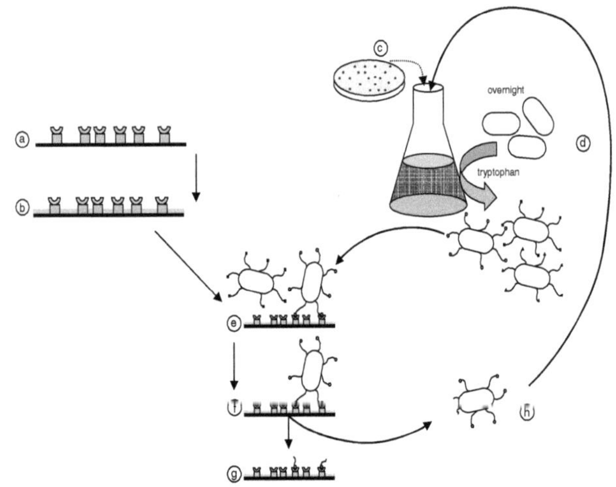

Fig. 16.4. Selection procedure of the "FLITRX^{TM} Random Peptide Display Library". Microtiter plates are coated with target molecule and carrier protein (**a**) and blocked with non-fat dry milk (**b**); *E. coli* carrying pFliTrx^{TM} are cultivated either on agar plates (**c**) or in liquid culture overnight (**d**) and FLITRX expression is induced upon inoculation of tryptophan medium with overnight culture; bacteria displaying target specific peptides are able to bind to microtiter plates (**e**), whereas non-binding bacteria are removed during washing steps (**f**); elution (**g**) is performed by vortexing and thereby breaking flagella by mechanical shearing; eluted bacteria (**h**) are used for inoculation of the next overnight culture or spread on agar plates for the next selection round.

the following properties: (i) it must not be a protein or homolog already existing in *E. coli*, and (ii) a specific antibody against it should be available. Possible peptide tags are, e.g., His- or FLAG-tags *(37)* and possible proteins are, e.g., GFP (green fluorescent protein) and GST (glutathione-*S*-transferase). For individual characterization western blots or sequencing of single clones are suitable.

1. If the selection procedure is carried out against molecules with small molecular weight, they should be conjugated to a carrier protein (like BSA, OVA, KLH) (*see* **Note 28**).
2. Proliferation of the FliTrx™ peptide library (*see* **Note 29**):
   a. Inoculate 50 ml IMC-Medium with tenfold of the peptide library (at least 1.8×10^9 clones).
   b. Incubate with shaking overnight at 25°C until saturation is reached.
3. Immobilization of the target molecule (*see* **Note 29**):
   a. Add 300 μl 20 μg/ml target molecule (in sterile water or buffer) (*see* **Note 30**) in each of 32 microtiter wells (*see* **Note 31**)
   b. Incubate overnight at 4°C.
4. Tryptophan induction:
   a. Inoculate 50 ml IMC-Medium (containing 100 μg/ml L-tryptophan) with 100-fold of the peptide library (detection by OD measurement).
   b. Incubate for 6 h at 25°C.
5. In the meanwhile, pour off the coating solution of the microtiter wells, rinse them with sterile water and block them with 1% (w/v) non-fat dry milk in 150 mM NaCl (300 μl/well for 1 h at room temperature and 50 rpm agitation) (*see* **Note 32**).
6. Panning:
   a. After 6 h induction, mix tryptophan culture (*see* **Notes 33** and **34**) with free-carrier protein (2% (w/v); final concentration) (*see* **Note 35**), 1% (w/v) non-fat dry milk and 300 μl 5 M NaCl (150 mM), add IMC-Medium to 10 ml final volume (*see* **Note 32**).
   b. Add 300 μl tryptophan culture mix per well, agitate them gently at 50 rpm (*see* **Note 36**) for 1 min and leave them bench-top for 1 h at room temperature.
   c. Remove the solution containing non-binding bacteria carefully (e.g., by pipetting).
   d. Add 300 μl IMC-medium to each well, incubate for 5 min at room temperature and gently agitate (50 rpm). Remove medium and repeat washing step another four times (*see* **Note 37**).

e. Decant the washing solution after the last step, add 300 μl IMC-medium to each well and vortex the whole plate for 30 s (*see* **Note 38**).

f. Inoculate 10 ml IMC-medium with the total eluate volume and cultivate it overnight at 25°C/180 rpm for the next selection round or centrifuge eluted cells for 2 min (2,000 × *g*), decant supernatant, resuspend the cells in 1/3 initial volume of IMC medium and spread them on RMG-Amp plates (overnight 25°C).

3.5. Purification of Peptide Aptamers

Once a protein or a peptide is identified from a library screen, it needs to be characterized in more detail and in complex with the target protein. The characterization might include measurements by ELISA, surface plasmon resonance (SPR), fluorescence or circular dichroism, depending on the nature of the selected interaction pair. For this purpose, the enriched binder needs to be present in purified form and sufficient amount. This can be achieved by both synthesizing the peptide and purifying it via reversed phase HPLC or by recombinant overexpression in *E. coli* and subsequent purification. For the latter, the binder is most often fused to a protein that can be purified by affinity chromatography, such as GST, chloramphenicol acetyl transferase (CAT) or maltose binding protein (MBP). Another possibility is to add a short peptide tag, e.g., six histidine residues, to the identified protein. By simultaneously inserting a proteolytic cleavage site, e.g., factor Xa (FXa, recognition site: Ile-Glu/Asp-Gly-Arg, cleavage after Arg), the enriched protein or peptide can be cleaved off from the fusion protein or tag.

The advantage of a fusion protein lies in a lower susceptibility to proteolytic degradation during expression in *E. coli* and easy purification. In addition, the tendency to form aggregates resulting in inclusion bodies is also reduced. However, it should be noted that each selected protein or peptide has to be treated individually with a defined expression and purification protocol.

Possibly the most common used fusion protein for the purification of recombinant proteins is GST. The native substrate of GST is glutathione (GSH), which binds to GST with high affinity. Consequently, during affinity chromatography, where GSH is covalently coupled to a carrier matrix (column or beads), the recombinant GST fusion protein binds to the GSH matrix. Elution of the GST fusion protein from the matrix is performed by adding soluble GSH, which competes with the immobilized GSH. The selected protein or peptide is subsequently isolated from the fusion protein by proteolytic degradation and purification via reversed-phase HPLC.

In the following protocol, the GST affinity purification of a helical peptide, followed by FXa cleavage and a reversed phase

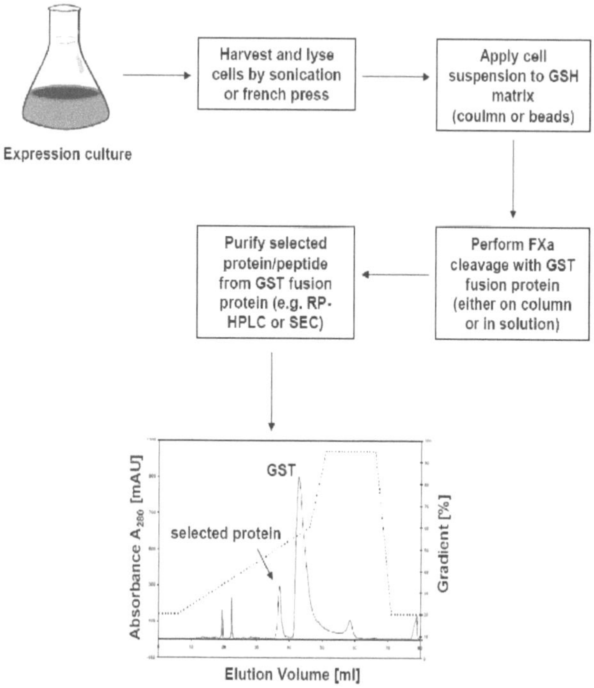

Fig. 16.5. Schematic flow chart for the purification of the peptide/protein identified from a library screen in fusion to GST. After binding and elution from a GSH matrix, the selected peptide or protein is cleaved from GST by incubation with FXa. The peptide or protein is recovered from the digest via a second purification on a further column, depending on the properties of the protein (in this case a reversed phase column).

high-performance liquid chromatography (RP-HPLC) is described. A schematic representation of the individual steps is described in **Fig. 16.5**.

1. Harvest *E. coli* cells from an expression culture and freeze them for 20 min at –80°C.
2. Resuspend the cells in GST binding buffer on ice (*see* **Note 39**) and transfer the lysate into a 50 ml polypropylene (PP) tube.
3. Lyse the cells either by sonication for 10 min on ice (with repeating 1 min interruptions to cool down the solution) or by french press.
4. Transfer the cell lysate into precooled SS34 centrifuge tubes and centrifuge for at least 30 min at 19,000 rpm (Sorval SS34 rotor; 41,000 *g*) at 4°C to spin down cell debris.
5. Transfer the supernatant into a pre-cooled 50 ml PP tube and sonicate again for 1 min on ice (*see* **Note 40**).

6. Load the supernatant via a superloop on the GST column (*see* **Note 41**) connected to a peristaltic pump or HPLC device that allows monitoring the absorbance of the loaded proteins by UV detection at 280 nm (*see* **Note 42**).
7. Wash the immobilized fusion protein with at least 60 column volumes of GST binding buffer to increase purity of the recombinant protein.
8. Elute the GST fusion protein from the column with elution buffer, containing 10 mM GSH. The protein should elute in the first 2–10 ml.
9. Transfer the elution fractions into dialysis tubes and exchange the buffer by extensive dialysis into FXa buffer at 4°C on a magnetic mixer (*see* **Note 43**).
10. Cleave the selected peptide or protein from GST by adding 1U of FXa enzyme per 10 μg of purified protein to the solution and incubate at 4°C or room temperature for an appropriate time (*see* **Note 44**).
11. Add 1 mM PMSF (phenylmethylsulfonyl fluoride) to stop the reaction. If appropriate, the cleavage efficiency can be checked on an SDS gel, where a clear shift from the upper GST fusion protein to the now remaining GST protein should be visible.
12. Load the FXa digest to a reversed phase column connected to an HPLC system that allows to monitor UV absorbance and to run gradients (*see* **Note 45**).
13. Separate the selected peptide or protein from GST using an appropriate gradient of ddH_2O 0.1% TFA to ACN 0.1% TFA (e.g., 1%/ min) and collect the fractions containing the peptide/ protein (*see* **Note 46**).
14. Concentrate the samples collected from the RP-HPLC run under vacuum (e.g., lyophilizer or speed-vac) and resuspend them in an appropriate buffer for determination of peptide/ protein concentration.

4. Notes

1. *E. coli* XL1-blue cells can also be used, but it should be made sure that the chosen cell line for amplification of the phage-library contains the F-episome, which is needed for F-pili formation. The F-episome contains a tetracycline resistance, which is why tetracycline is added to liquid cultures and plates.
2. X-Gal is light-sensitive. Wrap stock solutions and plates into aluminum foil.

3. Non-fat milk powder (approx. 2–5% (w/v)) can be used alternatively for blocking and/or washing. Due to different protein sizes in the milk, it might be more efficient – and cheaper – than BSA. Only in cases of streptavidin–biotin interactions, the biotin in the milk powder will interfere.
4. For easy preparation of M9 medium prepare a stock of 5 × M9 salts: 64 g $Na_2HPO_4 \times 7 H_2O$ (or 33.9 g Na_2HPO_4 anhydrous), 15 g KH_2PO_4, 2.5 g NaCl, 5.0 g NH_4Cl, add ddH_2O to 1 l and autoclave in aliquots of 100 or 200 ml. For 1 l minimal medium mix 200 ml sterile M9 salts, 2 ml sterile 1 M $MgSO_4$, 20 ml sterile 20% glucose, 1 ml sterile 0.1 M $CaCl_2$, ddH_20 ad 1 l. It is very important to use ultra-clean water and bottles to avoid any contamination with nutrients which would have an adverse effect on the mDHFR selection. All solutions should be cooled to at least 60°C before combining. To avoid precipitation, it is best to autoclave the water and $CaCl_2$ together first (for example, for 1 l of media, first autoclave approximately 750 ml UHP water and 1 ml 0.1 M CaCl2), then add the rest of the components, with gentle stirring or swirling. The final volume can be adjusted with sterile ddH_20.
5. It is important to decrease the temperature slowly so that the oligonucleotides can anneal properly.
6. For information on phage display vectors, also see the web page: http://www.biosci.missouri.edu/SmithGP.
7. Libraries can be transformed chemically or by electroporation. Usually electroporation yields higher transformation rates than chemical transformation. In either case, make sure to plate an amount of cells, which can easily be counted to determine the transformation rate and the library size (*see* **Section 2.1**).
8. It is important to work very accurate as phages are highly infectious and will infect any other *E. coli* culture susceptible for phage infection. Furthermore, if more than one person in the lab is working with phages, make sure to use completely different equipment and if possible have a spatial separation between respective lab members to avoid cross-contamination. Filter tip pipettes should be used for all phage-related work.
9. Shearing forces can cause damage to the pili of a considerable amount of cells, reducing the amount of infection events. Thus, e.g., Erlenmeyer flasks without baffles are recommended.
10. Directly shaking the culture upon infection of the phages might cause damage to a considerable amount of pili, due to immediate shearing forces.
11. Temperature and duration of the incubation can be varied, if the yield of phages is too low. For example let the cultures grow at 28°C and longer than 6 h if necessary.

12. To determine the phage concentration using a Photo-Spectrometer use the following equation *(38)*:

$$\frac{Pfu}{ml} = \frac{DF \times 6 \cdot 10^{16} \times (A_{269} - A_{320})}{\text{Size of genome in nt}}$$

with Pfu = phage forming units and DF = dilution factor.

13. Always check the plates for white plaques, as this means contamination with wild type phages, which are ubiquitous and contaminate easily.

14. Usually the washing steps become more stringent the more panning rounds are performed. That is, in panning round 1 wash the tube 10 x with TBS-T 0.05%, in panning round 2 wash 20 x with TBS-T 0.05%, and so on. Other parameters can be varied, like incubation time of washing steps (e.g., add washing buffer and incubate the immunotube for 5 min under gentle agitation).

15. This elution procedure is a very general method and can be applied to most libraries. Before starting, pre-titration of the amount of neutralization buffer to neutralize 1 ml of elution buffer (to pH 7.0) is recommended. There are many other possibilities to elute phages, depending on the target. Elution can be performed by adding high-ligand concentrations that compete with the phages, by adding trypsin, or high-salt concentrations *(25, 39–43)*.

16. The number of panning rounds can vary, depending on the target. It is useful to check the pools for enriched sequences to decide whether more panning rounds need to be performed or not. To make sure that no phages are enriched that bind to plastic or BSA, which is often used as blocking reagent, phages can first be incubated on immunotubes immobilized with BSA (subtractive panning). After ~30 min incubation, phages can be transferred to the immunotube containing the target. BSA binders will remain in the tube containing BSA.

17. Overnight growth is also possible for double-stranded plasmid preparation. However, in general it is not recommended to grow phage infected cells for prolonged times due to possible genetic instabilities.

18. If the signals in the phage ELISA are too high, less phages can be used (down to 1×10^{10} phages per well).

19. In the PCA, the murine mDHFR (UniProtKB/SwissProt entry P00375) is split into mDHFR fragment 1 (amino acids 1-105; also named DHFR1, F[*1, 2*] or DHFR[*1, 2*] in some publications) and mDHFR fragment 2 (amino acids 106–186; also named DHFR2, F[*3*] or DHFR[*3*]).

20. We typically clone the proteins of interest X and Y via NheI and AscI restriction sites into a pQE16 derivative (Qiagen) containing a G/S linker tagged to mDHFR fragment 1 (pAR230d-X-mDHFR1; ampicillin resistance) or fragment 2 (pAR300d-Y-mDHFR2; chloramphenicol resistance *(8, 13, 44)*.
21. For the CANDI procedure, a competing peptide is cloned via NheI and AscI in a third plasmid (pAR210d) from the same series as mentioned in **Note 20**, but lacking the mDHFR fragments and with tetracycline resistance *(13)*.
22. For stable propagation of expression constructs encoding toxic or hydrophobic proteins, the cells should express the *lacI* gene product that represses protein expression prior to IPTG induction. If *lacI* is not encoded in one of the vectors, it can be obtained by cotransforming the repressor plasmid pREP4 (Kan^R; Qiagen).
23. Cells should not be able to grow without IPTG in the presence of the *lacI* gene product. Therefore, they should also be plated on M9 minimal medium (+ Amp, Cm, Kan, TMP) without IPTG to confirm that there is no contamination and mDHFR activity is restricted to the screened interaction.
24. For weak interactions or toxic proteins, the use of many antibiotics might compromise growth. Consequently, Amp and Cm can be omitted as cell growth in the presence of TMP requires mDHFR complementation and thus selects for the presence of both mDHFR plasmids. Kan is required to maintain the *lacI* expressing pREP4, and Tet is essential in the case of CANDI for maintaining the plasmid with the competitor.
25. During later selection rounds, individual clones can be analyzed.
26. This may take up to 48 h for the first selection round. With each round, the bacterial growth rate will increase and the final OD_{600} will be reached faster.
27. The final OD_{600} should not be higher than 0.5 or the selection pressure will diminish, because a high-cell density results in the release of complex metabolites into the medium.
28. Conjugation to the carrier protein facilitates the adsorption to the plastic surface of microtiter plates by hydrophobic interactions and enables binding of library peptides with minimized steric inhibition. An additional tool to avoid selection of carrier protein specific peptides would be to change between two different carrier proteins.
29. **Steps 2** and **3** should be done in parallel.
30. The coating concentration of the target molecule can be reduced during several selection rounds to increase the selection pressure on single bacterial clones.

31. Lu et al. *(16)* performed panning in 60 mm dishes. 96-well microtiter plates provide a better surface to volume ratio and therefore an increased coating area.

32. Fimbriae at the *E. coli* surface are able to interact with oligosaccharides. This is important if the target molecules are glycosylated (like antibodies). In this case α-methyl mannoside (1%) (w/v) (*16*) should be also added.

33. Recommended: OD-measurement to detect bacterial concentration in tryptophan culture. Lu et al. (*36*) recommend OD_{550} = 0.8–1.2 and 10 ml induced cell culture for panning. A defined cell number helps to evaluate the enrichment of clones over the selection rounds. Due to the enrichment, the amount of cells can be reduced in later selection rounds. To be absolutely sure about the number of colony forming units (cfu) it is also possible to plate a small aliquot of tryptophan culture on RMG-Amp plates and incubate at 30°C overnight.

34. For characterization of different selection rounds, prepare glycerol stocks of overnight cultures before tryptophan induction.

35. The free carrier protein should be the same as used for coating, especially if the target molecule is conjugated to different carriers.

36. Flagella are sensitive to mechanical shearing. Every step in the panning procedure should be performed carefully.

37. Selection stringency can be increased by a higher number of washing steps or longer incubation for each washing step.

38. For observing recovery of eluted clones, it is recommended to remove 10 μl of eluate, plate it on RMG-agar plates and incubate overnight at 30°C.

39. Working on ice and fast handling as well as adding 1 mM PMSF or 1 mM EDTA decreases proteolysis and results in a higher yield of purified protein. Presence of lysozyme (0.2 mg/ml) enhances the lysis process.

40. Due to the shearing forces of the ultrasound, remaining DNA in the supernatant will be disrupted. Addition of DNAse also reduces the DNA amount in the cell lysate.

41. GSH sepharose beads can also be used for batch purification in a small reaction tube.

42. If no UV detector is available, the collected fractions during the elution can also be tested for enzymatic activity of GST.

43. The amount of dialysis volume depends on the volume of the eluted protein, but should at least be around six liters for 2 ml of eluted protein. Alternatively, cleavage on the column can be performed so that the proteins are not eluted from the

column by GSH. Instead, the column is equilibrated in the FXa buffer and the enzyme is injected manually with a syringe to the column. After incubation, the peptide or protein is eluted by washing the column.

44. The cleavage conditions need to be optimized separately for each protein and depend on the pH of the FXa buffer (optimum between pH 6.5 and pH 8) as well as on the incubation temperature and incubation time.
45. The target protein can be recovered from the FXa digest and separated from the GST fusion partner using either size exclusion chromatography, an ion exchange column or a reversed phase column, depending on the nature of the selected peptide or protein.
46. The elution volume of the identified peptide or protein depends on its hydophobicity. This elution volume should not overlay with the elution of the GST protein from the RP column. If this is the case, either the gradient needs to be changed or a FXa cleavage on the GST column needs to be performed.

References

1. Wolf, E., Kim, P.S. and Berger, B. (1997) MultiCoil: a program for predicting two- and three-stranded coiled coils. *Protein Sci* **6**, 1179–1189.
2. Lupas, A.N. and Gruber, M. (2005) The structure of alpha-helical coiled coils. *Adv Protein Chem* **70**, 37–78.
3. Mason, J.M. and Arndt, K.M. (2004) Coiled coil domains: stability, specificity, and biological implications. *Chembiochem* **5**, 170–176.
4. Müller, K.M., Arndt, K.M. and Alber, T. (2000) Protein fusions to coiled-coil domains. *Methods Enzymol* **328**, 261–282.
5. Arndt, K.M., Müller, K.M. and Plückthun, A. (2001) Helix-stabilized Fv (hsFv) antibody fragments: substituting the constant domains of a Fab fragment for a heterodimeric coiled-coil domain. *J. Mol. Biol.* **312**, 221–228.
6. Mason, J.M., Müller, K.M. and Arndt, K.M. (2007) Considerations in the design and optimization of coiled coil structures. *Methods Mol. Biol.* **352**, 35–70.
7. Straussman, R., Ben-Ya'acov, A., Woolfson, D.N. and Ravid, S. (2007) Kinking the coiled coil – negatively charged residues at the coiled-coil interface. *J. Mol. Biol.* **366**, 1232–1242.
8. Mason, J.M., Schmitz, M.A., Müller, K.M. and Arndt, K.M. (2006) Semirational design of Jun-Fos coiled coils with increased affinity: universal implications for leucine zipper prediction and design. *Proc. Natl. Acad. Sci. U.S.A.* **103**, 8989–8994.
9. Arndt, K.M., Pelletier, J.N., Müller, K.M., Alber, T., Michnick, S.W. and Plückthun, A. (2000) A heterodimeric coiled-coil peptide pair selected *in vivo* from a designed library-versus-library ensemble. *J. Mol. Biol.* **295**, 627–639.
10. Arndt, K.M., Pelletier, J.N., Müller, K.M., Plückthun, A. and Alber, T. (2002) Comparison of *in vivo* selection and rational design of heterodimeric coiled coils. *Structure* **10**, 1235–1248.
11. Hagemann, U.B., Mason, J.M., Müller, K.M. and Arndt, K.M. (2008) Selectional and mutational scope of peptides sequestering the Jun-Fos coiled coil domain. *J. Mol. Biol.* **381**, 73–88.
12. Pelletier, J.N., Arndt, K.M., Plückthun, A. and Michnick, S. W. (1999) An *in vivo* library-versus-library selection of optimized protein–protein interactions. *Nat. Biotechnol.* **17**, 683–690.
13. Mason, J.M., Müller, K.M. and Arndt, K.M. (2007) Positive aspects of negative design: simultaneous selection of specificity and interaction stability. *Biochemistry* **46**, 4804–4814.

14. Köhler, F. and Müller, K.M. (2003). Adaptation of the Ras-recruitment system to the analysis of interactions between membrane-associated proteins. *Nucleic Acids Res.* **31**, e28.
15. Benhar, I. (2001). Biotechnological applications of phage and cell display. *Biotechnol. Adv.* **19**, 1–33.
16. Lu, Z., Murray, K.S., Van Cleave, V., LaVallie, E.R., Stahl, M.L. and McCoy, J.M. (1995) Expression of thioredoxin random peptide libraries on the Escherichia coli cell surface as functional fusions to flagellin: a system designed for exploring protein–protein interactions. *Biotechnology (N Y)* **13**, 366–372.
17. Krebber, A., Bornhauser, S., Burmester, J., Honegger, A., Willuda, J., Bosshard, H.R. and Pluckthun, A. (1997) Reliable cloning of functional antibody variable domains from hybridomas and spleen cell repertoires employing a reengineered phage display system. *J. Immunol. Methods* **201**, 35–55.
18. Stebel, S.C., Arndt, K.M. and Muller, K.M. (2007) Versatile DNA fragmentation and directed evolution with nucleotide exchange and excision technology. *Methods Mol. Biol.* **352**, 167–190.
19. Müller, K.M., Stebel, S.C., Knall, S., Zipf, G., Bernauer, H.S. and Arndt, K.M. (2005) Nucleotide exchange and excision technology (NExT) DNA shuffling: a robust method for DNA fragmentation and directed evolution. *Nucleic Acids Res.* **33**, e117.
20. Denault, M. and Pelletier, J.N. (2007) Protein library design and screening: working out the probabilities. *Methods Mol. Biol.* **352**, 127–154.
21. Clark, J.M., Joyce, C.M. and Beardsley, G.P. (1987) Novel blunt-end addition reactions catalyzed by DNA polymerase I of Escherichia coli. *J. Mol. Biol.* **198**, 123–127.
22. Smith, G.P. (1985) Filamentous fusion phage: novel expression vectors that display cloned antigens on the virion surface. *Science* **228**, 1315–1317.
23. O'Brien, P.M. and Aitken, R. (2004) Antibody phage display: methods and protocols. *Methods Mol. Biol.* Humana Press **178**, 416.
24. Sidhu, S.S. (2005)Phage display in biotechnology and drug discovery. *Drug Discov. Ser.* Taylor & Francis Ltd, 768.
25. Russel, M., Lowman, H.B. and Clackson, T. (2004) *Introduction to Phage Biology and Phage Display. In Phage Display – A Practical Approach.* Oxford University Press: Oxford.
26. Willats, W.G. (2002) Phage display: practicalities and prospects. *Plant. Mol. Biol.* **50**, 837–854.
27. Smith, G.P. and Petrenko, V.A. (1997) Phage display. *Chem. Rev.* **97**, 391–410.
28. Martin, A., Schmid, F.X. and Sieber, V. (2003). Proside: a phage-based method for selecting thermostable proteins. *Methods Mol. Biol.* **230**, 57–70.
29. Arndt, K.M., Jung, S., Krebber, C. and Plückthun, A. (2000) Selectively infective phage technology. *Methods Enzymol* **328**, 364–388.
30. Jung, S., Arndt, K.M., Muller, K.M. and Pluckthun, A. (1999). Selectively infective phage (SIP) technology: scope and limitations. *J. Immunol. Methods* **231**, 93–104.
31. Schumacher, T.N., Mayr, L.M., Minor, D.L., Jr., Milhollen, M.A., Burgess, M.W. and Kim, P.S. (1996) Identification of D-peptide ligands through mirror-image phage display. *Science* **271**, 1854–1857.
32. Wiesehan, K. and Willbold, D. (2003) Mirror-image phage display: aiming at the mirror. *Chembiochem* **4**, 811–815.
33. Guichard, G., Benkirane, N., Zeder-Lutz, G., van Regenmortel, M.H., Briand, J.P. and Muller, S. (1994) Antigenic mimicry of natural L-peptides with retro-inverso-peptidomimetics. *Proc. Natl. Acad. Sci. U.S.A.* **91**, 9765–9769.
34. Remy, I. and Michnick, S.W. (1999) Clonal selection and *in vivo* quantitation of protein interactions with protein-fragment complementation assays. *Proc. Natl. Acad. Sci. U.S.A.* **96**, 5394–5399.
35. Subramaniam, R., Desveaux, D., Spickler, C., Michnick, S.W. and Brisson, N. (2001) Direct visualization of protein interactions in plant cells. *Nat. Biotechnol.* **19**, 769–772.
36. Lu, Z., LaVallie, E.R. and McCoy, J.M. (2003) Using bio-panning of FLITRX peptide libraries displayed on E. coli cell surface to study protein–protein interactions. *Methods Mol. Biol.* **205**, 267–280.
37. Terpe, K. (2003) Overview of tag protein fusions: from molecular and biochemical fundamentals to commercial systems.*Appl. Microbiol. Biotechnol.* **60**, 523–533.
38. Day, L.A. (1969) Conformations of single-stranded DNA and coat protein in fd

bacteriophage as revealed by ultraviolet absorption spectroscopy. *J. Mol. Biol.* **39**, 265–277.
39. Lunder, M., Bratkovic, T., Kreft, S. and Strukelj, B. (2005) Peptide inhibitor of pancreatic lipase selected by phage display using different elution strategies. *J. Lipid. Res.* **46**, 1512–1516.
40. Kazmin, D.A., Hoyt, T.R., Taubner, L., Teintze, M. and Starkey, J.R. (2000) Phage display mapping for peptide 11 sensitive sequences binding to laminin-1. *J Mol Biol* **298**, 431–445.
41. Goletz, S., Christensen, P.A., Kristensen, P., Blohm, D., Tomlinson, I., Winter, G. and Karsten, U. (2002). Selection of large diversities of antiidiotypic antibody fragments by phage display. *J. Mol. Biol.* **315**, 1087–1097.
42. D'Mello, F. and Howard, C.R. (2001) An improved selection procedure for the screening of phage display peptide libraries. *J. Immunol. Methods* **247**, 191–203.
43. Chames, P., Coulon, S. and Baty, D. (1998) Improving the affinity and the fine specificity of an anti-cortisol antibody by parsimonious mutagenesis and phage display. *J. Immunol.* **161**, 5421–5429.
44. Hecky, J., Mason, J.M., Arndt, K.M. and Müller, K.M. (2007) A general method of terminal truncation, evolution, and re-elongation to generate enzymes of enhanced stability. *Methods Mol. Biol.* **352**, 275–304.

Chapter 17

In Vitro Selection of Protein and Peptide Libraries Using mRNA Display

Terry T. Takahashi and Richard W. Roberts

Abstract

In vitro genetic approaches are powerful solutions to the protein design problem. mRNA display is an in vitro selection technique enabling the design of peptide and protein ligands ranging from one to over 100 amino acids. Libraries containing more than 10 trillion unique sequences can be synthesized and sieved with exquisite control over binding stringency and specificity.

Key words: mRNA display, in vitro selection, RNA–protein fusion, in vitro virus, PROfusion, combinatorial chemistry, ligand discovery, antibody mimetic, fibronectin, G protein.

1. Introduction

The discovery of functional ligands is important for applications such as imaging and potential drug leads. mRNA display is an in vitro selection technique that facilitates the discovery of functional ligands from trillion-member libraries. In mRNA display, a protein is fused to its encoding sequence, allowing the construction of an in vitro evolution cycle *(1)*. mRNA display selections have targeted RNA, proteins, and small molecules *(2)*, but can also isolate catalytic proteins *(3)*.

The following section describes a basic protocol for ligand discovery using mRNA display (*see* **Fig. 17.1**). The first round of selection and optimization can take 1–4 weeks and subsequent rounds of selection can take from 2 to 4 days. Depending on the complexity of the library, selections typically take from 1 to 12 rounds before converging.

Günter Mayer (ed.), *Nucleic Acid and Peptide Aptamers: Methods and Protocols, vol. 535*

DOI 10.1007/978-1-59745-557-2_17 Springerprotocols.com

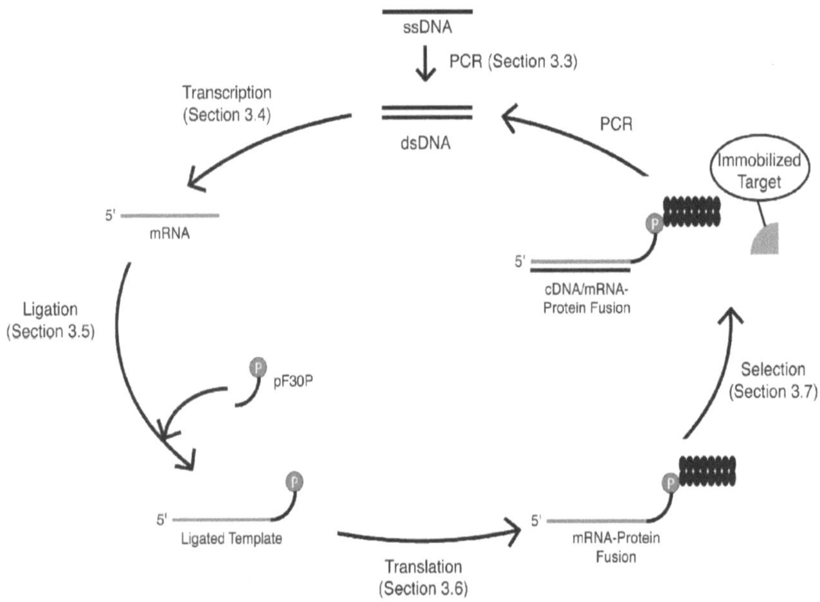

Fig. 17.1. An mRNA display selection cycle. ssDNA is PCR amplified (**Section 3.3**) into dsDNA. The dsDNA library is transcribed into mRNA by T7 RNA Polymerase (**Section 3.4**) and a pF30P linker is ligated to the 3' end of the mRNA (**Section 3.5**). This ligated template is translated in vitro (**Section 3.6**), purified, reverse transcribed and sieved against an immobilized target (**Section 3.7**). PCR regenerates a dsDNA library and the cycle is repeated.

A novice user may find the following protocol challenging as an error in an early step will affect the results of all downstream steps. We therefore recommend that a new user complete one selection cycle in its entirety on a small scale (as described in **Section 3.8**), making sure to include the controls described in **Section 3.7**, before prepping and attempting a selection using a large-scale library. Also consider performing a small complexity selection, where a known ligand is randomized at one or two positions. Such a selection will typically converge in 1–2 rounds and allow a quick diagnostic of the selection procedure. Lastly, a unique advantage of mRNA display is the potential expansion of the amino acid alphabet through the use of post translation modification *(4–6)* or through incorporation of unnatural amino acids *(7, 8)*. However, these methods introduce additional steps into the protocol and we recommend that a user perform several successful selections before attempting more difficult selections.

2. Materials

2.1. Urea PAGE Purification of Nucleic Acids

1. 2X Formamide loading buffer: 80% (v/v) ddH_2O, 20% formamide, add bromophenol blue until dark purple. Store at 4°C (*see* **Note 1**).
2. Twenty five percent acrylamide/bis solution (19:1) containing 7.5 M urea in ddH_2O, 7.5 M urea in ddH_2O, and 10X gel buffer (7.5 M urea, 890 mM Tris base, 890 mM boric acid, and 20 mM EDTA) (Sequagel® system, National Diagnostics, Atlanta GA). The solutions can be stored at room temperature for >12 months. Unpolymerized acrylamide is a neurotoxin; wear gloves and avoid exposure.
3. Ammonium persulfate (APS): prepare a 10% solution (w/v) in ddH_2O and store at 4°C for up to a month.
4. *N,N,N',N'*-Tetramethylethylenediamine (TEMED). Store at 4°C.
5. Vertical gel electrophoresis apparatus capable of running gels >15 cm long (Labrepco, Horsham, PA, or Owl Separation Systems, Portsmouth, NH) including 1.5 mm spacers, gel comb, casting boot, and glass plates.
6. 5X TBE Buffer: 445 mM Tris base, 445 mM boric acid, 10 mM EDTA. Filter through a 0.45 μm filter to prevent precipitation. Store at room temperature.
7. Elutrap Electroelution System (includes Elutrap chamber and BT1 and BT2 membranes) (Whatman, Florham Park, NJ). Store the membranes at 4°C (*see* **Note 2**).
8. Saran wrap
9. 20 × 20 cm glass-backed TLC plate with fluorescent dye excitable at 254 nm.

2.2. Phenol Extraction/Ethanol Precipitation

1. Phenol/chloroform/isoamyl alcohol (25:24:1 (v/v); *see* **Note 3**).
2. Sodium acetate: 3 M NaOAc, pH adjusted to 5.2. Store at room temperature.
3. One hundred percent ethanol. Store at –20°C.
4. Ninety five percent ethanol, five percent ddH_2O. Store at 4°C.
5. 1X TE buffer: 10 mM Tris–HCl, pH 8.0, 0.5 mM EDTA. Store at 4°C.

2.3. PCR

1. Non-water saturated *n*-butanol (various suppliers).
2. Taq DNA polymerase (various suppliers) or other thermostable polymerase. Store at –20°C

3. 10X PCR Buffer: 100 mM Tris–HCl, pH 9.0 at room temperature, 500 mM KCl, 20 mM $MgCl_2$, 1% (v/v) Triton X-100 (Sigma, St. Louis, MO). Aliquot and store at –20°C
4. 10X dNTP solution (various suppliers): 2 mM of each dNTP pH adjusted to 7–8. Aliquot and store at –20°C.
5. 20X SB Buffer: 200 mM NaOH, pH adjusted to 8.0 with boric acid.
6. Low melt agarose for small DNA fragments (various suppliers).
7. 5X SB Loading Buffer: Dilute the 20X SB to 10X with ddH_2O, add an equal volume of glycerol, and add bromophenol blue until dark purple (*see* **Note 1**).
8. 100 bp DNA ladder (New England Biolabs, Ipswich, MA).
9. Ethidium bromide solution (10 mg/mL). Store in a dark bottle. Ethidium Bromide is a mutagen. Wear gloves and avoid exposure. Dispose of ethidium bromide waste accordingly.
10. Agarose Gel Extraction Kit (Qiagen, Valencia, CA or other vendor).
11. TOPO TA Cloning Kit (Invitrogen, Carlsbad, CA).

2.4. Transcription

1. T7 RNA Polymerase (various suppliers). Store at –20°C.
2. 5X Transcription Buffer: 400 mM HEPES–KOH, pH 7.5, 10 mM spermidine, 200 mM DTT, 125 mM $MgCl_2$. Aliquot and store at –20°C
3. 5X NTP Solution: 20 mM each NTP, pH adjusted to 7–8 with NaOH (check pH with a pH strip). Aliquot and store at –20°C.
4. RNAsecure^{TM} Reagent (Ambion, Austin, TX): (*see* **Note 4**).
5. 0.5 M EDTA: Adjust pH to 8.0 with NaOH and sterile filter.

2.5. Ligation

1. T4 DNA Ligase (New England Biolabs or other supplier). Store at –20°C.
2. 10X Ligation buffer: 50 mM Tris–HCl, 10 mM $MgCl_2$, 1 mM ATP, 10 mM DTT. Store at –20°C.

2.6. Translation

1. Rabbit reticulocyte lysate kit (EMD Chemicals, San Diego, CA or other supplier). The kit contains potassium and magnesium salts as well as translation buffers and amino acid mixtures. Store at –80°C.
2. ^{35}S-methionine, in vitro translation grade, >800 Ci/mmol (Perkin Elmer, Waltham, MA, MP Biomedicals, Solon, OH, or other supplier). Store at 4°C.
3. 0.5 mM *L*-methionine (Sigma), pH adjusted to 7–8. Store at –20°C.

2.7. Selection

1. dT cellulose (New England Biolabs or GE Healthcare, Piscataway, NJ). Store at –20°C.
2. 1X Isolation Buffer: 100 mM Tris–HCl, pH 8.0, 1 M NaCl, 0.2% (v/v) Triton X-100. Sterile filter and store at 4°C.
3. Spin-X Filters, cellulose acetate, 0.45 μm (Corning, Corning, NY).
4. Centrex microfiltration system 0.45 μm (Whatman).
5. Linear acrylamide (5 mg/mL) (Ambion). Store at –20°C.
6. RNase H^- reverse transcriptase (Superscript II, Invitrogen): Supplied as a kit containing buffers and RT enzyme. Store at –20°C. The kit includes 5X first strand buffer (250 mM Tris–HCl pH 8.3, 375 mM KCl, 15 mM $MgCl_2$) and 0.1 M DTT.
7. 20 μM 3' primer.
8. Neutravidin agarose (Pierce, Rockford, IL) supplied as 50:50 (v/v) slurry. Store at 4°C (*see* **Note 5**).
9. Selection target (*see* **Note 6**).
10. Selection buffer (*see* **Note 7**).
11. SDS-out Precipitation Reagent (Pierce).

2.8. Additional Rounds of Selection

1. 2 mM Biotin solution: 2 mM D-biotin dissolved in 1X PBS (137 mM NaCl, 2.7 mM KCl, 4.3 mM Na_2HPO_4, 1.47 mM KH_2PO_4). Sterile filter and store at 4°C.
2. NAP-25 Columns (GE Healthcare).

3. Methods

Before beginning the protocol, a target and library must first be selected. The quality of the target is critical for success in an in vitro selection. Ideally, the target should be as pure and as homogeneous as possible to avoid the selection of ligands to impurities or inactive conformations. If possible, confirm the activity of the immobilized target through biochemical assays to insure that the immobilization does not inactivate the target. One milligram of a protein target is more than enough for an entire selection, however, the amount of target will depend on the complexity of the library used for selection.

When designing the random library, consider factors such as the length, degree of randomness, and the complexity to be used in the selection. Random peptide libraries are most often used in mRNA display and have yielded nanomolar ligands *(2)*. Longer protein libraries have been sieved successfully, however, these selections are more challenging due to the increased difficulties of constructing long open reading frames as well as the reduced efficiency of fusion between mRNA and peptide *(9–11)*.

Both natural and synthetic nucleic acid sequences can be the source of the library. cDNA libraries will yield natural protein sequences that interact with the target, while synthetic libraries may or may not yield sequences that resemble natural binding partners *(12, 13)*. It is also possible to randomize a natural sequence using mutagenic PCR *(14)* or to generate a library using DNA shuffling *(15)*. Synthetic libraries can be synthesized using NNS codons (*see* **Note 8**) or can be a doped sequence *(16, 17)*.

For the synthesis of random libraries, we recommend against having the synthesizer mix the random phosphoramidites as biases will result from the different reactivities of the phosphoramidites and different volumes of phosphoramidites delivered by the synthesizer; instead, a separate bottle should be included and used as a fifth base. In order to correct for the reactivities, the phosphoramidites should be mixed at a molar ratio of 3:3:2:2 for A:C:G:T, respectively. Peptide libraries shorter than ~37 amino acids can be synthesized as one oligo, however, for longer libraries multiple oligos must be synthesized and ligated together *(10, 18)* (*see* **Note 9**). The frequency of encountering a stop codon increases with library length and necessitates "preselection" of the library to increase full-length sequences *(18)*.

All libraries require a 5' constant region, a start codon, and a 3' constant region (*see* **Fig. 17.2a**). The most common 5' constant

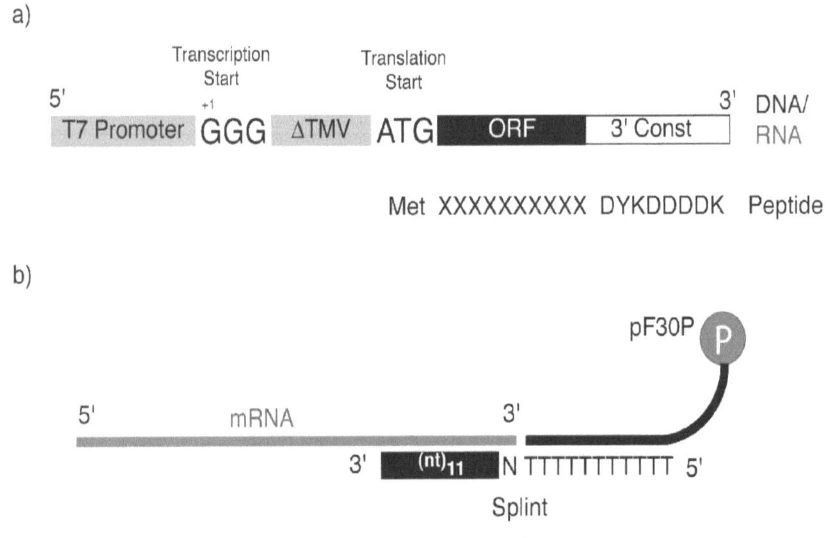

Fig. 17.2. (**a**) An mRNA display library. The 5' constant region contains a T7 promoter for transcription initiation, a GGG transcription start sequence, and a ΔTMV translation enhancer sequence. The translation begins at the initiator Met codon followed by an open reading frame (ORF) and the 3' constant region (in this case, a FLAG peptide). The 5' constant region and 3' constant region denote where the 5' primer and 3' primer anneal, respectively. (**b**) Splint-mediated ligation of mRNA and pF30P. The splint contains the sequence 5'-$(T)_{11}N(nt)_{11}$, where N is A/T/C/G and the $(nt)_{11}$ base pairs to the 3' end of the mRNA.

region is encompassed by the 5' primer 42.108 (5'-TAATACGACTCACTATAGGGACAATTACTATTTACAATTACA-3') and contains the T7 promoter sequence 5'-TAATACGACTCACTATA-3' necessary for in vitro transcription, a GGG transcription start sequence preferred by T7 RNA Polymerase, and a translation enhancer sequence derived from Tobacco Mosaic Virus (ΔTMV). The start codon ATG follows the 5' constant region and is necessary for translation initiation. Fusion formation does not seem to depend on the 3' end of the mRNA and many 3' end sequences have been successfully used in mRNA display selections. For example, innocuous sequences such as $(Gly)_n$ or $(Gly\text{-}Ser)_n$ or one that allows peptide-based purification (e.g., FLAG (DYKDDDDK) or His_6) have been used.

Lastly, the complexity of the library used for selection should be considered. The increased technical difficulties of performing a more complex selection must be weighed against the increased probability of discovering ligands from more complex libraries. To reduce the risk of selection failure due to mishandling of large volumes of reagents, consider performing the first round multiple times with smaller volumes and combining the resulting pools for the second round of selection.

3.1. Purification of Nucleic Acids by Urea-PAGE

1. Obtain an appropriate DNA library. Calculate the volume of ddH_2O required to resuspend the DNA based on 50–100 μL of water per 100 nmol of DNA. Add an equal volume of 2X Formamide loading buffer.
2. These instructions are for the V16 vertical system from Labrepco, but can easily be adapted to other vertical gel systems. Assemble the gel system and run the gel at constant power according to the manufacturer's recommendations.
3. Based on the size of the DNA fragment, prepare 50 mL of an appropriate percentage 1.5 mm-thick urea–polyacrylamide gel according to the manufacturer's recommendations. Add 500 μL of 10% APS and 50 μL of TEMED and mix quickly (*see* **Note 10**). Insert the comb diagonally into the gel to avoid trapping air bubbles and allow the gel to polymerize.
4. Prepare 1 L of 1X TBE by dilution of the 5X TBE stock with ddH_2O.
5. Assemble the gel in the gel system. Wash the wells with buffer to remove any unpolymerized acrylamide. Pre-run the gel at 25 W for 15–30 min (*see* **Note 11**).
6. Wash the wells with buffer again to remove any urea that diffused into the wells then load 100 L of sample per lane (*see* **Note 12**). Run the gel at 25 W until the bromophenol blue dye reaches the bottom of the gel. Typically, this will take 60–75 min, depending on the percentage gel that was poured.

7. Remove the glass plates and place the gel on a piece of Saran wrap. Place on a fluorescent TLC plate and use a handheld UV light to UV shadow the nucleic acid, taking care to minimize the UV exposure to prevent damage. The nucleic acid will show up as dark purple/black bands against the fluorescent plate. Excise the bands with a clean razor or glass cover slide (*see* **Note 13**). The gel slices can be stored at –20°C overnight if desired.
8. Assemble the elutrap device according to the manufacturer's specifications (*see* **Note 14**). Place the gel slices in the elutrap device and prepare 2 L of 0.5X TBE. Fill the inner chamber so that the buffer completely covers the gel slices. Add buffer to the outer chamber so that the buffer level on the outside is the same height or higher than the buffer level in the inner chamber.
9. Elute the DNA for 1–2 h at 200 V. Most of the DNA will be eluted in the first hour, which is usually sufficient for most applications. Reverse the polarity of the device and run for 15 s. Collect the eluent using a P200, taking care not to poke a hole in the membranes (*see* **Note 15**).
10. Estimate the volume of the eluent by backpipetting, then ethanol precipitate the DNA as described in **Section 3.2**.

3.2. Phenol Extraction/Ethanol Precipitation of Nucleic Acids

If only ethanol precipitating, skip **Step 1**.

1. Add an equal volume of phenol/chloroform/isoamyl alcohol and vortex. Spin for 30–60 s at top speed in a centrifuge to separate the organic and aqueous layers. Remove the top aqueous layer (*see* **Note 16**).
2. Add 1/10 volume of 3 M NaOAc, pH 5.2, then add 2–3 volumes of –20°C 100% ethanol. Split the sample into multiple tubes if necessary.
3. Allow the DNA to incubate at –20°C for 15–30 min. Spin at top speed (14–15 K RPM) in a 4°C centrifuge (*see* **Note 17**).
4. Remove the supernatant with an aspirator or pipette. Do not to disturb the pellet. Add 1 mL of 95% ethanol, invert the tube and remove the supernatant without disturbing the pellet.
5. Dry the pellet by placing in a Speedvac or by allowing to air dry upside down on a Kimwipe.
6. Resuspend the pellet in water or 10 mM Tris–HCl, pH 8.0. (*see* **Note 18**).
7. Determine the DNA concentration by UV absorbance.

3.3. PCR Amplification of DNA

3.3.1. Optimization of the PCR Reaction

1. Obtain a 5' and 3' primer for amplification of the desired template. Resuspend the primer in 100 μL of ddH_2O and add 1 mL of *n*-butanol. Vortex, and spin at top speed in a 4°C centrifuge (14–15 K RPM). Wash the pellet with 1 mL of 95% ethanol and allow to dry in a Speedvac. Resuspend the pellet in water and determine the DNA concentration by UV

absorbance. Adjust the primer concentration to 100–200 μM. Aliquot the primer into 50–100 μL aliquots and store at −20°C (*see* **Note 19**).

2. Prepare a 1 μM stock of ssDNA template.
3. Prepare a series of PCR reaction mixes in 0.65 mL tubes containing 1–8 pmol of ssDNA template (1–8 μL of 1 μM), 100–200 pmol of each primer (1 μL of 100–200 μM primer), 10 μL of 10X dNTP mix, 10 μL of 10X PCR mix, and add ddH_2O to 100 μL (*see* **Note 20**).
4. Place the tubes on a PCR block and preheat the reaction to 94°C. Once the reactions reach 94°C, add 1 μL of Taq to each tube. Perform 4–6 cycles of PCR of 94°C for 30 s, 55°C for 30 s, 72°C for 45 s.
5. While the reactions are cycling, prepare a 2% agarose gel with the low melting point agarose. This protocol is for the Bio-rad mini-sub cell GT but can be adapted for other systems. Weigh 1 g of LMP agarose in a 125 mL Erlenmeyer flask, then add 50 mL of 1X SB buffer. Microwave until the agarose is dissolved, stirring occasionally. Caution is advised since the solution can easily boil over. Cool the flask under running tap water until it can comfortably be touched, then add 2 μL of ethidium bromide. Pour the solution into the casting and place the comb and allow to harden.
6. Mix 1 μL of 5X SB Loading buffer with 4 μL of PCR product and run on the gel. (*see* **Note 21**). On an adjacent lane, run 1 μL of 100 bp DNA ladder. Run the gel for 4 min at 300 V then image the gel, though longer run times will be required for longer templates (*see* **Note 22**).

3.3.2. Cloning of the Rd 0 Library

1. Pour a preparative agarose gel and gel purify a 100 μL PCR reaction (*see* **Note 23**). Visualize the band using a handheld UV lamp and excise with a clean razor.
2. Extract the DNA from the gel slice using a gel extraction kit according to the manufacturer's specifications.
3. Clone the purified DNA using the TOPO-TA cloning kit according to the manufacturer's specifications.
4. Sequence >10 clones to insure that the library was synthesized as designed and there is no strong bias in the nucleotide composition. If there is an undesigned bias, consider resynthesizing the library.

3.3.3. Preparation of a Large-Scale Library

1. Determine the size of the library required for selection; typical libraries range from 10^{12} to 10^{14} molecules. Assuming that each DNA sequence entering the PCR is unique, the pool complexity will be the number of molecules entering the PCR multiplied by the fraction of molecules extendable (*see* **Note**

24), also called a pool equivalent. Calculate the number of 100 μL reactions needed to obtain the desired pool complexity (e.g. a single 100 μL reaction with 4 pmol template that is 10% extendable contains ~2×10^{11} molecules; a pool of 1×10^{13} molecules would require ~4 mL of PCR).

2. Perform a large scale PCR reaction. If more than ten tubes are loaded onto the PCR block, change the cyclic reaction to 2 min at 94°C, 2 min at 55°C, and 4 min at 72°C to adjust for the increased mass that must be cycled (*see* **Note 25**).
3. Run an agarose gel to check the quality of the PCR product.
4. Phenol extract and ethanol precipitate the library as described in **Section 3.2**.
5. Resuspend the dsDNA in 1X TE or 10 mM Tris–HCl (*see* **Note 18**).
6. Run a dilution series of the dsDNA pool (e.g., 1/4, 1/8, 1/16, etc.) on an agarose gel and compare the dsDNA product to the nearest-sized band of the 100 bp ladder to determine the concentration of dsDNA (in ng/μL; *see* **Note 26**).
7. Determine the number of pool equivalents that have been synthesized (i.e., the number of times the initial pool has been amplified). For a 6 cycle PCR, theoretically 32 pool equivalents should have been synthesized, however, the actual number can range from ~5 to 20 (*see* **Note 27**). The dsDNA pool can be stored at −20°C indefinitely.

3.4. In Vitro Transcription

1. In order to preserve the complexity of the library, at least three pool equivalents must be carried through each subsequent step until the affinity selection step. Calculate the volume of the transcription reaction required to retain the pool complexity, assuming a final template DNA concentration of 100 nM. For example, a pool of 1×10^{13} molecules would require ~50 pmol to be carried through in a 0.5 mL transcription (*see* **Note 28**).
2. Prepare the transcription reaction; per mL add the dsDNA PCR product from **Section 3.3.3** to a final concentration of 100 nM, 200 μL of 5X NTP buffer, 200 μL of 5X Transcription buffer, 40 μL of 25X RNAsecure, and ddH_2O to 1 mL (*see* **Note 29**). Heat the reaction to 65°C for 5–10 min, then cool on ice. Once cool, add 10 μL of T7 RNA polymerase, mix, and incubate at 37°C for 2–4 h to overnight. Generally, most of the product is generated within 1–2 h, however, longer incubations will yield more product.
3. Add 1/10 volume of 0.5 M EDTA, pH 8.0 and vortex until the white precipitate is dissolved.
4. Phenol extract and ethanol precipitate as in **Section 3.2** (*see* **Note 30**).

5. Gel purify the mRNA on an appropriate percentage urea–PAGE. Up to 2 mL of transcription reaction can be loaded on a 1.5 mm gel, depending on the quality of the mRNA. Electroelute the mRNA as in **Section 3.1**, and ethanol precipitate and quantitate the mRNA as in **Section 3.2**. The mRNA can be stored at –20°C indefinitely (*see* **Note 31**)

3.5. Ligation of the Puromycin Linker

1. Synthesize a DNA splint that contains the sequence 5'-$(T)_{11}N(nt)_{11}$-3', where N is a random base and the $(nt)_{11}$ region base pairs to the 3' end of the mRNA (*see* **Note 32**).
2. Synthesize the pF30P linker. The linker may be ordered from a number of commercial sources (e.g., IDT DNA, Coralville, IA, or the Keck Biotechnology Resource Laboratory, New Haven, CT). Order a 1 μmol scale synthesis of the sequence $8(A)_{21}777ACC6$, where 8 is chemical phosphorylation reagent I, 7 is spacer phosphoramidite 9, and 6 is puromycin CPG. Purify the pF30P via urea–PAGE, electroelute, and ethanol precipitate as in **Section 3.1**.
3. Determine the volume of translation required to carry three pool equivalents through the translation and purification steps to just before the affinity selection step. Assume an overall yield of 3% through all steps from translation to right before the affinity selection step (*see* **Note 33**). For example, for a pool of 1×10^{13} molecules, ~1,700 pmol should be translated in order for 3×10^{13} molecules to enter the selective step.
4. Calculate the volume of ligation reaction based on the number of pmol required for translation, assuming a conservative yield of 20% of input mRNA (e.g., 1,000 pmol of ligated product per 500 μL of ligation reaction). For a pool of 1×10^{13}, roughly 1 mL of ligation would be required.
5. Prepare the large scale ligation reaction; a 0.5 mL ligation contains 10 μM mRNA, 15 μM splint, 10 μM pF30P, 50 μL 10X ligase buffer, and ddH_2O to 487.5 μL (*see* **Note 34**). Add 12.5 μL of T4 DNA ligase and incubate at room temperature for 30–60 min.
6. Add an equal volume of 2X Formamide loading buffer and gel purify on an appropriate percentage gel as in **Section 3.1** (*see* **Notes 35 and 36**).
7. Excise the ligated product and electroelute as in **Section 3.1**.
8. Determine the concentration of the mRNA-pF30P ligated template by UV (*see* **Note 37**). Adjust the concentration to 10 μM.

3.6. In Vitro Translation

This section details the preparation of a translation mix using the Novagen Red Nova Lysate, though other translation mixes can be used (*see* **Note 38**).

1. Prepare a translation reaction equal to the volume calculated in **Section 3.1, Step 4**; each cold 25 μL reaction contains 1 μL of 10 μM ligated template from **Section 3.5**, 1 μL of 2.5 M KOAc, 0.5 μL 25 mM $MgCl_2$, 2 μL 12.5 X translation mix-Methionine, 0.5 μL of 0.5 mM L-methionine, 10 μL of ddH_2O, and 10 μL of red nova lysate. In a separate tube, prepare at least one hot reaction, where 2 μL in vitro translation grade ^{35}S-labeled methionine is added in place of the cold methionine in a final volume of 25 μL (*see* **Note 39**). Incubate at 30°C for 1–2 h.
2. Add 2 μL of 1 M $MgCl_2$and 7 μL of 2.5 M KCl per 25 μL of translation. Incubate at room temperature for 5–15 min. Alternatively, store at –20°C overnight.

3.7. Selection

1. The following protocol is for a translation volume of 2.5 mL. Adjust the reagent volumes accordingly for larger or smaller translations. Weigh 75 mg of dT cellulose (3 mg per 100 μL of translation) and transfer to a 15 mL conical tube. Wash the beads with 10 mL of isolation buffer (IB) and spin to pellet the beads. Remove the supernatant and repeat twice for a total of three washes.
2. Combine the hot and cold reactions, then dilute to 25 mL (a tenfold dilution) with IB and add the washed dT cellulose. Rotate at 4°C for 1 h.
3. Before beginning the dT washing steps, fill two 1.7 mL tubes with water and heat at 65°C.
4. Transfer ~ 5 mL of the slurry to a Centrex filter, then spin at 1,500 × *g* (*see* **Note 40**). Remove the filtrate and repeat until all of the solution is filtered.
5. Add 5 mL of IB to the top of the Centrex filter, insuring that the beads are resuspended. Spin at 1,500 × *g* to wash the beads, and repeat four times for a total of five washes. Count 1 mL of the last wash and ensure that it is <2,000 cpm. If not, continue washing until the cpm are below 2,000.
6. Transfer the filter to a new collection tube and elute the fusions with 2 × 1 mL of 65°C water (*see* **Note 41**). Count 50 μL of the eluent to determine the total cpm for the entire reaction.
7. Ethanol precipitate the fusions as in **Section 3.2**, except add 6 μL of linear acrylamide per 500 μL of eluent (*see* **Note 42**).
8. For every 50 μL of translation, prepare 19 μL of RT mix (4 μL of 1X First strand buffer, 2 μL 0.1 M DTT, 1 μL of 20 μM 3' primer, 2.5 μL of 10X dNTPs, and 8.5 μL of ddH_2O) (*see* **Note 43**).
9. Resuspend the pellet with the RT mix, then heat the reaction to 65°C for 5–10 min, and place on ice. Once cooled, remove

1 μL for a no RT control reaction. Place the cooled RT mixture at 42°C for 2 min then add 1 μL of Superscript II enzyme per 19 μL of RT mix, and incubate at 42°C for 50 min. Heat at 65°C for 5–10 min to kill the RT enzyme, then cool on ice. Remove 1 μL for a +RT control reaction. Count 1/100 (10 μL for a 2.5 mL translation; count at least 1 μL if the entire RT reaction is <100 μL) of the cooled RT reaction and compare the cpm to that obtained in **Step 6** to determine the yield.

10. While the reverse transcription reaction is incubating, prepare the target beads (*see* **Note 44**). Place 100 μL of 50/50 slurry of neutravidin agarose in a Spin-X filter and wash with 3 × 700 μL of selection buffer. Add the target and incubate at 4°C for 1 h. Transfer to a new Spin-X filter and wash 3 × 700 μL with selection buffer. Transfer to a 1.7 mL tube and add buffer to make a 50/50 slurry.
11. Combine the RT reaction, target beads, and 940 μL of selection buffer (*see* **Notes 45 and 46**). Incubate at 4°C for 1 h (*see* **Note 47**).
12. Transfer to a Spin-X filter and wash with 5 × 700 μL of selection buffer (with BSA and tRNA) and once with 700 μL selection buffer without tRNA or BSA. Add 700 μL of selection buffer without tRNA or BSA and transfer to a new Spin-X filter for the final wash (*see* **Note 48**).
13. Elute the fusions by adding 100 μL of 0.15% SDS, incubating for 30 s, and spinning through the Spin-X filter (*see* **Note 49**). Repeat the elution for a total of 200 μL of elution.
14. Add 10 μL of SDS-out reagent and incubate on ice for 20 min.
15. Spin at top speed in a 4°C centrifuge then transfer the supernatant to a Spin-X filter. Spin and transfer the eluent to a new 1.7 mL tube.
16. Add 4 μL of linear acrylamide and ethanol precipitate the fusions as in **Section 3.2**.
17. Prepare a PCR mix with 13 μL of each primer, 130 μL of 10X dNTP mix, 130 μL of 10X PCR mix, and add ddH_2O to 1,287 μL. Resuspend the pellet with 990 μL of the mix and aliquot into 10 × 0.65 mL tubes. Aliquot 99 μL of the mix into a 0.65 mL tube as a no template control reaction (*see* **Note 50**). Lastly, add 99 μL of PCR mix into the –RT and +RT reactions.
18. Calculate the initial template concentration in the PCR reaction. For example, for 2.5 mL of translation, then the initial template concentration will be approximately 0.3 nM assuming a binding efficiency of ~1% (*see* **Note 51**). Based on the

initial template concentration, calculate the number of cycles needed in order to amplify the library, assuming a final template concentration of 0.1 μM. Similarly, calculate the initial concentration and cycles required for the +RT reaction.

19. Cycle the PCR reaction according to **Section 3.3**, but stop the reaction three cycles before the number calculated in **Step 18**. Run 4 μL of the PCR product on a 2% SB agarose gel to determine if more cycles are necessary (*see* **Note 52**). If there is no product, perform three additional cycles, and check again. If a light band is seen, perform 1–2 more cycles. Continue until the product band is similar in density to the nearest-sized band of the 100 bp ladder (*see* **Note 53**). The +RT and –RT controls will provide confirmation that the reverse transcription was successful. There should be a >4 round difference between the +RT and –RT control reactions.

3.8. Additional Selection Rounds

Once the first round of selection is completed in **Section 3.9**, repeat the following cycle until binding is observed.

1. Phenol extract and ethanol precipitate the enriched pool. Resuspend the DNA pellet in 200 μL of 1X TE buffer or 10 mM Tris–HCl, pH 7.5.
2. Use 100 μL of the resuspended DNA in a 1 mL transcription as in **Section 3.4** (*see* **Note 54**).
3. Add EDTA, phenol extract and ethanol precipitate.
4. While the mRNA is ethanol precipitating, remove the cap from a NAP-25 column, pour out the solution, and cut off the tip with a clean razor blade. Equilibrate the column with 5 × 5 mL of ddH_2O (*see* **Note 55**).
5. Resuspend the mRNA in 1 mL of ddH_2O, and add to the equilibrated NAP-25 column. Once the flow stops, add 1.5 mL of ddH_2O and allow to flow through the column. Add 0.5 mL and collect the first fraction, then repeat five times for a total of 6 × 0.5 mL fractions.
6. Ethanol precipitate the fractions. Generally, the product elutes in fractions 2–4, though some product will elute in fractions 5 and 6. If higher yield is desired, precipitate fractions 2–6, however, if higher quality is desired, precipitate fractions 2–4.
7. Perform a ligation reaction as in **Section 3.5**, however, increase the mRNA template concentration to 30 μM if the mRNA was desalted using the NAP-25 column (*see* **Note 56**). Gel purify the product as in **Section 3.1**.
8. Perform a 100 μL cold translation and a 25 μL hot translation as in **Section 3.6**.

9. Add 100 μL of 25% dT cellulose slurry (*see* **Note 57**) and 1 mL of IB to each translation reaction. Incubate at 4°C for 30–60 min.
10. Spin the tubes containing the fusions and cellulose at 1,500 × *g* and remove ~ 500 μL of supernatant. Resuspend the beads and transfer to a Spin-X filter.
11. Spin at 1,500 × *g* for 30 s. Remove the filtrate and add 700 μL of IB, making sure that the beads are resuspended. Repeat the washing and filtration steps for a total of five washes. Save the last wash and count in a scintillation counter. The last wash should be below ~2,000 cpm.
12. Elute the fusions twice with 250 μL of 65°C ddH_2O.
13. Count 1/10 the volume of the hot sample. Typically, for a 25 μL reaction, we routinely get 100–500,000 cpm of material for the whole reaction (*see* **Note 58**)
14. Prepare a binding reaction of 100,000 cpm of hot fusions, 20 μL of 50/50 slurry of target, and add selection buffer to 1 mL. Similarly, prepare a no target control reaction that only omits the target. Incubate at 4°C for 1 h, transfer the solutions to a Spin-X filter and wash 3X with 700 μL of selection buffer. Count the supernatant, the washes, and the beads in a scintillation counter. Determine the percent binding by dividing the cpm left on the beads by the total number of cpm in the reaction (supernatant + washes + beads) (*see* **Note 59**).
15. Precipitate the cold reaction in the presence of linear acrylamide, and reverse transcribe the reaction as in **Section 3.7**. Combine 40 μL of the RT reaction, 20 μL of 50/50 slurry of target, 5 μL of 2 mM biotin, and 935 μL of selection buffer. Incubate, wash, elute and PCR amplify the reaction as in **Section 3.7**.

3.9. Analysis of Selected Sequences

1. Clone and sequence the library as described in **Section 3.3.2**.
2. Once individual sequences are obtained, they can be aligned manually or using web-based alignment programs (e.g., www.expasy.org).
3. Select a few sequences for further study. We routinely make fusions with these sequences and test for binding (as described in **Section 3.8**). The clones with the highest binding affinity are then selected for further study.
4. Synthesize the peptide sequences using synthetic methods or by cloning the sequences into expression vectors. It is not uncommon for the 3' constant region to add to the affinity of the peptides, thus several constructs can be synthesized.

4. Notes

1. The amount of bromphenol blue added is not critical as it acts as an electrophoresis marker. Xylene cyanol can also be added if desired. This is particularly useful when gels must be run for hours as in the case of longer mRNAs.
2. Other gel elution methods can be used, such as crush and soak, however, we prefer electroelution for its relatively high speed and yield.
3. The phenol/chloroform/isoamyl solution can be purchased premixed or made in house. Pink solutions of phenol indicate oxidation and should be discarded to avoid nucleic acid damage.
4. RNAsecure can be omitted if RNA degradation is minimal.
5. The target may be immobilized by other methods such as neutravidin acrylamide (Ultralink, Pierce), streptavidin agarose, or other biotin-binding matrix. Covalent methods of immobilization include NHS-activated sepharose or agarose, epoxy-activated sepharose, or cyanogen bromide-activated sepharose.
6. Roughly 1–10 μg of a 30 kDa protein will be needed per round of selection (33–330 pmol). However, we recommend at least 1 mg of protein since more protein is required for the first round of selection as well as optimization, monitoring selection progress, and analysis of functional sequences.
7. The selection buffer will depend on the particular requirements of the target. We have successfully used buffers such as 25 mM Hepes pH 7.5, 150 mM NaCl, 0.05% (v/v) Tween 20, 0.05% (w/v) BSA, and 1 μg/mL tRNA. Buffers should have a buffering agent (e.g., Hepes, Tris, or phosphate), a monovalent salt (NaCl or KCl), a non-ionic detergent (Tween-20, NP-40, or Triton X-100), Bsa, and tRNA. Magnesium, reducing agents (DTT or β-mercaptoethanol), or protein specific requirements (such as GDP for G proteins *(13)*) can also be added, as needed.
8. NNS codons (where N=A/T/G/C and S=G/C) are the best random codon for unbiased access to all 20 amino acids for a library translated in the rabbit reticulocyte lysate (from *Oryctolagus cuniculus*). NNN codons contain a higher frequency of stop codons and NNK codons (K=G/T) are less frequently used than NNS in *O. cuniculus*.
9. While it is possible to construct a random library via PCR of two random oligos, ligation is the method of choice to retain complexity. Synthetic DNA is often 10–20% extendable by Taq and PCR of these oligos will result in a significant loss of complexity since only 1–4% of all sequences will be fully extended.

10. Do not vortex the gel solution as introducing oxygen will prevent gel polymerization.
11. We find that placing an aluminum plate (~0.17 cm thick) on the glass plates helps to equalize the heat across the gel and reduces smiling.
12. For templates >200 nt, it is sometimes necessary to boil the sample at 95°C for 1–2 min before loading to completely denature the nucleic acid. The sample should be loaded directly after heating.
13. Since a DNA synthesis produces much more DNA that is required for downstream steps, it is often better to sacrifice yield by cutting more conservatively to obtain a higher quality product.
14. We have noticed that recent batches of BT1 and BT2 membranes contain a salt that precipitates with nucleic acid. Washing the membranes with ddH_2O or with running buffer prevents the unwanted precipitation of this salt.
15. The clear BT1 membrane is much tougher than the BT2 membrane. It is easier to insert the pipette tip closer to the BT1 membrane.
16. Avoid any white precipitate at the organic/aqueous interface. Most of the liquid can be removed by setting a pipette for a smaller volume and backpipetting to remove the last amounts of the aqueous layer. Alternatively, products such as the Phase Lock Gel (Eppendorf, Westbury, NY) can also be used.
17. It is good practice to place the tubes in the centrifuge so that the pellet will be known in orientation. For example, place the tubes in the centrifuge so that the tabs of the 1.7 mL tubes face inward.
18. For dsDNA libraries, be sure to use 10 mM Tris–HCl or 1X TE to resuspend the DNA as pure water will denature the DNA and prevent transcription of the pool.
19. It is good practice to aliquot and store primers in small volumes since primers can easily be contaminated with template DNA. We butanol precipitate the primers instead of gel purifying since it is easy to contaminate the primers with template DNA during gel purification.
20. Filter tips should be used to avoid potential PCR contamination, though other precautions should also be used such as separating pre- and post-PCR rooms, or preparing reactions in laminar flow PCR workstations.
21. The gel can be more quickly loaded if 1 μL spots of 5X loading buffer (one spot per sample) are placed on a piece of parafilm. The PCR samples can then be mixed on the parafilm and loaded onto the gel.

22. The optimal template concentration is the sample that yields the largest amount of a PCR product as a single band with very little smearing present. If no bands are seen, optimize the PCR reaction by varying the annealing temperature or the concentrations of template, $MgCl_2$, primer, or Taq.
23. The gel purification and elution steps can be omitted if there is no primer dimer.
24. Due to errors in synthesis, not all input synthetic DNA will be extendable. Typically, 10–20% of a 100-mer oligo can be extended by Taq polymerase. The actual fraction can be determined by labeling the complementary primer with ^{32}P, performing one cycle of PCR, and running the products on a denaturing gel.
25. It is possible to load 200 μL into a 0.65 mL tube with a slight loss in product yield. If this is done, the cycling times must be extended and a heated lid on the thermocycler must not be used.
26. Densitometry must be used instead of UV absorption to determine the product yield since some of the primers and dNTPs will also precipitate and will contribute to the UV absorbance of the stock DNA solution.
27. The first cycle of PCR will make the input DNA double stranded, thus only five cycles of doubling will theoretically occur ($2^5 = 32$).
28. More material can be carried through a step if required. For example, we would perform at least 1 mL of transcription in this example.
29. Working with RNA requires several basic precautions including wearing gloves and using disposable materials as much as possible. Glassware can be washed with RNase removal solutions (e.g., RNase Zap (Ambion), and others). Tips and tubes can be purchased RNase/DNase free and used without autoclaving. Water can be DEPC treated, however, we use a commercially available water purification system capable of producing RNase-free water (Barnstead, Dubuque, IA or Millipore, Billerica, MA).
30. No −20°C incubation step is needed after adding the ethanol since the concentration of mRNA and NTPs is very high.
31. The RNA can be DNase treated at this step (*see (13)*). It is possible to omit this step since any ligated DNA will not be translated and will be removed at the selective step.
32. Including the N base allows ligation of N and N+1 products *(19)*. It is possible to use different chemistries to ligate the puromycin on the C-terminus *(20, 21)*. Although the psoralen method *(20)* may be a quicker ligation procedure that

does not require gel purification, we have found (data not shown) that the psoralen ligation strategy inhibits the reverse transcription reaction.

33. Ideally, the efficiency of each step should be determined for each selection. This is a conservative estimate of 0.1 for fusion formation, 0.75 for dT purification, 0.9 for ethanol precipitation, 0.9 for reverse transcription, and 0.5 for loss of material to tubes, tips, etc. This will require a ^{32}P-labeled template and require running the products of each step on a urea–PAGE gel.
34. The reaction can be annealed by heating to 95°C for 1–2 min, then cooling on ice. Annealing does not seem to increase the ligation efficiency for short (<200 nt) templates, however, in some cases has increased the ligation efficiency for longer (>400 nt) templates. The ratio of mRNA:splint:pF30P also does not seem to make a difference in the amount of ligated product obtained. RNase inhibitors can be also added to the ligation reaction, however, in our hands are not necessary.
35. Gel purification of the splint-mediated ligation is necessary to avoid the RNase H activity found in the rabbit reticulocyte lysate *(22)*.
36. For templates longer than 400 nt, it is often difficult to separate the ligated and unligated products. In this case, dT-cellulose chromatography can be used after gel purification to separate the ligated and unligated products.
37. The $\varepsilon_{260\ nm}$ for puromycin in water = 11,790 $M^{-1}\ cm^{-1}$ *(23)*.
38. For preparation of large libraries, several milliliters of translation must be performed. In these cases, it is more economical to prepare the lysate according to the procedure of Jackson and Hunt *(24)*. Reticulocyte-rich rabbit blood can be purchased from Pel Freeze (Rogers, AR).
39. The entire large-scale reaction can be performed hot, though we tend to avoid this as it is more difficult to work with a large volume of very radioactive material and generates a large volume of radioactive waste. The most direct way of measuring and following the material through the multiple selection steps is to spike in a ^{32}P-labeled template, however, this is requires working behind a shield.
40. If a larger volume than 2.5 mL of translation is performed, the reaction can be divided among additional filters.
41. At this point, the fusions can be purified with a peptide-based purification scheme (e.g., FLAG, His_6 *(18)*, or thiopropyl sepharose *(22)*), which can increase enrichment. This step is especially necessary if the fusion formation efficiency is low, as in the case for selections of proteins larger than 100 amino acids.

42. Linear acrylamide is necessary as a co-precipitant to improve the efficiency of the ethanol precipitation.
43. DTT can be omitted in the reverse transcription reaction, which is important when disulfide-cyclized libraries are used or if the target contains disulfide bonds (as in the case of an antibody) and no buffer exchange step is desired.
44. We find that a fresh preparation of beads with target prevents loss of activity of proteins, however, if the immobilized target remains active after storage at 4°C, that can be used also.
45. In some cases, a buffer exchange is required between reverse transcription and the affinity selection step. Either ethanol precipitation in the presence of linear acrylamide (*see* **Section 3.7, Step 7**) or spin desalting columns (Princeton separations, GE Healthcare, or Biorad) can be used.
46. It is also possible to perform a negative selection step (preclear) in an attempt to reduce isolation of matrix-binding sequences. To do this, incubate the RT reaction as described, except add beads containing no target. Up to five consecutive negative selection steps may be necessary since even in the best circumstances, only 80% of material is bound (as measured by the radioactive binding assay in **Section 3.8, Step 14**). Alternatively, perform the preclear in a column format which we have observed to be slightly more efficient than a batch preclear.
47. In the later rounds of selection, the incubation temperature can be increased to increase the stringency of the selection.
48. Transferring the beads to a new Spin-X filter prevents any nonspecific sequences stuck to the original filter from being eluted by SDS in the next step. If the library is sticking to the Spin-X filter (which can be determined by counting the filter in a scintillation counter), then the washing can be performed in a 1.7 mL tube.
49. Besides SDS, NaOH and RNase are other potential nonspecific eluents. It is also possible to PCR the pool directly from the beads. If enough target can be purified, addition of non-biotinylated target into the supernatant will specifically elute the pool.
50. It is important to perform a no template control reaction every round to detect contamination of PCR solutions. If a band is observed in the no template control, discard the primers and PCR solutions, retranslate the pool, and repeat the selection.
51. A 2.5 mL translation at 400 nM yields 1,000 pmol input into the translation. A 3% overall yield and 1% binding yields 0.3 pmol of material on beads after affinity selection. The 0.3 pmol

is diluted to 1,000 for PCR giving an initial concentration of ~0.3 nM.

52. It is not uncommon for the actual number of PCR cycles to exceed those calculated by 3–10 rounds since the efficiency of the binding is not known. Thus, the number of cycles calculated represents a lower bound. As long as the negative control contains no band, the selection round was likely successful. Cycles for the first round of selection will depend on the size of the translation reaction while later rounds typically take 9–24 cycles. If >30 cycles are required for amplification, consider retranslating and reselecting.
53. Avoid over PCR of the library. Performing too many cycles can cause a shift in the size of the library but also has the effect of normalizing the relative proportions of different pool members, reducing the selective enrichment due to binding.
54. The volume of transcription can be decreased in later rounds to 250–500 μL.
55. The RNA can be gel purified instead of loaded onto a NAP-25 column.
56. The volume of ligation can be decreased to 125–250 μL.
57. A 25% slurry can be prepared by washing 100 mg of dT cellulose beads three times with 10 mL of IB, aliquoting into four, 1.7 mL tubes and adjusting the buffer volume until a 25% slurry is obtained. The slurry can be stored at 4°C for > 1 month.
58. If less than 100,000 cpm are purified from a 25 μL translation, consider retranslating and repurifying the library. Incomplete desalting of the mRNA-F30P template can also cause low translation yields.
59. The no target control is an essential control since selection of ligands to the matrix or neutravidin is possible. If nonspecific binding is seen, we recommend going back 2–3 rounds and switching the immobilization matrix as described in **Note 5**. Once binding does not increase with additional rounds of selection, increase the selection pressure by increasing temperature, salt, or adding a specific competitor to isolate higher affinity sequences. Alternatively, the pool can be cloned and sequenced.

Acknowledgments

We would like to thank Dr. Ryan Austin, Stephen Fiacco, and Shannon Howell for helpful comments on the manuscript. This work supported by NIH grant NIH GM R01 60416 and NIH GM R21 76678.

References

1. Roberts, R.W. and Szostak, J.W. (1997) RNA-peptide fusions for the in vitro selection of peptides and proteins. *Proc. Natl. Acad. Sci. U.S.A.* **94**, 12297–12302.
2. Takahashi, T.T., Austin, R.J. and Roberts, R.W. (2003) mRNA display: ligand discovery, interaction analysis and beyond. *Trends Biochem. Sci.* **28**, 159–165.
3. Seelig, B. and Szostak, J.W. (2007) Selection and evolution of enzymes from a partially randomized non-catalytic scaffold. *Nature* **448**, 828–831.
4. Li, S. and Roberts, R.W. (2003) A novel strategy for in vitro selection of peptide-drug conjugates. *Chem. Biol.* **10**, 233–239.
5. Millward, S.W., Takahashi, T.T. and Roberts, R.W. (2005) A general route for post-translational cyclization of mRNA display libraries. *J. Am. Chem. Soc.* **127**, 14142–14143.
6. Millward, S.W., Fiacco, S., Austin, R.J. and Roberts, R.W. (2007) Design of cyclic peptides that bind protein surfaces with antibody-like affinity. *ACS Chem. Biol.* **2**, 625–634.
7. Josephson, K., Hartman, M.C. and Szostak, J.W. (2005) Ribosomal synthesis of unnatural peptides. *J. Am. Chem. Soc.* **127**, 11727–11735.
8. Li, S., Millward, S. and Roberts, R. (2002) *J. Am. Chem. Soc.* **124**, 9972–9973.
9. Keefe, A.D. and Szostak, J.W. (2001) Functional proteins from a random-sequence library. *Nature* **410**, 715–718.
10. Olson, C.A. and Roberts, R.W. (2007) Design, expression, and stability of a diverse protein library based on the human fibronectin type III domain. *Protein Sci.* **16**, 476–484.
11. Xu, L., Aha, P., Gu, K., Kuimelis, R.G., Kurz, M., Lam, T., Lim, A.C., Liu, H., Lohse, P.A., Sun, L., Weng, S., Wagner, R.W. and Lipovsek, D. (2002) Directed evolution of high-affinity antibody mimics using mRNA display. *Chem. Biol* .**9**, 933–942.
12. Huang, B.C. and Liu, R. (2007) Comparison of mRNA-display-based selections using synthetic peptide and natural protein libraries. *Biochemistry* **46**, 10102–10112.
13. Ja, W.W. and Roberts, R.W. (2004) In vitro selection of state-specific peptide modulators of G protein signaling using mRNA display. *Biochemistry* **43**, 9265–9275.
14. Cadwell, R.C. and Joyce, G.F. (1992) Randomization of genes by PCR mutagenesis. *PCR Methods Appl.* **2**, 28–33.
15. Stemmer, W.P. (1994) Rapid evolution of a protein in vitro by DNA shuffling. *Nature* **370**, 389–391.
16. LaBean, T.H. and Kauffman, S.A. (1993) Design of synthetic gene libraries encoding random sequence proteins with desired ensemble characteristics. *Protein Sci.* **2**, 1249–1254.
17. Ja, W.W., Wiser, O., Austin, R.J., Jan, L.Y. and Roberts, R.W. (2006) Turning G proteins on and off using peptide ligands. *ACS Chem. Biol.* **1**, 570–574.
18. Cho, G., Keefe, A.D., Liu, R., Wilson, D.S. and Szostak, J.W. (2000) Constructing high complexity synthetic libraries of long ORFs using in vitro selection. *J. Mol. Biol.* **297**, 309–319.
19. Liu, R., Barrick, J.E., Szostak, J.W. and Roberts, R.W. (2000) Optimized synthesis of RNA-protein fusions for in vitro protein selection. *Methods Enzymol.* **318**, 268–293.
20. Kurz, M., Gu, K. and Lohse, P.A. (2000) Psoralen photo-crosslinked mRNA-puromycin conjugates: a novel template for the rapid and facile preparation of mRNA-protein fusions. *Nucleic Acids Res.* **28**, E83.
21. Tabuchi, I., Soramoto, S., Suzuki, M., Nishigaki, K., Nemoto, N. and Husimi, Y. (2002) An efficient ligation method in the making of an in vitro virus for in vitro protein evolution. *Biol. Proc. Online* **4**, 49–54.
22. Barrick, J.E., Takahashi, T.T., Balakin, A. and Roberts, R.W. (2001) Selection of RNA-binding peptides using mRNA-peptide fusions. *Methods* **23**, 287–293.
23. Starck, S.R., Green, H.M., Alberola-Ila, J. and Roberts, R.W. (2004) A general approach to detect protein expression in vivo using fluorescent puromycin conjugates. *Chem. Biol.* **11**, 999–1008.
24. Jackson, R.J. and Hunt, T. (1983) Preparation and use of nuclease-treated rabbit reticulocyte lysates for the translation of eukaryotic messenger RNA. *Methods Enzymol.* **96**, 50–74.

Chapter 18

Ligand-Regulated Peptide Aptamers

Russell A. Miller

Abstract

The peptide aptamer approach employs high-throughput selection to identify members of a randomized peptide library displayed from a scaffold protein by virtue of their interaction with a target molecule. Extending this approach, we have developed a peptide aptamer scaffold protein that can impart small-molecule control over the aptamer–target interaction. This ligand-regulated peptide (LiRP) scaffold, consisting of the protein domains FKBP12, FRB, and GST, binds to the cell-permeable small-molecule rapamycin and the binding of this molecule can prevent the interaction of the randomizable linker region connecting FKBP12 with FRB. Here we present a detailed protocol for the creation of a peptide aptamer plasmid library, selection of peptide aptamers using the LiRP scaffold in a yeast two-hybrid system, and the screening of those peptide aptamers for a ligand-regulated interaction.

Key words: Peptide aptamers, yeast two-hybrid, scaffold, LiRP, ligand-regulation, selection.

1. Introduction

From its conception and early application the peptide aptamer approach has held promise as a route for the rapid generation of transdominant inhibitors of protein function *(1)*. In this approach an individual member of a library of random peptide sequences is selected for its ability to interact with a target protein. Both the relative ease with which peptide aptamers can be identified using selection methods such as the yeast two-hybrid and the high likelihood that a selected aptamer will bind in a functionally relevant site of a target protein suggest that this method will continue to prove useful. Traditionally the approach has utilized small inert scaffold proteins, including thioredoxin *(1)*, GFP *(2)*, staphylococcus nuclease *(3)*, and SteA *(4)* to display the randomizable peptide aptamer region. We have worked toward extending the

Günter Mayer (ed.), *Nucleic Acid and Peptide Aptamers: Methods and Protocols, vol. 535*

DOI 10.1007/978-1-59745-557-2_18 Springerprotocols.com

peptide aptamer approach by developing a scaffold that can control the presentation of the randomized peptide aptamer region with a cell-permeable, small-molecule ligand *(5, 6)*, creating a two-component system comprised of a small molecule effector and a transdominant protein modulator. This approach merges the benefits of small molecules, including the inhibition of single domains, exquisite temporal control of protein activity, and doseability of effects, with the powers of the genetically encoded, selection-based peptide aptamer approach, including tissue specific expression of a protein inhibitor and potential for the rapid generation of protein-targeting agents.

Our ligand-regulated peptide (LiRP) scaffold protein enables the cell permeable small molecule rapamycin to regulate the binding interactions of a peptide linker region with target proteins. The overall scaffold protein contains three protein domains: FK506 binding protein 12 (FKBP) *(7)*; FKBP-Rapamycin Binding domain (FRB) from the mammalian Target of Rapamycin protein (mTOR) *(7)*; and glutathione-S-transferase (GST). Peptides presented in the context of the FKBP-peptide-FRB-GST scaffold protein are free to interact with target proteins in the absence of rapamycin. In the presence of rapamycin, a complex forms between FKBP, rapamycin, and FRB that limits the conformational freedom of the peptide that links the c-terminus of FKBP with the n-terminus of FRB. In addition, the large GST protein domain acts to sterically occlude access to the peptide aptamer linker. Functionally, this architecture allows for the inhibition of aptamer–target interactions with cell permeable and biologically orthogonal rapamycin derivatives.

One of the great strengths of the peptide aptamer approach is its utilization of the yeast two-hybrid system and the wealth of highly optimized reagents and protocols for high-efficiency transformation and selection that exist for the technique. The yeast two-hybrid strain PJ69-4a *(8, 9)* was used as the parent strain for our studies and was chosen because it is optimized for selection experiments and enables the use of two independent growth selection markers (His6 and Ade2 for growth on media lacking histidine or adenine, respectively) allowing for a reduction in false positive hits. For the ligand-regulated peptide aptamer approach it is necessary to ensure that the ligand used is functionally inert at the concentrations at which it is delivered; for selection of rapamycin-regulated peptide aptamers we modified the PJ69-4a strain by introducing a genomic mutation in the endogenous TOR protein, TOR2 *(10, 11)*, which renders the strain resistant to the high concentrations of rapamycin necessary for the characterization of ligand-regulated phenotypes. This produced the yeast strain sBFB4 *(5)*, which was shown to grow in concentrations of Rapamycin greater than 1 μM.

The production of a good peptide aptamer plasmid library is one of the most crucial aspects of our approach, as the complexity and ratio of viable/nonviable members contribute greatly to the likelihood of successfully identifying peptide aptamers. For our peptide aptamer plasmid libraries we used very traditional methods, employing an oligonucleotide with degenerate nucleotides (corresponding to seven consecutive random aminoacids) that were flanked by restriction enzyme recognition sites (*see* **Fig. 18.1**). Using this oligo we could create double-stranded DNA using a complementary primer and DNA polymerase, cut the restriction sites, and ligate the cut, random insert in the plasmid backbone. This procedure required optimization, but

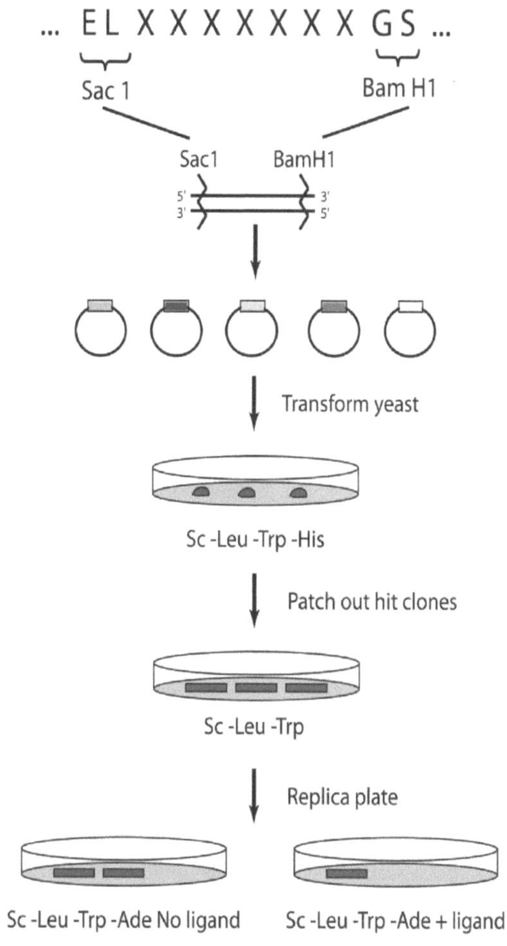

Fig. 18.1. This flow chart details the major steps in the selection of ligand-regulated peptide aptamers.

with a lot of care taken to prevent over or under digestion of the insert and plasmid we obtained libraries in which over 80% of the members had a single insert.

The key modification of our approach for the identification of ligand-regulated peptide aptamers is the combination selection/screen procedure that allows the rapid identification of peptide aptamers and characterization of the ability of the target–aptamer interaction to be modulated by rapamycin. This is accomplished by first performing a standard peptide aptamer selection by transforming the peptide aptamer library in to a yeast two-hybrid strain expressing the target protein of choice by the PEG/lithium acetate yeast transformation method (*see* **Note 1**) *(12)*. This leads to the auxotrophic selection of individual library members that interact with the target protein. These individual colonies are then regrown on new plates from which they can be screened for the ability to interact with the target protein in the presence of various concentrations of the scaffold-binding small molecule rapamycin (*see* **Fig. 18.2**) (*see* **Note 1**).

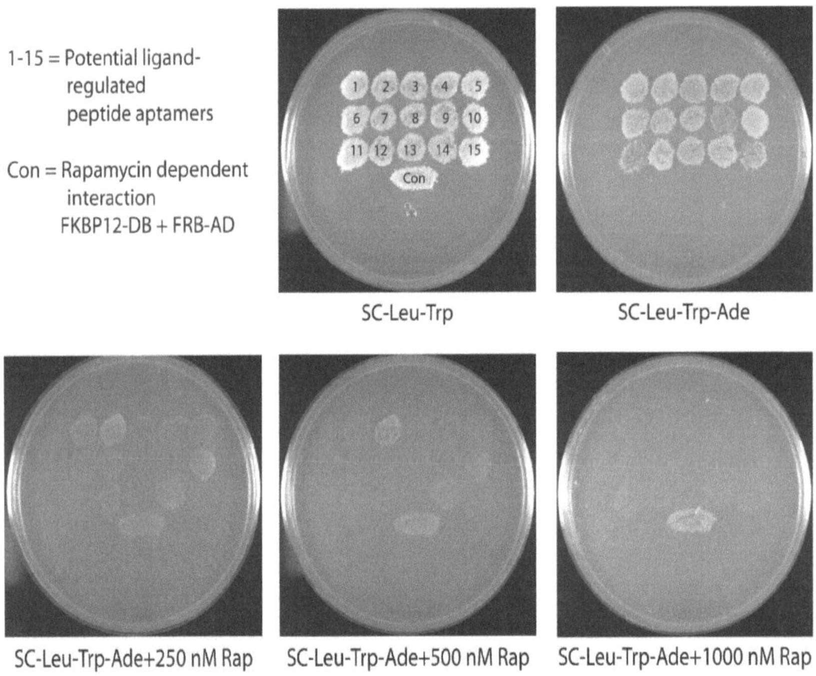

Fig. 18.2. 15 LiRP proteins targeting the AMPK $\alpha 2$ protein were transformed into yeast strain sBFB4. Yeast colonies were replica plated onto SC-Leu-Trp (plasmid retention), SC-Leu-Trp-Ade (selection media), and SC-Leu-Trp-Ade with increasing concentrations of rapamycin. Growth of LiRP strains and the rapamycin-dependent control strain is shown after 3 days of growth.

2. Materials

2.1. Degenerate Insert Synthesis

1. The degenerate oligo for a BamHI–SacI cloning reaction, with the sequence GTG CAT CGG GTT GAG CTC $(NNS)_7$ GGA TCC ACT GTA GGT CAC, where N equals a 1:1:1:1 mixture of A:G:C:T and S equals a 50/50 mixture of G or C (*see* **Note 2**), should be ordered with HPLC and/or PAGE purification to minimize the number of truncated clones.
2. The antisense oligo for the creation of a double-stranded degenerate DNA insert, with the sequence GTG ACC TAC AGT GGA TCC, should be ordered.
3. Klenow DNA polymerase, 10X reaction buffer, and 10 μM each dNTP's (Promega)
4. Phenol/chloroform/isoamyl alcohol (25:24:1) premixed solution and 100% ethanol (Sigma)
5. 3 M Sodium acetate (Sigma) dissolved in sterile water at pH 5.2 and filter sterilized

2.2. Restriction Digest of Insert and Vector (see Note 2)

1. BamHI (high concentration, 40 U/μL), SacI (high concentration, 40 U/μL), and Promega Buffer E (Promega) for restriction enzyme digests.
2. Electrophoresis grade agarose (Sigma), and TAE running buffer (40 mM Tris, 20 mM acetic acid, and 1 mM EDTA) and ethidium bromide (0.5 μg/mL final concentration) are used for the gel electrophoresis. Take care when handling the ethidium bromide, as it is a known mutagen.
3. QIAquick gel extraction kit (QIagen) for purification of cut DNA fragments.

2.3. Large-Scale Library Ligation and Transformation

1. T4 DNA ligase (high concentration, 20 U/μL) and 10x T4 DNA ligase buffer (Promega) for ligation reaction.
2. G-25 microspin desalting columns can be purchased from GE-Healthcare.
3. High-electroporation-efficiency bacterial Strain MC1061 can be obtained from the American Type Cuture Collection.
4. The ampicilin stock solution is made by dissolving ampicilin at a concentration of 50 mg/mL in water. Sterile filter this by passing through a 0.2 μm syringe filter into microcentrifuge tubes that can be stored at –20°C as a 1,000x stock solution.
5. LB broth is made by mixing 5 g yeast extract, 10 g bacto tryptone, and 10 g sodium chloride (all from Sigma) in 1 L of water and sterilized in an autoclave. LB agar is made by including 17 g of Agar (Sigma) in the autoclaved mixture, cooling the mixture in a 50°C water bath, adding 1 mL of the appropriate

1,000x antibiotic stock solution and pouring into sterile, plastic 10 cm petri dishes.

6. 1 L of SOC electroporation recovery media is made by adding 20 g bacto tryptone, 5 g yeast extract, 2 mL of 5 M sodium chloride, 2.5 mL of 1 M potassium chloride, 10 mL of 1 M magnesium chloride, 10 mL of 1 M magnesium sulfate, and 20 mL of 1 M glucose to 950 mL of water, and this is sterilized in an autoclave.
7. 0.1 cm gap electroporation cuvettes from Bio-Rad are used for the electroporation of bacteria. Store these at –20°C prior to electroporation to minimize heat shock of bacteria during electroporation.
8. Plasmid DNA Mega prep from Promega for the purification of library plasmid DNA.
9. For purification of plasmid DNA from individual library clones use a Promega Mini Prep kit.
10. For sequencing of plasmid DNA, use sequencing primer with the sequence CCTCCCGGGACAGAAACA, purchased from IDT DNA and submit to a DNA sequencing facility using the facilities concentration requirements.

2.4. Yeast Two-Hybrid Validation of Target Protein (see Note 3)

1. Yeast YPAD liquid medium is made by mixing 10 g yeast extract, 20 g bacto peptone, 24 mg adenine hemisulfate, and 50 mL of 40% glucose (20 g) mixed in a final concentration of 1 L with sterile water. This mixture is autoclaved for sterilization.
2. Make a 50% PEG solution by dissolving 50 g of PEG molecular weight 3,350 (Sigma) in a beaker with 35 mL of water and mix with a magnetic stir bar until all the PEG has gone into solution (*see* **Note 3**). Measure the amount of PEG solution and bring the volume up to 100 mL with water. Filter sterilize this solution to ensure sterility using a 0.2 μm vacuum filter apparatus.
3. Carrier DNA for the yeast two-hybrid transformation: dissolve 100 mg of salmon sperm DNA (sigma) in 50 mL TE buffer (20 mM Tris–HCl pH 8.0 and 1 mM EDTA) to make a final concentration of 2 mg/mL ssDNA. Ensure the DNA is fully dissolved by mixing with a stir bar overnight at 4°C. Sterile filter the ssDNA by filtration through a 0.2 μm vacuum filter and aliquot this solution in 1 mL aliquots that can be stored at –20°C.
4. 1 M Lithium acetate (Sigma): mix solid in milli-pure water and sterilize the solution by autoclaving.
5. 50x Stock solution of amino acid master mix (AAMM) for synthetic complete medium: mix 0.4 g each of uracil, arginine, methionine and inositol, 1.2 g of lysine, 1.6 g of isoleucine, 1.0 g of phenylalanine, 2.0 g each of glutamic acid

and aspartic acid, 3.0 g of valine, 4.0 g of threonine, and 7.5 g of serine in 400 mL of sterile water. Autoclave this mixture and store for less than 6 months at 4°C.

6. 50x Stock solutions of histidine (His), adenine (Ade), leucine (Leu), and tryptophan (Trp) (synthetic complete media): mix 1.0 g of histidine, 1.0 g of adenine, 4.0 g of leucine, and 8.0 g of tryptophan each with 1 L of sterile water in separate autoclavable bottles. Autoclave these mixtures and store them for less than 6 months at 4°C.
7. 50x Stock solution of tyrosine (Tyr) (synthetic complete media): mix 1.5 g of tyrosine and 16.67 mL of 6 N NaOH (the sodium hydroxide helps to solubilize the tyrosine) in 1 L of sterile water, autoclave this mixture, and store for less than 6 months at 4°C.
8. 20x Stock solution of glucose (40% solution): mix 200 g of glucose in 500 mL of sterile water and autoclaving for sterilization.
9. 1 L Liquid synthetic complete base medium (SC): mix 1.7 g yeast nitrogen base without amino acids and 5 g ammonium sulfate (both from Sigma) in 900 mL of sterile water. Autoclave this mixture and store at room temperature for less than 6 months. As needed, take a desired amount of SC base media and add glucose and the appropriate amino acids to a 1 × final concentration (for 100 mL SC-leu-trp media add 89 mL SC base, 2 mL AAMM, 2 mL His, 2 mL Ade, 2 mL Tyr, and 5 mL of 20x glucose solution).
10. 1 L of SC agar for selection plates: add 1.7 g yeast nitrogen base and 5 g ammonium sulfate (both from Sigma) in 400 mL of sterile water into a 1 L erlenmeyer flask. In a separate 2 L flask resuspend 20 g of agar (Sigma) in 500 mL of sterile water with a stir bar. Autoclave both flasks and then cool to 50°C. While stirring on a magnetic stir plate, add the SC base media into the 2 L flask containing the agar and add glucose solution and the appropriate amino acid mixtures to a 1 × final concentration. Pour this media into petri dishes and allow them to cool on the bench top. Wrap dishes in parafilm and store at 4°C for less than 6 months. For this section you will need both SC-trp and SC-leu-trp petri dishes.
11. 3 M Stock solution of 3-amino-1,2,4-triazole (3-AT): add 25.2 g of 3-AT to 100 mL of sterile water. Sterilize the solution by filtering through a 0.2 μm syringe filter. Store this stock solution in 5 mL aliquots at –20°C.
12. For selection experiments prepare SC-leu-trp-his+3 mM 3-AT as described above, with the addition of 1 mL of a 3 M stock solution of 3-AT per liter of media, added after cooling to 50°C.

13. For replica plating we used a Scienceware replica plater (Sigma #Z363391) with velvet squares cut from low pile velvet cloth. The velvet cloths were cleaned and reused by rinsing thoroughly in hot water with agitation, autoclaving submersed in water, spreading out on the benchtop to dry, piling and wrapping in tinfoil, followed by a final autoclave step with an extended drying cycle to ensure sterility and dryness.

2.5. Large-Scale Selection of Peptide Aptamers and Screen for Ligand-Regulated Phenotype

1. A 1 mM rapamycin stock solution (rapamycin can be purchased from Sigma) can be made by dissolving 1 mg of rapamycin in 1.09 mL of 100% ethanol. This solution can be stored at –20°C for 2 months.
2. For selection plates containing rapamycin we prepared the agar containing SC media as above. Immediately prior to pouring the plates the appropriate amount of a 1 mM stock solution of rapamycin was added to the media while mixing on a stir plate. These plates were then allowed to cool over night, wrapped in parafilm, and stored at 4°C until use (not more than 1 month). Due to the high cost of rapamycin, these plates should be made as necessary.

2.6. Isolation of Ligand-Regulated Peptide Aptamer Plasmid DNA

1. Yeast lysis buffer consists of 50 mM Tris–HCl, 50 mM EDTA, 10 mM β-mercaptoethanol, and 1.2 M sorbitol.
2. The enzyme lyticase was purchased from Sigma.

3. Methods

3.1. Library Creation

3.1.1. Degenerate Insert Synthesis

1. Mix the degenerate antisense oligonucleotide (4 nmols), the complementary annealing oligonucleotide (4 nmol), 8 μL of 10x Klenow DNA polymerase buffer, 8 μL of 10 mg/mL BSA, and 40 μL of 10 mM dNTPs in a final volume of 795 μL. Anneal the oligonucleotides by heating the reaction mixture to 95°C for 5 min followed by an incubation at 40°C for 30 min.
2. Begin the DNA extension reaction by adding 25 U (5 μL) of Klenow DNA polymerase to the tube containing annealed oligonucleotides from **Step 1.1** and incubate at room temperature for 2 h.
3. Purify the DNA from the extension step by adding an equal volume of phenol/chloroform/isoamyl alcohol (25:24:1) to the reaction. Vortex this mixture and then centrifuge in a microcentrifuge to separate the aqueous and organic layers and collect the aqueous layer.

4. Precipitate the extended DNA insert by adding 0.1 volumes sodium acetate (pH 5.2) and 2.5 volumes 100% ethanol, and incubate this mixture at −20°C for 30 min. Pellet the precipitated DNA by centrifugation at top speed in microcentrifuge at 4°C for 10 min. Decant the liquid and wash the DNA pellet three times with 70% ethanol. Air dry the pellet and redissolve the DNA in 100 μL sterile nuclease-free water.

3.1.2. Restriction Digest of Insert and Vector

1. Perform a restriction digest of the double-stranded library insert by mixing 50 μL of purified insert DNA, 10 μL of Promega buffer E, 10 μL of 10 mg/mL BSA, 20 μL of high-concentration BamHI (200 U), and 20 μL of high-concentration SacI (200 U) in a total volume of 1,000 μL. Incubate this reaction at 37°C for 5 h (*see* **Note 4**).
2. Purify the cut double-stranded DNA from this digestion with a phenol/chloroform extraction and ethanol precipitation, as described in **Section 3.1.1, Step 3–4**.
3. To selectively purify twice cut insert from singly cut or undigested insert, resolve the DNA insert fragments on a 4% agarose/TAE gel, in the presence of EtBr. Using approximately 1 cm combs in the agarose gel, run no more than ~5 μg of cut insert per well, to ensure that the resolution between uncut, singly cut, and doubly cut bands is adequate. Run this agarose gel at 50 V until sufficient separation exists between various DNA species (at least 1 h), as observed under long wavelength UV light (300–315 nm) (*see* **Note 5**). Excise the proper twice-cut DNA insert band from the gel and purify it using multiple QIAquick gel purification columns (follow the suggested binding capacity in QIAgen product manual, 10 μg). The DNA insert should finally be eluted in sterile nuclease-free water.
4. To generate the cut LiRP scaffold plasmid, assemble a restriction digest by mixing 200 μg of plasmid DNA, 10 μL of Promega buffer E, 10 μL of 10 mg/mL BSA, 4 μL of high-concentration BamHI (40 U), and 4 μL of high-concentration SacI (40 U) in a total volume of 1,000 μL. Incubate this reaction at 37°C for 8 h followed by an incubation at 65°C for 15 min to inactivate the enzymes.
5. Selectively purify cut LiRP plasmid by first resolving the DNA on a 2% agarose/TAE gel in the presence of EtBr. As with the insert purification, to ensure proper resolution of DNA species, load only 5 μg cut plasmid DNA per 1 cm well. Run the agarose gel at 50 V until the different DNA species can be resolved (at least 1 h) and cut out the doubly digested DNA band. Purify this cut LiRP plasmid using QIAquick gel purification columns, and elute the DNA in sterile, nuclease-free water (*see* **Note 5**).

6. Determine the concentration of both cut LiRP plasmid and cut DNA insert by absorbance at 260 nm. If the concentration of either the insert or plasmid is bellow 150 ng/μL, concentrate using a speed-vac to obtain sufficiently concentrated DNA for the subsequent ligation step.

3.1.3. Large-Scale Library Ligation and Transformation

1. Assemble a large-scale LiRP library ligation by mixing 30 μg of the cut LiRP plasmid, 3 μg of the cut DNA insert, 25 μL of 10x T4 DNA ligase buffer, and 10 μL of high-concentration T4 DNA ligase in a final reaction volume of 250 μL. Incubate the ligation reaction overnight at 16°C followed by an incubation at 65°C for 15 min to inactivate the enzymes. Desalt the ligation reactions by gel-filtration using GE-Healthcare G-25 microspin columns pre-equilibrated in sterile water (*see* **Note 6**).
2. Prepare electrocompetent *E. coli* strain MC1061 by first inoculating 1 L of LB medium with a 5 mL saturated overnight culture of the bacteria. Allow this culture to grow at 37°C to an OD_{600} nm of 0.8. Pellet this culture by centrifugation at 15,000 × *g* for 10 min at 4°C. Wash the pellet twice by resuspending it in 1L of sterile, ice-cold water followed by centrifugation at 15,000 × *g* for 10 min at 4°C. Resuspend the pellet in 1 mL of sterile, ice-cold water and centrifuge at 15,000 × *g* in a microcentrifuge for 10 min at 4°C. Resuspend the bacterial pellet in an equal volume of sterile, ice-cold water and place on ice.
3. Set up multiple electroporations (10–30 per ligation depending on bacterial yield and ligation volume) by mixing 85 μL electrocompetent bacterial suspension with 15 μL of desalted ligation mixture on ice. Immediately prior to electroporation, add bacteria/DNA mixture to a precooled (–20°C) electroporation cuvette and electroporate using a Gene Pulser II apparatus (Bio-Rad) with the following parameters: field strength 20 kV/cm; resistance 200 Ω; capacitance 25 μF. Immediately following electroporation, resuspend cells in cuvette by adding 1 mL of pre-warmed, 37°C SOC media. Transfer cell suspension from the electroporation cuvette to a 10 mL test tube and incubate in a 37°C shaker for 1 h (*see* **Note 7**).
4. Combine all independent ligations in 1 L of LB media containing 50 μg/mL of Ampicilin and incubate in a shaking 37°C incubator. Allow 10 min of shaking to sufficiently mix the culture and take a 1 mL aliquot that can be used for library size and complexity measurements. Perform serial dilutions from this aliquot, diluting as far as 10^{-10} of the total culture. From these dilutions plate 100 μL onto two LB–Ampicilin petri dishes and allow them to grow overnight at 37°C. Shake the 1 L LB flask with the electroporated cells overnight at 37°C to amplify the plasmid library.

5. After overnight amplification of the plasmid library, harvest the library DNA with a Mega prep kit, using two columns to ensure total recovery of plasmid DNA.
6. Determine the library size (independent library members) by counting the number of colonies on serially diluted LB–Ampicilin plates, multiplying the number of colonies by the fold dilution of the total culture that was plated on that dish and averaging the library size value from the multiple LB–Ampicilin plates. Isolate 40 individual library members from these plates and grow them overnight in LB–Ampicilin liquid media in a shaker at 37°C. Isolate plasmid DNA from these clones using a mini prep kit (Promega Wizard Mini Prep Kit). Using the sequencing primer submit all 40 clones for sequencing to assess the insert efficiency for the library ligation (*see* **Note 8**).

3.2. Target Validation

3.2.1. Yeast Two-Hybrid Validation of Target Protein (*see* **Note 4**)

To validate a chosen target protein for the peptide aptamer approach it is important to verify that the target protein can function in the yeast two-hybrid system; this requires that the selected target protein does not independently activate transcription when fused to a DNA binding domain and, ideally, that a functional interaction can be demonstrated in the yeast two-hybrid with a known binding protein as the activation domain fusion. As an example here, we will show the validation and screening for the AMPK $\alpha 2$ target protein. This protein has previously been shown to interact with the AMPK$\beta 1$ protein by yeast two-hybrid *(13)* and this interaction will serve as the control interaction verifying the functionality of the target protein in the yeast two-hybrid.

1. One day prior to small-scale transformations, inoculate 100 mL of YPAD media with the yeast two-hybrid strain sBFB4 and grow in an incubated shaker at 30°C.
2. After overnight growth split the culture into two 50 mL conical tubes and centrifuge at 3,000 × *g* for 5 min. Decant the media and resuspend both pellets in 20 mL of sterile water and combine the tubes. Pellet the yeast again by centrifugation at 3,000 × *g* for 5 min. Decant the water and resuspend the pellet in 20 mL of sterile water. Centrifuge the yeast again at 3,000 × *g* for 5 min. Decant the water and resuspend the pellet in 1 mL of sterile water and move this yeast solution into a sterile microcentrifuge tube. Centrifuge the yeast again at 3,000 × *g* for 5 min, decant the water, and resuspend the yeast pellet in an equal volume of sterile water.
3. For each individual transformation (*see* **Table 18.1**), add 100 μL of the yeast suspension to a microcentrifuge tube. Centrifuge the yeast in these tubes at 3,000 × *g* for 5 min and decant the water leaving only the yeast pellet.

Table 18.1
Small-scale transformations

Transformation	Bait plasmid	Prey plasmid
1	Empty vector (pGBKT7)	Empty vector (pGAD-GH)
2	AMPK alpha 2 – pGBKT7	Empty vector (pGAD-GH)
3	Empty vector (pGBKT7)	AMPK beta 1 – pGAD-GH
4	AMPK alpha 2 – pGBKT7	AMPK beta 1 – pGAD-GH
5	FKBP12 – pGBKT7	FRB – pGAD-GH
6	AMPK alpha 2 – pGBKT7	Nothing

4. Each small-scale yeast transformation will contain 270 μL of 50% PEG 3350 solution, 30 μL of 1 M lithium acetate, 50 μL of 2 mg/mL salmon sperm carrier DNA, and 5 μL of a bait and prey plasmid DNA. To set up six different transformations you should premix enough PEG 3350, lithium acetate, and carrier DNA for 10 reactions (2.7 mL, 300, and 500 μL, respectively) mix thoroughly, and add 350 μL of this solution to each transformation tube containing 50 μL of washed yeast. To these transformation tubes you can then add 5 μL of bait and prey plasmids (DNA from a mini prep is sufficiently concentrated).
5. Mix each transformation by vortexing (*see* **Note 9**) and incubate for 1 h at 42°C.
6. After the heat shock, centrifuge the transformations at 3,000 × *g* for 10 min, decant the supernatant, and resuspend yeast pellet in 500 μL sterile water. Plate 100 μL of transformations 1–5 onto SC-leu-trp plates and 100 μL of transformation 6 onto a SC-trp plate. Incubate the plates at 30°C until colonies have formed (2–3 days). Once colonies have grown to sufficient size (1–3 mm in diameter) the plates can be taken out of the incubator and stored at 4°C for use in subsequent experiments.
7. Using an inoculation loop, patch individual colonies onto a new SC-leu-trp plate, ultimately moving 2–4 colonies from transformations 1–5 onto the master SC-leu-trp plate. Allow patches to grow for 2–3 days in a 30°C incubator.
8. Using a sterile velvet cloth and a replica plating apparatus transfer yeast patches from the master SC-leu-trp plate

onto: a new SC-leu-trp plate, an SC-leu-trp-ade selection plate, and an SC-leu-trp-his+3 mM 3-AT selection plate. Following replica plating, use a new sterile velvet cloth and the replica plating apparatus to remove excess transferred yeast from replica plates (*see* **Note 10**). Allow the replica plates to grow for 2–5 days at 30°C, taking images or noting the timing of growth of particular strains.

3.3. Large-Scale LiRP Selection/Screen

3.3.1. Large-Scale Selection of Aptamers and Screen for Ligand-Regulated Phenotype

1. For the large-scale transformation of the LiRP library you will use single yeast sBFB4 colonies expressing AMPK alpha 2 (transformation 5 from **Section 2.1**) as the recipient strain. One day prior to the large-scale transformation inoculate an overnight culture of yeast strain 5 into 100 mL of SC-trp liquid culture in a 250 mL erlenmeyer flask. Incubate this culture overnight at 30°C in a shaking incubator.
2. The following morning, determine the OD_{600} of the overnight culture and inoculate a 300 mL culture of pre-warmed 2X YPAD liquid media to a starting OD_{600} of 0.4. Allow this culture to grow to a final OD_{600} of 1.0 (approximately 6–8 h).
3. Pellet this yeast culture by centrifugation at 3,000 × *g* and resuspend the yeast pellet in 50 mL of sterile water. Repeat this washing step 1 additional time.
4. Prepare a the PEG–lithium acetate transformation mix by combining 7.2 mL 50% w/v PEG 3350, 1.08 mL 1 M lithium acetate, 1.08 mL carrier DNA, and 150 μg of library plasmid DNA. Resuspend the washed yeast pellet with this transformation mixture and aliquot 1 mL of the resuspended solution into sterile microcentrifuge tubes. Incubate transformations in a 42°C water bath for 1 h, keeping the yeast in suspension by frequently inverting the tubes (*see* **Note 11**).
5. Recombine the individual transformation tubes into a 50 mL conical tube and pellet the yeast by centrifugation at 3,000 × *g* for 15 min. Resuspend this yeast pellet in 30 mL of 1x YPAD media and incubate at 30°C in a shaking incubator for 30 min. Following this recovery period, pellet the yeast by centrifugation at 3,000 × *g* for 5 min and wash this pellet with 30 mL of sterile water. Centrifuge the yeast again and resuspend the pellet in 12 mL of sterile water.
6. Take a small aliquot of the resuspended, transformed yeast and serially dilute them into sterile water, plating two plates each that represent 10^{-4}, 10^{-5}, and 10^{-6} of the library onto SC-leu-trp plates. These plates will allow you to assess the size of the transformation.
7. With the rest of the 12 mL of transformed yeast, plate 200 μL of the suspension onto 60 individual 10 cm petri dishes containing SC-leu-trp-his+3 mM 3-AT agar media. Spread

the transformed cell solution on the plates using sterilized 1 mm glass beads (approximately six beads per plate), moving the plates back and forth on the benchtop to ensure complete coverage of the plate surface. Incubate these plates in a 30°C incubator for 3–14 days (colonies should form by day 4–5).

8. Once individual colonies form on the selection plates move them from the selection plates with a sterile inoculation loop and plate them in patches onto plasmid retention SC-leu-trp plates. In addition to selected colonies, on each plate include yeast strains 4 and 5 from **Section 2.1**, controls for, respectively, ligand-independent interaction and a ligand-dependent interaction. Incubate these plates in a 30°C incubator for 3 days or until robust yeast patches are present for ligand-selection replica plating.
9. Replica plate selected peptide aptamers onto SC-leu-trp-ade, SC-leu-trp-ade+1 μM Rapamycin, and SC-leu-trp plates using sterile velvet cloth and a replica plating apparatus. Replica clean plates with sterile velvet clothes until no visible yeast debris remain. Incubate these plates in a 30°C incubator, imaging with digital camera to document the growth of the various yeast patches in the presence or absence of rapamycin.

3.3.2. Isolation of Ligand-Regulated Peptide Aptamer Plasmid DNA

1. To isolate the plasmid DNA from yeast strains, first inoculate 5 mL of SC-leu-trp liquid cultures with the potential hit strains identified in **Section 3.1**. Grow these cultures to saturation in a shaking incubator at 30°C overnight.
2. Pellet 2 mL of the overnight culture by centrifugation at 3,000 × g and resuspend the yeast in yeast lysis buffer containing 200 U of Lyticase and incubate at 37°C in a rotary shaker overnight to digest yeast cell walls and form yeast protoplasts.
3. Fully lyse these cells and purify the plasmid DNA using a promega DNA mini prep kit. Using this purified DNA, transform chemically competent DH5α bacteria and plate on LB–Amp plates to select for peptide aptamer plasmids. Amplify individual colonies and isolate DNA from 5 mL cultures using a Promega mini prep kit.
4. Submit DNA for sequencing with the insert sequencing primer from 2 to 4 individual clones from each putative ligand-regulated peptide aptamer yeast patch.

3.4. Hit Confirmation and Characterization

1. Using the same small-scale yeast transformation protocol outlined in **Section 3.2.1, Step 1–6**, transform all potential ligand-regulated peptide aptamers into the sBFB4 yeast strain with either the target protein–DNA binding domain fusion or an empty pGBKT7 vector as a negative control.

2. Using an inoculation loop, patch individual colonies onto a new SC-leu-trp plate, ultimately moving 2–4 colonies from each transformation onto master SC-leu-trp plate (multiple plates may be necessary to include all hits). In addition to the putative hit strains and negative controls, patch transformation strains 4 and 5 from **Section 2.1** onto each master SC-leu-trp plate as ligand-independent and ligand-dependent control interaction strains. Allow the patches to grow for 2–3 days.
3. Using a sterile velvet cloth and a replica plating apparatus transfer yeast patches from the master SC-leu-trp plate onto: a new SC-leu-trp plate, an SC-leu-trp-ade selection plate, and SC-leu-trp-ade plates containing 250, 500, and 1,000 nM Rapamycin. Following replica plating, use a new sterile velvet cloth and the replica plating apparatus to remove excess transferred yeast from the replica plates. Allow the replica plates to grow for 2–5 days at 30°C, taking images or noting the timing of growth of particular strains (*see* **Fig. 18.2**, for an example of growth after 3 days).

4. Notes

1. In addition to the cited reference, the Gietz Lab has a very useful web site (http://home.cc.umanitoba.ca/~gietz/) that describes in detail many of the aspects of the yeast transformation protocols described here, which are adapted from the Gietz protocols.
2. The use of G as the third wobble codon in the degenerate oligonucleotide is essential, as it reduces the number of potential stop codons from 3 to 1, while still maintaining the possibility for all 20 aminoacids.
3. This will take a while; make sure to set the bar at a slow speed to minimize air bubble formation, as this will make it difficult to determine if the PEG is in solution.
4. The timing of this reaction is crucial due to the detrimental effects of having uncut or inappropriately cut restriction sites, as these inappropriate events will decrease the effective number of clones in the library. One aspect of this is maintaining the appropriate concentration of reaction buffer throughout the reaction. This can be done using a PCR incubator or other device that employs a heated lid, preventing the concentration of the reaction due to condensation. This concentration can also increase the star activity (cutting of non-consensus sites) of the enzymes by increasing the effective concentration of glycerol in the reaction.

5. It is important in the purification of digested DNA fragments that a low amount of DNA is run in each lane; if too much DNA is loaded it leads to a streaky pattern that will prevent the precise excision of a single band. Also, care should be taken to expose the DNA to as little UV light as possible to reduce the amount of degradation.
6. Purification of the excess salts and buffers from the ligation mixture is essential for the use of the ligated DNA produced in the subsequent electroporation step. Without this purification the electroporation will fail due to arcing within the cuvette, leading to overheating and a non-uniform electrical charge.
7. Electroporation of DNA into the cell is extremely harsh on bacteria; to minimize the number of cells lost following electroporation the pre-warmed SOC media should be added immediately after electroporation.
8. While the use of restriction digest analysis can be used to assess the insert efficiency in the library, we prefer the use of sequencing as it allows a more detailed view of not only the randomness of the inserts but a view of the potential errors that create aberrant library members that can be addressed in future libraries. In early libraries we identified that a non-consensus BamHI site with five of the six nucleotides similar with the GGATCC recognized by BamHI was being cut in a small percentage of clones. This allowed us to pay careful attention to preventing this star activity.
9. It is very important to thoroughly mix the yeast into the PEG/LiOAc transformation mixture. Once it is completely mixed no yeast clumps or particulates will be visible. If vortexing does not work for this you can pipette up and down using a 1 mL micropipettor.
10. There is a fine line between the removal of excess yeast debris and the complete removal of the transferred yeast. A good rule should be that no yeast is visible when observing the plate from the bottom side, but that upon close inspection of the surface of the agar plate the individual patches are just visible. It is important to practice this technique with known interacting partners and negative controls to make certain patches behave as expected prior to observing experimental clones.
11. Ensuring that the yeast transformation tubes are evenly mixed throughout the 1 h heat shock step will help to increase your transformation efficiency. The yeast will settle out of solution considerably in 10 min, so a simple inversion or two every 5–10 min is necessary to maintain their even distribution.

Acknowledgments

I would like to acknowledge my research advisor Dr. Peter Belshaw, who conceived of the ligand-regulated peptide aptamer approach and provided me with continued support during the development and implementation of the approach. I would also like to thank Dr. Brock Binkowski, who worked collaboratively with me throughout the development and implementation of the LiRP approach and provided advice with the preparation of this manuscript. I would also like to thank the NIH (GM065406) for funding to Dr. Belshaw.

References

1. Colas, P., Cohen, B., Jessen, T., Grishina, I., McCoy, J. and Brent, R. (1996) Genetic selection of peptide aptamers that recognize and inhibit cyclin-dependent kinase 2. *Nature* **380**, 548–550.
2. Abedi, M.R., Caponigro, G. and Kamb, A. (1998) Green fluorescent protein as a scaffold for intracellular presentation of peptides. *Nucleic Acids Res.* **26**, 623–630.
3. Norman, T.C., Smith, D.L., Sorger, P.K., Drees, B.L., O'Rourke, S.M., Hughes, T.R., Roberts, C.J., Friend, S.H., Fields, S. and Murray, A.W. (1999) Genetic selection of peptide inhibitors of biological pathways. *Science* **285**, 591–595.
4. Woodman, R., Yeh, J.T., Laurenson, S. and Ko Ferrigno, P. (2005) Design and validation of a neutral protein scaffold for the presentation of peptide aptamers. *J. Mol. Biol.* **352**, 1118–1133.
5. Binkowski, B.F., Miller, R.A. and Belshaw, P.J. (2005) Ligand-regulated peptides: a general approach for modulating protein-peptide interactions with small molecules. *Chem. Biol.* **12**, 847–855.
6. Miller, R.A., Binkowski, B.F. and Belshaw, P.J. (2007) Ligand-regulated peptide aptamers that inhibit the 5'-AMP-activated protein kinase. *J. Mol. Biol.* **365**, 945–957.
7. Choi, J., Chen, J., Schreiber, S.L. and Clardy, J. (1996) Structure of the FKBP12-rapamycin complex interacting with the binding domain of human FRAP. *Science* **273**, 239–242.
8. James, P. (2001) Yeast two-hybrid vectors and strains. *Methods Mol. Biol.* **177**, 41–84.
9. James, P., Halladay, J. and Craig, E.A. (1996) Genomic libraries and a host strain designed for highly efficient two-hybrid selection in yeast. *Genetics* **144**, 1425–1436.
10. Heitman, J., Movva, N.R. and Hall, M.N. (1991) Targets for cell-cycle arrest by the immunosuppressant rapamycin in yeast. *Science* **253**, 905–909.
11. Lorenz, M.C. and Heitman, J. (1995) TOR mutations confer rapamycin resistance by preventing interaction with FKBP12-rapamycin. *J. Biol. Chem.* **270**, 27531–27537.
12. Gietz, R.D. and Woods, R.A. (2002) Transformation of yeast by lithium acetate/single-stranded carrier DNA/polyethylene glycol method. *Methods Enzymol* **350**, 87–96.
13. Thornton, C., Snowden, M.A. and Carling, D. (1998) Identification of a novel AMP-activated protein kinase beta subunit isoform that is highly expressed in skeletal muscle. *J. Biol. Chem.* **273**, 12443–12450.

Chapter 19

Isolation of Peptide Aptamers to Target Protein Function

Luisa Lopez-Ochoa, Tara E. Nash, Jorge Ramirez-Prado, and Linda Hanley-Bowdoin

Abstract

Peptide aptamers are small recombinant proteins typically inserted into a supportive protein scaffold. These short peptide domains can bind to their target proteins with high specificity and affinity, often resulting in an altered target protein. We describe high-throughput protocols that facilitate the selection and characterization of peptide aptamers from yeast dihybrid libraries. These protocols include the preparation and evaluation of the bait fusion and the peptide aptamer screen. They also include confirmation of interaction specificity as well as isolation and sequencing of peptide inserts. Once the amino acid sequence is determined, we describe a protocol for aligning and comparing short peptide sequences and assessing the statistical significance of the alignments.

Key words: Peptide aptamer, protein–protein interaction, yeast two-hybrid assay, library screen, high-throughput protocol.

1. Introduction

Peptide aptamers are recombinant proteins selected for specific binding to a target protein *(1)*. They have been applied to basic research, therapeutics and drug development, and used to modify protein function in bacterial, yeast, animal and plant systems *(2)*. Peptide aptamers generally consist of a short peptide domain inserted into a supporting protein scaffold that enhances specificity and affinity by conformationally constraining the peptide. Constrained aptamers can bind to their targets with 10^2–10^3 fold higher affinity than unconstrained peptides *(3, 4)*. Peptide aptamers are typically identified in vivo using stringent yeast dihybrid conditions. The in vivo selection process enhances the probability that the peptide aptamers will be stably expressed and correctly folded and will interact with their targets in an

Günter Mayer (ed.), *Nucleic Acid and Peptide Aptamers: Methods and Protocols, vol. 535*

DOI 10.1007/978-1-59745-557-2_19 Springerprotocols.com

intracellular context *(5)*. In this chapter, we describe high-throughput protocols that greatly facilitate the selection and characterization of peptide aptamer isolated in yeast dihybrid screens (*see* **Fig. 19.1**).

The first step in a peptide aptamer screen is to choose a peptide library that best meets the needs of the project. Peptide aptamer libraries vary with respect to the choice of scaffold, peptide length, selection stringency, and the number of selectable markers *(6, 7)*. Some are available upon request *(4)* and other can be purchased (Clontech). The choice of library and plasmids used in the screen will dictate the selection media for plasmids and

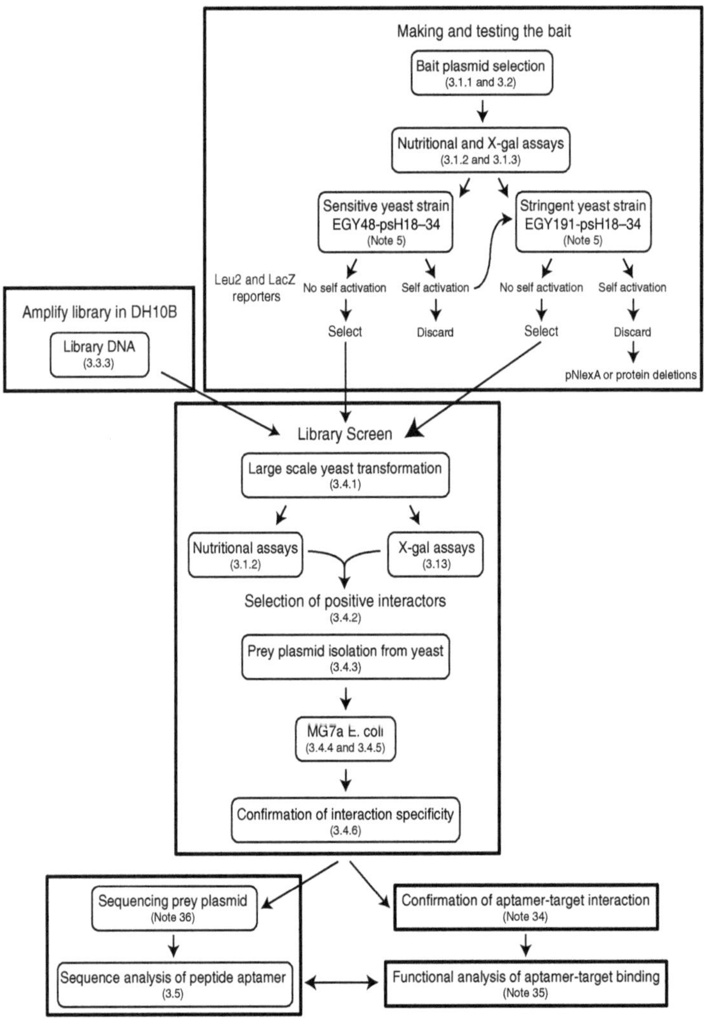

Fig. 19.1. Flow chart of peptide aptamer screen.

protein interactions. The protocols described here were developed using the pJM-1 library, which is based on the LexA dihybrid system *(4)*. The pJM-1 library encodes 20 amino acid peptides inserted into the active site of *E. coli* Thioredoxin (TrxA) *(8)*. The TrxA-peptides are fused to the SV40 nuclear localization signal, the *E. coli* B42 activation domain (AD) and the hemagglutinin (HA) epitope tag *(4)*. The peptide aptamers are expressed from the yeast GAL1 promoter, which is repressed by glucose and activated by galactose. The bait plasmids contain the LexA DNA binding domain (DBD). Bait strains such as EGY48, EGY42 and EGY191 carry the *Leu2* mutation and a *LEU2* reporter cassette under the control of the LexA promoter. The bait, prey (or library) and *LacZ* plasmids are selected in medium lacking histidine (-H), tryptophan (-W) or uracil (-U), respectively. Yeast colonies expressing peptide aptamers that interact with the bait protein grow on medium supplemented with galactose and lacking leucine (-L). The *LacZ* gene encoding β-galactosidase carried on the pSH18-34 *(8)* plasmid serves as a second reporter for interactions *(9)*. Yeast cells expressing an interacting peptide aptamer and the bait protein can be monitored for β-galactosidase activity in the presence of a substrate that produces a blue color upon cleavage *(10)*.

Typically, a large number of putative colonies are recovered in a peptide aptamer screen, and it is essential to use a high-throughput strategy to identify those that bind specifically to the target protein. We describe here the adaptation of yeast two-hybrid protocols to the 96-well format for this purpose. These protocols include preparation and evaluation of the bait fusion and the peptide aptamer screen. They also include confirmation of interaction specificity as well as isolation and sequencing of peptide inserts. Once the amino acid sequence is determined, we describe a protocol for aligning and comparing short peptide sequences and assessing the statistical significance of the alignments. There is often a sequence redundancy among a subset of the peptide aptamers that reflects "hot spots" in the target. Some proteins contain regions better suited for protein–protein interactions and are highly "aptamerogenic" *(6, 11)*.

In some instances, proteins that are positive for interaction in yeast dihybrid assays do not interact directly and, instead, a third protein acts as a bridge between the bait and the prey. Hence, it is advisable to validate binding of the peptide aptamer to the target protein by an independent method, such as copurification or coimmunoprecipitation. Once the interaction is confirmed, the next step is to evaluate the effect of the peptide aptamer on the function of the target protein in a relevant biological assay. Generally, functional screening of large numbers of candidate aptamers is facilitated by the use of in vitro or transient assays, if available. Stable transformants that express the peptide aptamers can also be evaluated, but this approach is more labor intensive and is better suited as secondary screen. The functional information can be used to classify the peptides and facilitate the identification of conserved motifs. Given the diverse nature of biological systems, specific description of functional assays is beyond the scope of this chapter.

In the absence of functional information about the target protein, peptide aptamers can be used as "mutagens" to study protein function. Their interaction sites on the target protein can be mapped and correlated with phenotypic effects. For example, the peptide aptamers might mimic authentic interactions of the target protein and their sequences can be used to aid in the identification of bona fide partners.

2. Materials

2.1. Making and Testing the Bait

1. Yeast strains: *Saccharomyces cerevisiae* strains EGY48 (*MAT his3 trp1 ura3-52 leu2::LexA6op-*) and EGY191 (*MAT his3 trp1 ura3-52 leu2::LexA2op-LEU2*) (Invitrogen, *(5)*).
2. Yeast plasmids; Bait plasmids pEG202 *(12)*, pNLexA (Origin, *(13)*). For fusions in the N-terminus/C-terminus of LexA, respectively.
3. Drop out aminoacid 20X stock solution (-HLWU): Made from the amino acids in **Table 19.1**. Weigh the indicated amount and dissolve in distilled water (dH_2O). You might

Table 19. 1
Drop out medium amino acid stocks

	20X (mg/L)	Sigma Cat #
L-Adenine hemisulfate salt	400	A-9126
L-Valine	3,000	V-0500
L-Arginine HCl	400	A-8094
L-Isoleucine	600	I-7403
L-Lysine HCl	600	L-8662
L-Methionine	400	M-9625
L-Phenylalanine	1,000	P-2126
L-Threonine	4,000	T-8625
L-Tyrosine	600	T-3754

Table 19.2.
Individual nutrient solutions

	100X (mg/L)	Sigma Cat #
L-Uracil	2,000	U-0750
L-Histidine HCl monohydrate	2,000	H-5669
L-Tryptophan	2,000	T-0254
L-Leucine	10,000	L-8912

need to warm the solution to completely dissolve the amino acids. Adjust volume to 1 L and autoclave.

4. Individual aminoacid 100X stock solutions for Histidine (H), Tryptophan (W), Uracil (U) and Leucine (L): Weigh the amount of the particular aminoacid indicated in **Table 19.2** and dissolve in dH_2O. Autoclave to sterilize (*see* **Note 1**).
5. Stock solutions to supplement synthetic drop out mediums: 40% (w/v) glucose (dextrose, Fisher, D-16-3), 40% (w/v) galactose (D-galactose 99% minimum, Sigma, G0750-550G), 40% (w/v) raffinose (Raff. Difco). Dissolve required amount in dH_2O. Autoclave at 121°C for 15 min.
6. Synthetic drop out medium (DO): 6.7 g yeast nitrogen base without amino acids (Difco), 50 mL drop out aminoacid 20X stock solution (-HLWU), 600 mL dH_2O. Mix the ingredients (except agar), Adjust pH to 5.6 with 10 N NaOH and volume to 1 L with dH_2O. Add 20 g Bacto Agar (Difco) for solid medium. Autoclave. When ready to use, supplement with the required aminoacid (100X aminoacid stock solution) and the appropriate carbon source to a final concentration of 2% glucose (v/v) or 2% galactose/1% raffinose (v/v) (*see* **Note 2** and **Table 19.3**).
7. YPD medium: 10 g Bacto-yeast extract, 10 g Bacto-peptone, 900 mL dH_2O. Mix ingredients, adjust volume to 1 L with dH_2O. Autoclave. Cool at 55°C and supplement with 50 ml 40% glucose.

2.2. Testing the Bait for Self-Activation

1. Yeast strains EGY42 or EGY191 (Invitrogen, *(5)*).
2. *LacZ* reporter plasmid pSH18-34 (Invitrogen, *(8)*).
3. DO media: Glu-U; Glu-HWU.
4. *See* **Section 2** for small-scale yeast transformation.

2.2.1. Nutritional Assays

1. DO media (agar): Glu-HWU, Gal/Raff-HWU, Glu-HLWU, Gal/Raff -HLWU.
2. Nunc omnitrays.

Table 19. 3
Drop out media

Supplement	Gal/Raff-HU	Gal/Raff-HWU	Gal/Raff-HLWU	Glu-HU	Glu-HWU	Glu-HLWU
Galactose	+	+	+	–	–	–
Raffinose	+	+	+	–	–	–
Glucose	–	–	–	+	+	+
Uracil	–	–	–	–	–	–
Histidine	–	–	–	–	–	–
Tryptophan	+	–	–	+	–	–
Leucine	+	+	–	+	+	–

3. Flat toothpicks.
4. Sterile water.
5. 96-well round bottom culture plates with lid (Costar Cat. No. 3799).
6. Multi-channel pipettes (10 and 200 μL, 1 mL).
7. Numbered well orienter for 96-well plates (USA Scientific).

2.2.2. β-Galactosidase/ Filter Lift X-Gal Assays

1. Z-buffer: To prepare 1 L mix the following; 16.1 g Na_2HPO_4-7 H_2O, 5.5 g NaH_2PO_4-H_2O, 0.75 g KCl, 0.246 g $MgSO_4$-7 H_2O. Adjust pH to 7 and autoclave.
2. X-gal stock: 5-bromo-4-chloro-3-indolyl-b-D-galactopyranoside. Dissolve X-gal in N,N-dimethylformamide (DMF) at a concentration of 20 mg/mL. Store at –20°C protected from the light.
3. X-gal/Z-buffer solution: Prepare at the moment of use: 1.67 mL X-gal stock, 100 mL Z buffer, 0.27 mL β-mercaptoethanol.
4. Whatman qualitative circle (12.5 cm grade 5, Cat. No. 1005 125).
5. Forceps.
6. Liquid nitrogen.

2.3. Small-Scale Yeast Transformation with LiAc

1. Yeast strain (EGY48 or EGY191).
2. DO medium or YPD.
3. 15 and 50 mL Conical centrifuge tubes.
4. 10X TE: 100 mM Tris–HCl, 10 mM EDTA, pH 8.0. Autoclave.
5. 10X LiAc stock solution: 1 M lithium acetate; Adjust to pH 7.5 with diluted acetic acid. Autoclave.

6. 50% (w/v) PEG 4000 (Merk-Schuchardt). Autoclave.
7. PEG/LiAC solution: 40% PEG 4000, 1X TE buffer, 1X LiAc. Prepare prior to use.
8. Salmon sperm carrier DNA Stock (Sigma, D-1626). Extract with phenol–chloroform, quantify and store at −20°C in 1 mL aliquots (approximately 2 mg/mL) *(14)*. Denature before use by boiling for 5 min and cooling on ice.
9. Thermo Scientific Barnstead Labquake® Rotisserie or Nutator mixer.

2.4. Library Amplification

1. Yeast two-hybrid peptide aptamer library DNA: In this book chapter "pJM-1" *(4)*.
2. Electro-competent cells: *E. coli* strain DH10B (10 × 10^{10}efficiency, Invitrogen).
3. SOC medium (1 L): Add 20 g Bacto Tryptone, 5.5 g Bacto yeast extract, 2 mL 5 M NaCl, 2.5 mL 1 M KCl, 10 mL 1 M $MgCl_2$, 10 mL 1 M MgSO4, and 20 mL 1 M glucose, to 900 mL distilled water (dH_2O). Adjust volume to 1 L with dH_2O. Autoclave 15 min at 120°C.
4. TB medium: Prepare two solutions for 1 L media: (a) Mix 12 g Bacto Tryptone, 24 g Bacto yeast extract and 4 mL glycerol. Adjust volume to 900 mL with dH_2O. Autoclave 15 min at 120°C. (b) Phosphate solution: 2.3 g KH_2PO_4 and 12.5 g K_2HPO_4. Adjust volume to 100 mL with dH_2O and autoclave. Cool media to 55°C and mix with phosphate solution.
5. LB medium: Add the following to 800 ml water: 10 g peptone, 5 g yeast extract and 10 g NaCl. Adjust pH to 7.5 with NaOH. Adjust volume to 1 L with dH_2O. Add 20 g Bacto agar to solid media. Autoclave 15 min at 120°C.
6. Carbenicillin: Prepare a 100 mg/mL stock. Sterilize by filtration. Store at −20°C.
7. Electroporation cuvettes (0.1 cm gap).
8. Electroporator.
9. QIAfilter plasmid maxi kit (Qiagen).

2.5. Screening

2.5.1. Large-Scale Yeast Transformation

1. DO liquid media: Glu-HU, Gal/Raff–HU.
2. DO solid media: Gal/Raff -HLWU and Glu-HWU.
3. 200 mm Petri dishes (Becton Dickinson Labware).
4. 15 and 50 mL conical centrifuge tubes.
5. DMSO (Dimethyl sulfoxide).

2.5.2. Selection of Positive Interactors

1. Nunc omnitrays.
2. DO media: Glu-HWU, Gal/Raff-HWU, Glu-HLWU, Gal/Raff-HLWU.

3. Flat toothpicks.
4. Glycerol/$MgSO_4$solution: 65% (v/v) glycerol, 100 mM $MgSO_4$, 25 mM Tris–HCl, pH 8. Autoclave.
5. Numbered well orienter for 96-well plates (USA Scientific, 9193-5500).
6. Foil PCR adhesive film.
7. Multi-channel pipettes (10 and 200 μL, 1 mL).
8. Sterile water.
9. 96-well round bottom culture plates with lid (Costar Cat. No. 3799).

2.5.3. Yeast Plasmid Isolation

1. Costar 2 mL 96-well assay block.
2. DO Medium: Glu-HWU.
3. Microplate replicator (Boekel, Cat. No. 140500).
4. 95% Ethanol.
5. Airpore tape sheet (Qiagen).
6. Multi-channel pipettes (10 and 200 μL, 1 mL).
7. Yeast lysis buffer: 1.2 M sorbitol, 100 mM $NaHPO_4$. Prepare 20 mL, filter sterilize, divide into 1 mL aliquots and store at –20°C. Just before use add 5.0 units/μL lyticase.
8. 20% (w/v) SDS.
9. R.E.A.L. Prep 96 Plasmid Kit (Qiagen).
10. Glass beads 425–600 μm (Sigma, G-9268).
11. Tape pads (Qiagen).
12. Isopropanol.
13. 75% Ethanol.
14. 200 μL 96-well polypropylene V-bottom plates (Greiner Bio-one Cat. No. 5665-1201).

2.5.4. Transformation of Yeast Plasmid DNA into MG7α E. coli and Prey Plasmid Selection

1. *E. coli* strain MG7α (ATCC Ref. No. MBA-78, *(15)*).
2. 96-well PCR plate.
3. LB broth and agar.
4. Carbenicillin 50 μg/mL.
5. Nunc 24-well multidish culture plates (Cat. No. 144530).
6. Nutator mixer (BD diagnostics).
7. 5X Minimal medium 9 (M9) salts: 210 mM Na_2HPO_4, 120 mM KH_2PO_4, 45 mM NaCl, 95 mM NH_4Cl.
8. M9-W/M9-H medium: 750 mL dH_2O, 50 mL 20X DO stock (**Section 2.1**), agar 20 g for solid medium. Autoclave and store at 25°C. Just before use, add 200 mL 5X M9 salts,

2 mL 1 M $MgSO_4$, 10 mL 40% (w/v) glucose, 100 µL 1 M $CaCl_2$, 1 mL 1 M Thiamine–HCl. For**M9-W** 10 mL of each aminoacid 100X stock solutions: Uracil, Leucine, and Histidine (**Section 2.1**). For**M9-H**10 mL of each aminoacid 100X stock solutions: Uracil, Leucine, and Tryptophan.

9. 50% Glycerol.
10. 200 µL 96-well polypropylene V-bottom plates (Greiner Bio-one Cat. No. 5665-1201).

2.5.5. 96-Well E. coli Prey Plasmid Isolation

1. Costar 2 mL 96-well assay block.
2. LB media.
3. Carbenicillin (50 µg/mL).
4. Microplate replicator (Boekel, Cat. No. 140500).
5. 95% Ethanol.
6. Airpore tape sheet (Qiagen).
7. Multi-channel pipettes (8-channel pipettes: 10 and 200 µL, 1 mL).
8. R.E.A.L. Prep 96 Plasmid Kit (Qiagen).
9. Tape Pads (Qiagen).
10. Isopropanol.
11. 75% Ethanol.
12. 200 µL 96-well polypropylene V-bottom plates (Greiner Bio-one Cat. No. 5665-1201).

2.5.6. Confirming Specificity of Interaction

1. Materials for LiAc yeast transformation (*see* **Section 2.1.4**).
2. Plates/Racks of 12 microtube strips of 8 1.1 mL (ISC Bioexpress, Cat. No. p-8705-2).
3. 8-caps strips (ISC Bioexpress, Cat. No. p-8705-c).
4. Nunc 12-well multidish culture plates (Cat. No. 150200).
5. DO agar medium: Glu-HWU.
6. Nutator mixer (BD diagnostics).
7. Materials for nutritional selection and *LacZ* assays.

3. Methods

3.1. Making and Testing the Bait

The bait plasmid pEG202, also called pLexA, contains a polylinker for cloning the protein of interest fused to the *E. coli* LexA DNA binding domain (DBD) *(9)*. The polylinker is 3' of the LexA sequence so it will produce fusions with the protein of interest to the LexA C-terminus. To characterize the bait strain, it is recommended to test for reporter self-activation, as described below. In addition, immunoblotting can be used to verify expression of a full-length, stable bait protein (Anti-LexA antibodies are commercially available). There is also a pEG202-related plasmid, pNLexA, for making fusions to the LexA N-terminus (Origin, *(13)*, *see* **Note 3**). This vector can be used in those cases, where it is suspected that the N-terminal region of a protein may be important for interaction *(16)*, or when the fusion in pEG202 self activates transcription (*see* **Note 4**). Whenever possible, use controls to verify the bait fusion. If partners of the protein of interest are already known, fuse them to the prey plasmid and confirm that they interact with the bait. This would indicate that your LexA fusion is active.

Using standard cloning methods, construct the bait plasmid by inserting a cDNA encoding the protein of interest into the bait vector on frame with LexA (pEG202 and/or pNLexA). DNA sequencing of the resulting plasmid is recommended to confirm that coding regions of the bait and LexA are in frame. The bait plasmids carry ampicillin resistance for selection in *E. coli* and the histidine marker for auxotrophic selection in yeast.

3.1.1. Testing the Bait for Self-Activation

1. To test whether the bait self activates the *Leu2* and *LacZ* reporters, transform the bait strain EGY42 or EGY191 (*see* **Note 5**) with the *LacZ* reporter plasmid pSH18-34 *(8)*. Because the pSH18-34 plasmid has the URA3 marker, select transformants in DO Glu-U medium.
2. Cotransform EGY48/pSH18-34 or EGY191/pSH18-34 with the bait plasmid and an empty library vector or a prey that does not interact with your bait. Also, transform an unrelated bait (*see* **Note 6**) with the same prey plasmids. If a prey control positive for interaction is available, transform it with both baits. Select the colonies in DO Glu-HWU plates. Transformations can be made using the lithium acetate method (*see* **Section 3.2**). Incubate the plates for 2–3 days at 30°C. Proceed with next protocol *(16)*.

3.1.2. Nutritional Assays

1. Prepare the following DO agar plates in Nunc omnitrays: Glu-HWU *(3)*, Gal/Raff-HWU *(2)*, Glu-HLWU (1), Gal/Raff-HLWU (1) (*see* **Notes** 7 and **8**, *see* **Table 19.3**).
2. Select four colonies of similar size from each transformation (after 2–3 days incubation, *see* **Section 3.1.1**), streak them gently with a flat toothpick in 1 cm squares or patches on Glu-HWU plates. Incubate them overnight at 30°C.
3. Perform the drop assay as follows: With a multi-channel pipette add 100 μL of sterile water to each well of the first four columns of a sterile 96-well round bottom culture plate. Using a flat toothpick, take similar amounts of the yeast

colonies (amount equivalent to a 2 mm colony) from the patches on Glu-HWU plates. Dissolve the yeast in the water by rolling the toothpick around thoroughly in each well. Mix the yeast suspensions with a 200 μL multi-channel pipette. Alternatively, you can use a plate shaker in lower motion to maintain the yeast in solution. Perform a dilution series (1×10^{-1} to 1×10^{-4}) of the yeast solution in a fresh section of the 96-well round bottom culture plate. Make sure to mix the yeast suspensions before plating to ensure quantitative results. With a 10 μL multi-channel pipette, plate 4 μL drops of each yeast solution in the omnitrays containing the following media: Glu-HLWU, Glu-HWU (two plates), Gal-HLWU, Gal-HWU (two plates) (*see* **Note 9**). Attach the numbered well orienter for 96-well plates to each plate bottom to use as a guide (*see* **Note 10**). Incubate the Glu-HWU and Gal/Raff-HWU plates at 30°C for 1–2 days. Retrieve the plates and proceed with X-Gal assays. For nutritional assays incubate the Glu-HLWU and Gal/Raff-HLWU plates for 3–4 days at 30°C.

4. Yeast should grow well on the Glu-HWU and Gal/Raff-HWU plates after 2 days. After 3–5 days of incubation, no growth should be observed on the Glu-HLWU plates. The bait plus the negative control bait plasmid should not grow on Gal/Raff–HLWU plates. However, yeast corresponding to the bait plus the positive control/known partner should appear between days 3 and 5 on Gal/Raff–HLWU plates. The unrelated bait should not grow on medium lacking leucine. Gal/Raff–HLWU plates should be monitored daily for growth of yeast corresponding to bait and negative prey control plasmids.

3.1.3. β-Galactosidase/Filter Lift X-Gal Assays (After the Method Developed by Stern and Col)

1. Perform the X-gal and nutritional assays simultaneously. Retrieve the following plates from the incubator after 1–2 days: Glu-HWU and Gal/Raff-HWU from previous section (*see* **Note 11**) *(10)*.
2. Prepare fresh X-gal/Z-buffer solution (see Section 2). About 15 mL should be enough for two plates. Place 5–7 mL of X-gal/Z-buffer solution in the center of two clean Nunc omnitrays.
3. Cut two Whatman filters to fit a 96-well plate (11.5 × 7.5 cm). Using forceps, place a filter on top of the yeast spots. With the side of the forceps, press down and rub the filter until the yeast attach to it. To orient the filter to the agar, poke holes with a needle in asymmetric locations.
4. Using forceps, remove the filter and immediately immerse it in liquid nitrogen (colonies faced up), keep it submerged for 10 s.

5. Remove the filter from the liquid nitrogen and place it (colonies side up) on top of an absorbent towel, allowing the filter to thaw. Repeat the freezing/thawing one more time.
6. With the forceps place the filter on top of the X-gal/Z-buffer solution (colonies side up). Make sure the X-gal solution does not go to the upper part of the filter (*see* **Note 12**).
7. Incubate the filter at 30°C. Check periodically for the appearance of blue colonies (*see* **Note 13**).
8. Select colonies positive in both the nutritional and *LacZ* assays. Discard any colonies that grow on Glu-HLWU plates.

3.2. Small-Scale Yeast Transformation with LiAc

Perform all procedures at room temperature and under sterile conditions. A variation of this protocol in the 96-well format is described in **Section 3.4.6** *(16)*.

1. Inoculate the yeast strain from a 1–2 day old patch into 3 mL YPD or DO media, as required (*see* **Note 14**). Incubate overnight at 30°C with shaking at 200 rpm.
2. Add 1 mL of the overnight culture to a flask containing 10 mL liquid media. Incubate 3–4 h at 30°C with shaking at 200 rpm. OD_{600} should reach 0.3–0.6.
3. Transfer the culture to a 15 mL conical tube and pellet by spinning at 3,000 rpm for 5 min Resuspend the cells with 10 mL sterile water. Spin to pellet cells.
4. Resuspend cells in 1.5 mL of freshly prepared 1X LiAc/TE solution.
5. Prepare the following transformation mix (multiply the amounts/volumes described here, per the number of samples to be transformed): (a) 240 μL freshly prepared PEG/LiAc/TE solution, (b) 20 μg boiled salmon sperm carrier DNA, (c) 50 μL of the yeast suspension in 1X LiAc/TE.
6. Add 360 μL of this mix to individual tubes containing 0.1–1 μg plasmid DNA. Gently mix tube content by inversion.
7. Incubate tubes at 30°C for 30 min with inversion mixing (use a Labquake rotisserie or Nutator mixer inside the incubator).
8. Heat shock by transferring tubes to a water bath at 42°C for 15 min.
9. Centrifuge tubes at 3,000 rpm for 5 min and remove the supernatant with a micropipette tip.
10. Add 500 μL sterile water and resuspend the cells by mixing with a micropipette tip.
11. Plate 50 and 150 μL samples onto plates of appropriate DO selection medium. Incubate at 30°C for 3–4 days and isolate transformants.

3.3. Library Amplification

If you have the pJM1-library as plasmid DNA, you will need to first introduce it in *E. coli*. Transform 5 μg into high efficiency (10×10^{10}) electro-competent DH10B*E. coli* cells. The pJM-1 library consists of 3×10^9 members *(4)*. To have a representative library, you would need to obtain at least 3×10^9CFU on your initial transformation. These transformants will be grown to amplify the number of independent clones contained in the library (the goal is to amplify the library 5–10 fold). The library titer will then be determined by plating a dilution series. Aliquots can be frozen for long-term storage (*see* **Note 15).** If you start with the library already in *E. coli*, determine the library titer by going directly to **Step 6** in this protocol.

1. Perform a DNA dialysis before electroporation as follows: Add 20 mL dH_2O to a dish. With forceps, place one Millipore 0.022 μm filter on top of the water (floating). Do not allow the liquid to go into the surface. Dilute 5 μg of library DNA in 50 μL water and place 10 μL drops on top of the filter. Allow dialysis for 20 min. Recover the DNA drops into a fresh tube. Adjust DNA volume to 100 μl.
2. Thaw electro-competent cells on ice and place cuvettes (0.1-cm gap) on ice. Mix the 100 μL library DNA with 1,000 μL competent cells (always on ice). Electroporate at 2.5 kV, 200 Ω, 25 μF. Incubate the cells in 50 mL SOC medium for 1 h at 37°C to allow recovery.
3. Determine the efficiency of transformation by performing a dilution series starting with 100 μL of the transformation diluted into 1 mL SOC medium. Plate dilutions from 10^{-3} to 10^{-7} on LB plates with carbenicillin (50 μg/mL) (or proper antibiotic).
4. Add the rest of the cells to 1 L TB medium with carbenicillin (50 μg/mL) and incubate overnight at 37°C with shaking.
5. Isolate DNA from 500 mL of the overnight culture using a good quality method (we recommend the large-scale Qiagen Maxiprep kit).
6. To determine the library titer, perform serial dilutions with 1 mL of cells and plate on LB carbenicillin 50 μg/mL (Plate 1×10^{-4} to 1×10^{-7}dilutions only). Incubate at 37°C. Chill the rest of the cells on ice and store overnight at 4°C.
7. Count colonies and calculate the number of Colony Forming Units (CFU) per mL. Make aliquots of the amplified library at a density of at least 1×10^8CFU/mL (*see* **Note 15**). For storage, add glycerol to 20% and freeze at –80°C.

3.4. Screening

3.4.1. Large-Scale Yeast Transformation

To have a representative number of peptide aptamer clones, perform a large-scale sequential transformation with 150 μg of the peptide aptamer library DNA and the corresponding yeast strain containing the bait and*LacZ* plasmids (*see* **Notes 16–18**). Use sterile technique *(16)*.

1. Inoculate a fresh bait strain transformed with the *LacZ* and the bait plasmids (*see* **Note 19**) into 50 mL of Glu-HU medium in a culture tube. Incubate overnight at 30°C with continuous shaking at 230 rpm. Grow to stationary phase (OD_{600}> 1.5).
2. Inoculate a 1 L flask containing 300 mL Glu-HU medium with the overnight culture. Add sufficient culture to reach an OD_{600}= 0.2–0.3.
3. Incubate at 30°C with continuous shaking (230 rpm) until OD_{600} reaches 0.5 (3–6 h).
4. Pellet the cells by centrifugation at 3,000 rpm for 5 min in 50 mL conical tubes at room temperature.
5. Wash the cells with sterile water as follows: Discard the supernatant and resuspend the yeast in 10 mL water by vortexing gently. Once the cells are resuspended, add 30 mL sterile water to each tube and mix by inversion. Centrifuge at 3,000 rpm for 5 min.
6. Resuspend cells in 5 mL sterile water and pool them into one conical tube. Centrifuge as indicated above.
7. Discard supernatant and resuspend in 1.5 mL 1X TE/LiAc (prepared just before use, see Section 2)
8. Mix together 1 mL of 10 mg/mL denatured salmon sperm DNA and 150 μg library DNA.
9. Add DNA mixture to cell suspension from **Step** 7.
10. Add 40 mL 1X LiAc/8X PEG/1X TE solution. Swirl to mix and incubate at 30°C for 30 min.
11. Add 17.6 mL DMSO and mix well by inversion or swirling.
12. Heat shock at 42°C for 15 min with occasional swirling to facilitate heat transfer.
13. Chill cells on ice for 5 min.
14. Centrifuge at 3,000 rpm for 5 min.
15. Remove supernatant and resuspend cells in 40 mL Gal/Raff-HU medium.
16. Incubate with gentle shaking for 4 h at 30°C to induce library expression.
17. Make serial dilutions from 1 mL of the transformation culture using the DO Gal/Raff -HWU medium. Plate on GLU-HWU plates and incubate 2–4 days at 30°C until colonies are visible. Count the colonies and determine the number of CFU of transformed yeast.
18. Centrifuge the remaining transformation culture at 3,000 rpm for 5 min.

19. Resuspend in 20 mL Gal/Raff-HU medium. Plate transformed yeast onto twenty 200-mm Petrie dishes containing Gal/Raff-HLWU medium (ca. 200 mL per plate). Make sure to spread the yeast over the entire plate. Incubate at 30°C.

3.4.2. Selection of Positive Interactors

A peptide aptamer screen can yield a large number of putative interactors. On average, each transformation will have as many as 200 interacting colonies *(6)*. It is recommended to perform two or three large-scale transformations. Hence, it is beneficial to employ protocols that allow easy and efficient handling of large number of samples. It is also necessary to eliminate false positives. The following protocol describes how to select colonies that have positive interactions. At all times, negative and positive controls should be included in parallel for both the *LacZ* and *Leu2* assays to confirm that the selection is working properly.

1. Day 1. Harvest the colonies growing on the screening plates (Gal/Raff -HLWU) 4–7 days after transformation (*see* **Note 20**) and patch them on Glu-HWU omnitray plates to repress the activity of the *GAL1* promoter. Incubate overnight at 30°C. These are your "master replica plates".
2. Day 2. The yeast on the "master replica plates" can be stored at 4°C for up to 1 month. As an alternative, the colonies can be frozen for long-term storage. To freeze, add 100 µL liquid Glu–HWU medium to each well of the 96-well V-bottom plates. Inoculate each colony into a well using the replicator and grow overnight at 30°C. On Day 3, add 100 µL 65% glycerol/$MgSO_4$ solution to each well, cover with foil PCR adhesive film and freeze at –80°C.
3. Day 2. Set up assays to confirm the interaction of bait and preys with both *LacZ* and *Leu2* reporters as follows. With a multi-channel pipette, add 200 µL sterile water to each well of a 96-well round bottom culture plate. Using a flat toothpick, scrape the yeast colonies from the master replica plate making sure to take the same amount for each colony (equivalent to a 2 mm yeast colony). Dissolve each colony in the water by rolling the toothpick thoroughly. Mix yeast suspensions with 200 µL multi-channel pipette and immediately perform a 1×10^{-1} dilution of the yeast solution into a fresh 96-well round bottom culture plates (alternatively, you can use a plate shaker set on low). With a 10 µL multi-channel pipette, plate 4 µL drops of each yeast solution on the orienter-attached omnitrays containing the following media: Glu-HLWU, Glu-HWU (two plates), Gal-HLWU, Gal-HWU (two plates). Allow plates to dry under laminar hood, cover them with saran wrap and incubate them at 30°C.
4. Day 3. Take one Glu-HWU and one Gal/Raff-HWU plate from the incubator and perform filter lift assays as indicated

above (*see* **Section 3.1.3**). Store the additional Glu-HWU and Gal/Raff-HWU plates at 4°C, which can be used as controls for growth to show that an even quantity of yeast was placed on all plates.

5. Days 3–8. Retrieve the Glu-HLWU, and Gal/Raff-HLWU plates after 3–6 days. Verify your control plates are working properly. Select colonies that are positive for β-galactosidase assays/X-gal assays, grow well on Gal/Raff–HLWU plates and do not grow on Glu–HLWU plates (*see* **Note 21**).
6. Patch selected interactors on a fresh Glu–HWU omnitray and incubate overnight at 30°C. You can now discard the initial "master replica plates" from **Step 1** (of this section) and replace them with these new "master replica plates".

3.4.3. Yeast Plasmid Isolation (see Note 22)

1. Inoculate fresh yeast from the master replica plate described in the previous section (**Step 6**) into liquid media (*see* **Note 23**). Fill a 2 mL 96-well assay block with 1.5 mL Glu -HWU medium using a multi-channel pipette. Sterilize the 96-pinned replicator by submerging it into 95% ethanol and burning off the ethanol with a flame. Using the sterilized microplate replicator, touch each of the replicator pins to an individual patch on the plate being careful not to mix or touch more than one yeast patch. Place the replicator into 96-well block plate containing the DO medium making sure to submerge each of the replicator pins into the medium. Cover the plate with an air-pore tape sheet and incubate overnight shaking at 30°C.
2. Pellet yeast cells by spinning the plate at 2,000 rpm for 5 min using a 96-well adapted centrifuge.
3. Carefully pour off the supernatant dabbing excess medium on paper towels. Resuspend the pelleted cells in residual medium (ca. 50 μL) using a 200 μL multi-channel pipette and vortexing.
4. Add 10 μL yeast lysis buffer to each well and mix by vortexing. Incubate the plate at 37°C overnight with no shaking.
5. Add 10 μL 20% SDS to each well and vortex.
6. Add 300 μL Solution R1 from R.E.A.L. Qiaprep kit and vortex (*see* **Note 24**).
7. Add ca. 50 μL glass beads and vortex.
8. Centrifuge the mixture for 10 min at 2,500 rpm.
9. Remove the supernatant and transfer into new 2 mL 96-well assay block making sure that no glass beads are present.
10. Add 300 μL Solution R2. Cover with a tape pad making sure that it is sealed and mix by inverting several times. Let the mixture incubate at room temperature for 5 min.

11. Add 300 μL Solution R3 to halt lysis. Seal with a new tape pad and mix by inverting.
12. Pellet precipitated material by spinning at 4,000 rpm for 30 min.
13. Place the 96-well filter plate (in R.E.A.L. Prep 96 Plasmid Kit) on a clean 96-well block plate making sure the wells line up. Transfer the supernatant to a 96-well filter plate/block plate being sure to remove as little precipitant as possible. Centrifuge for 5 min at 2,500 rpm.
14. Add 750 μL room-temperature isopropanol to each well of the square-well block containing the cleared lysate. Tape the block carefully pressing the pad with a tissue, and mix immediately by inverting a few times. Incubate the plate on ice for 20 min.
15. Spin the plate for 30 min at 4,000 rpm. Pour off the supernatant and blot the excess on paper towels. Add 300 μL 75% ethanol and spin at 4,000 rpm for 5 min. Remove the ethanol and blot with paper towels (*see* **Note 25**).
16. Dry the DNA pellet in a speed-vac for 30 min or allow it to dry for 2 h in a 37°C incubator. Resuspend the pellet in 40 μL water or TE.

3.4.4. Transformation of Yeast Plasmid into MG7α E. coli and Prey Plasmid Selection

1. Yields of plasmid DNA isolation from yeast are usually low. In addition, the obtained DNA is a mixture of bait, prey and *LacZ* plasmids. Therefore, is strongly recommended to transform *E. coli* MG7α (ATCC Ref. No. MBA-78, *(15)*) with the yeast plasmid obtained in **Section 3.4.3** and then to select *E. coli* cells carrying the prey plasmid (but not the bait plasmid) in selective M9-W medium.
2. Day 1. Prepare or thaw chemically competent *(10)* MG7α *E. coli* cells on ice (*see* **Notes 26–29**).
3. Add 10 μL of each yeast plasmid to a prechilled 96-well PCR plate (*see* **Note 30**).
4. Add 100 μL of competent cells to each well.
5. Incubate the plate on ice for 30 min.
6. Heat shock the cells in a 42°C water bath for 90 s.
7. Chill the plate on ice for 5 min.
8. Add 100 μL LB without antibiotic and shake for 1 h at 37°C.
9. Centrifuge for 5 min at 2,500 rpm to pellet the cells.
10. Pour off excess LB and resuspend the pellet in residual liquid.
11. Melt LB agar and add carbenicillin to 50 μg/mL. Add ca. 1 mL LB–carbenicillin to each well of the 24-well multidish culture plate. Allow the media to solidify and dry completely (under laminar hood).

12. Add all the resuspended cells into each well of LB agar 24-well culture plates being careful to mark the plates to correspond to the original 96-well transformation plate. To spread cells evenly, place the plates on the Nutator mixer and allow the cells to dry before placing at 37°C overnight.
13. Prepare omnitray plates with M9-W, M9-H and LB carbenicillin (50 μg/mL) agar media.
14. Day 2. Select *E. coli* transformants from the 24-well culture plate and streak onto M9-W and M9-H plates with a toothpick. Incubate at 37°C for 36–48 h.
15. Select transformants that grow well on M9-W plates but do not grow on M9-H plates. Streak selected colonies on LB carbenicillin (50 μg/mL) plates and store at 4°C (*see* **Note 31**). For long-term storage, add 100 μL 50% glycerol to each well of a 96-well V-bottom plate. Add 100 μL of MG7α cells from an overnight culture and store at –80°C.

3.4.5. 96-Well *E. coli* Prey Plasmid Isolation

1. Inoculate MG7α *E. coli* transformants from stored plates (*see* **Section 3.4.4. Step 14**): Fill a 96-well block plate with 1.5 mL of liquid LB carbenicillin (50 μg/mL) using a multichannel pipette. Sterilize the microplate replicator by submerging it into 95% ethanol and burning off the ethanol with a flame. Using the sterilized replicator, press each of the replicator's pins to each well of frozen –80°C stock *E. coli* plate. Place the replicator in the LB carbenicillin filled 96-well block plate, submerging each replicator pin into the medium. Cover the plate with airpore tape sheets and shake overnight at 37°C.
2. Pellet cells by spinning at 4,000 rpm for 5 min.
3. Carefully pour off supernatant dabbing excess medium on paper towels. Resuspend pelleted cells in residual medium (ca. 50 μL) using a multi-channel pipette and vortexing.
4. Add 300 μL Solution R1 from R.E.A.L. Qiaprep kit and vortex (*see* **Note 32**).
5. Add 300 μL Solution R2. Seal with block tape and mix by inverting several times. Let mixture incubate at room temperature for 5 min.
6. Add 300 μL of Solution R3 to halt lysis. Seal with a new block tape and mix by inverting.
7. Pellet precipitated material by spinning at 4,000 rpm for 30 min.
8. Tape the 96-well filter plate (included in R.E.A.L. Prep 96 Plasmid Kit) to a clean 96-well block plate making sure the wells line up accordingly. Transfer supernatant to 96-well

filter plate/block plate being sure to remove as little precipitant as possible.

9. Centrifuge for 5 min at 4,000 rpm.
10. Add 750 μL of room-temperature isopropanol to each well of the square-well block containing the cleared lysates and seal the block with Foil PCR adhesive film (rub top of plate with a tissue to help adhesion of the foil film). Mix by inverting several times and incubate on ice for 20 min.
11. Centrifuge the plate for 30 min at 4,000 rpm.
12. Pour off the supernatant and blot the excess on paper towels.
13. Add 300 μL of 75% ethanol and spin at 4,000 rpm for 5 min. Remove the ethanol and blot with paper towels.
14. Dry the DNA pellets in a speed vac or allow pellet to dry overnight at room temperature.
15. Resuspend the pellets in 100 μL water or TE (*see* **Note 33**).

3.4.6. Confirming Specificity of Interaction

Once the library plasmids have been isolated from yeast, it is important to confirm that the selected peptide aptamers specifically interact with the bait of interest and not to unrelated proteins (*see* **Notes 34 and 35**). This is achieved by retransforming yeast with the bait strain and the new isolated peptide aptamer plasmids. These same peptide aptamer plasmids are also introduced into yeast with an unrelated bait. Two methods commonly employed to introduce large numbers of independent plasmids into yeast are mating *(17)* and direct transformation *(16)*. In our hands, the direct transformation method has worked better, and it is feasible to perform 192 transformations in a single day (see below).

1. Using the LiAc transformation protocol (*see* **Section 3.1.4**), transform EGY191-pSH1834 or EGY48-pSH1834 with the plasmids containing the bait of interest and the unrelated bait. These bait strains are then transformed with the prey plasmids to be evaluated. You can perform up to 192 transformations simultaneously by taking advantage of the 96-well format. Follow the protocol in **Section 3.1.4** with the following variations: (a) Start with 200 mL of the yeast culture. (b) Place the plasmid DNA to be transformed in each tube of a plate with microtube strips of eight 1.1 mL. (c) Add 300 μL of the transformation mix and cover with 8-strip caps. (d) Resuspend yeast in 200 μL sterile water with a multichannel pipette and plate 100 μL of the transforming yeast into 12-well culture plates filled with 3 mL selective DO Glu-HWU media. (e) Place the 12-well culture plates with the yeast on the nutator until the yeast have spread over the well and have dried (for about 30–40 min). Cover plates with plastic cling wrap and incubate at 30°C for 3–4 days.

2. Confirm protein–protein interaction using nutritional selection and X-gal assays. Plate drops as indicated in **Section 3.1.2** on Gal/Raff-HLWU, GLU-HLWU, Gal/Raff-HWU and GLU-HWU plates. Those library plasmids that interact with the bait of interest and not with the unrelated bait are verified interactors. These clones can be sequenced for peptide identification (*see* **Note 36**).

3.5. Analysis of Peptide Sequences

Many of the selected peptide aptamers will have different sequences that reflect differences in the ways they bind to the target protein. The random library, the testing process, and the selection method can also enrich for aptamers with similar amino acid motifs. Because of their small size, aligning the peptide aptamers using programs (ClustalW defaults) designed for larger proteins will fail to find any significant alignment hits. The aptamers will have only a few conserved amino acids that could be separated by gaps of random residues which will result in low alignment scores (bits). Alignment programs can be modified for the small peptides by decreasing the stringency of the search. This can be done by increasing the *E*-value cut off (low score bits) and lowering the gap penalties. Reducing the stringency will allow for positive alignment hits to be observed when comparing the aptamers, but this change can also increase the possibility of finding alignments produced by chance. To address this issue, it is necessary to create a database of random peptide sequences of similar size and amino acid content to those aptamers isolated in the screen. These will be used in the following protocol to determine if the aptamer motifs uncovered under the lower stringency alignment are significantly different to those found by chance in a randomly generated peptide population.

3.5.1. Generate Databases of Aptamers

1. If functional information is available, classify aptamers by their activity, e.g., “interfering” or “noninterfering”.
2. Create two fasta-formatted files including either the “interfering” or “noninterfering” aptamers. Create a third fasta file containing all of the aptamers. These are the three Experimental Databases (EXP-db).
3. Determine the total amino acid content for each of the EXP-db, e.g., “interfering”, “noninterfering” and “all”. [The amino-counter.pl[1] script, coded using BioPeri modules *(11, 18)*, can be used for this and is available at http://biochem.ncsu.edu/faculty/hanley-bowdoin/Pages/homepage.html under tools].
4. Using the amino acids proportions found for each of the three categories in the EXP-db, generate a 100 sets of Random databases (RAN-db) for each EXP-db (“interfering”, “noninterfering” and “all”). Each of the RAN-db should have the same number of aptamers as its corresponding EXP-db and the random aptamers should have the same residue length as

the experimental ones [The ranPEP.pl[1] script, coded using BioPerl modules *(11, 18)*, can be used and is available at http://biochem.ncsu.edu/faculty/hanley-bowdoin/Pages/homepage.html under tools].

5. Generate BLAST-searchable databases of all the EXP-db and RAN-db using NCBI formatdb. [Executables for various platforms and documentation is available at http://www.ncbi.nlm.nih.gov/blast/download.shtml].

3.5.2. BLASTP Pairwise Alignments to Search for Probable Interfering/Noninterfering Motifs

1. Perform pairwise BLASTP searches of each database (EXP-db and RAN-db) to itself. Original fasta files can be used as queries to perform batch searches. For example, the fasta file containing the experimental "interacting" aptamers used as a batch query against the "interacting" EXP-db. Score values in bits are calculated from the number and type of positive hits to consecutive amino acid residues with small or no gaps. To decrease the stringency of the search, increase the expected *E*-value cutoff (>20) and decrease the gap penalties in small step increments.

3.5.3. Statistical Significance of Pairwise Alignments: Expected Versus Observed Positive Hits on the Databases

1. For each EXP-db and RAN-db, parse the resulting reports discarding hits of any aptamer to itself and record:
   a. the number of aptamers with at least one hit for any given database (count every aptamer only once even if it has more than one hit)
   b. the total number of hits for each database (include every aptamer as many times as it has a hit)
2. These two totals for each EXP-db ("interfering", "noninterfering" and "all") are the "Observed Values". Graph the number of aptamers with hits and the total number of hits for each of the three RAN-db sets as a frequency histogram. The totals should fall under a normal distribution. The means calculated from each graph are the "Expected Values" of the experiment. The standard deviation for the RAN-db sets can also be calculated from the graph
3. Determine if the "Observed Values" are significantly different from the "Expected Values" using a one-sample Student's *t*-Test. Calculate the *t*-statistic using the equation:

$$t = \frac{(\overline{x} - \mu)}{S/\sqrt{n}}$$

where $\overline{x}$ is the sample mean (Observed Value), μ is the population mean (Expected Value), S is the standard deviation of the RAN-db sets, and n is the sample size (n=100). Compare the obtained *t*-statistic to a table of *t*-values (for n–1 degrees of freedom) to determine if it reaches the threshold of statistical significance (reject null hypothesis).

3.5.4. Inspect Aptamer Alignments to Find Probable Interfering/Noninterfering Motifs

The pairwise alignments of the EXP-db can be further analyzed to infer probable motifs for interfering and noninterfering aptamers. There are many free and commercial packages available to create and edit alignments like those included in Invitrogen's Vector NTI Advance^{TM} (free license for nonprofit researchers) and Tom Hall's BioEdit, as well as many online alignment tools.

4. Notes

1. 100X Tryptophan (W) stock solution should be protected from the light.
2. Galactose is not as good of a carbon source as glucose. Hence, raffinose is added to the medium in a final concentration of 1% to supplement yeast carbon requirements.
3. Testing your bait is essential for knowing if the protein of interest is working well in the yeast two-hybrid system. In some cases, the bait may be inactive when fused to the LexA DBD but may work well when fused to the Gal4 DBD. A bait could self activate the expression of the *Leu2* and *LacZ* reporters or could produce a fusion in which domains important for interaction are blocked. In these cases, it may be helpful to create N-terminal LexA fusions using the pNLexA plasmid. In most cases, bait proteins will translocate into the yeast nucleus and will occupy LexA operators *(12)* (to evaluate that your bait is going to the nucleus you can perform a repression assay with plasmid pJK101 *(19)*). In rare instances, baits are excluded from the nucleus, for these cases, the use of a close relative of pEG202, which carries the SV40 T-nuclear localization signal between the LexA coding sequence and the polylinker is recommended (this plasmid could be obtained from Bert Vogelstein and his coworkers).
4. In pEG202, LexA DBD self activates the expression of both reporters. However, this activity generally disappears when another protein is fused to its C-terminus *(9)*.
5. Depending on the background activity of the bait, it may be necessary to use a different yeast strain. EGY48 contains six LexA operators, which make it more sensitive, while EGY191 contains only one LexA operator, therefore, is a more stringent strain. If background/self activation is observed with the bait protein (protein of interest) in EGY48, it is recommended to use EGY191, where background could be

reduced. However, keep in mind that only stronger interactions will be detected in EGY191, due to its astringency (*see* **Note 3**).

6. You can construct your own unrelated bait plasmid or request one, most researchers are willing to share their yeast constructs.
7. The Nunc omnitrays fit up to 96 patches. To facilitate the patching process, use the numbered well orienter.
8. To avoid colony confluence, omnitrays with media should be allowed to dry under the laminar hood for 20–30 min without lids.
9. You can also streak equivalent amounts of yeast on Glu-HWU, Gal/Raff-HWU, Glu-HLWU, and Gal/Raff -HLWU plates instead of doing the drop assay. However, because four different media is tested, you will need to streak each colony four times. The drop assay is recommended because it is less time consuming and is more quantitative than the streaks.
10. Electric multi-channel pipettes provide a more accurate delivery of small volumes and are especially useful when large numbers of samples have to be plated.
11. It is very important that the yeast colonies are fresh (recently transformed) when performing the *LacZ* reporter assay.
12. Avoid excess liquid going over the filter surface to prevent colony confluence.
13. Prolonged incubation (> 8 h) may give false positives.
14. This protocol can be used with cultures that have been stored at room temperature or in a refrigerator. The yield will be reduced with older cultures but will generally be sufficient to isolate a number of transformants of the desired genotype.
15. The library can be kept for up to 1 year with out losing significant viability when stored at a density of 1×10^8 cells/mL.
16. A different strategy is also commonly used to introduce the library plasmids into the bait strain. This approach is based on mating two yeast strains of different mating types, one carrying the bait plasmid and the other carrying the peptide aptamer library *(17, 20)*.
17. At difference with a cDNA library, in a peptide aptamer screen is not necessary to exhaust the library. It is excessive to attempt to plate the members to the level 3×10^9. Therefore, when 3×10^7 colonies have been transformed and analyzed, the screen can be considered complete.
18. Higher efficiencies are achieved with sequential transformation.
19. It is recommended to a generate growth curve of your bait under selective conditions (Glu-HU medium) before

performing the actual large-scale transformation. This will confirm the bait is not toxic to the yeast strain and will provide you with information regarding the time frame necessary to reach the required OD_{600}.

20. During selection for protein–protein interactions, the time of yeast growth may vary due to bait expression, strength of interaction, concentration of yeast plated, incubators, etc. Therefore, it is important to note the growth of positive controls and any background growth of negative controls. You can keep harvesting "positive" colonies daily until colonies begin to appear on plates transformed with the library and the negative control bait plasmids. It is also helpful to know how yeast containing the bait plasmid grows on DO-HU media as used in the bait selection (*see* **Section 3.1.1**). After 2–3 days, some positive colonies will begin to grow, but 5 days may be necessary to observe slower growing colonies.
21. It is possible that more than one library plasmid will be present in the positive colonies. Streaking colonies on Glu–HWU media allows segregation of plasmids. Hence, it is a good idea to start plasmid isolation from single colonies.
22. Several protocols suggest that is possible to cure plasmids from yeast by removing the selective pressure and this method can be used to separate the bait plasmid from the library plasmid. Positive interactors containing all three plasmids (bait, library, and *LacZ* reporter) can be grown in liquid DO Glu–W for 2–3 days at 30°C. The bait plasmid should be randomly lost from some of the transformants due to lack of selective pressure. However, we have tried this technique without success, i.e., a high proportion of yeast maintain the bait plasmid. As a consequence, we recommend isolating the yeast plasmid (miniprep protocol) followed by transformation into *E. coli*.
23. Plasmid DNA can be isolated from yeast grown as a solid patch or a liquid culture. Liquid culture may result in more yeast cells and higher yield of plasmid DNA. For a solid patch, scrape yeast from fresh patches and resuspend in water.
24. As an alternative to the Qiagen R.E.A.L. prep. kit, plasmid isolation can be done with a phenol chloroform extraction and ethanol precipitation.
25. Two additional 75% ethanol washes are typically recommended. However, the yield of yeast DNA from this prep. is low, and a single ethanol wash will provide the maximum yield of yeast DNA.
26. MG7α is an *E.coli* strain that carries the *trpC*, *leuB*, and *hisB* mutations that can be complemented by the yeast TRP1, LEU2, and HIS3 genes. MG7α also contains a deletion of

the *endA* gene that removes any residual *endA* endonuclease activity. It is an improved version of the *E.coli* KC8 strain (*see* **Note 27**, *(21)*) but with superior plasmid DNA production characteristics. This strain saves time by eliminating the need to transform plasmid DNA into DH5α. To achieve higher plasmid yields because it provides the selection of KC8 and the plasmid yield of DH5α *E. coli* strains *(15, 21)*.

27. Competent KC8 *E. coli* cells (Clontech) can also be used to selectively rescue the library plasmid or the bait plasmid from positive yeast cotransformants identified in a two-hybrid library screen. KC8 cells carry *trpC, leuB*, and *hisB* mutations that can be complemented by the yeast *TRP1, LEU2*, and *HIS3* wild-type genes *(16)*. KC8 cells have Kanamycin resistance. It may be necessary to transform the plasmids recovered from KC8 to a strain such has DH5α, which for better DNA quality production (*see* **Note 26**).
28. Transformation of electro-competent *E.coli* is often recommended due to the high transformation efficiency that can be obtained with these cells and the interference of yeast genomic DNA with transformation. You must have an electroporator and a cuvette with 0.1 cm gap.
29. Chemically competent cells will also work cells with a high efficiency of transformation are used (10^7 cfu/μg of pUC19 DNA).
30. If no colonies are obtained, it may be necessary to add more plasmid DNA to the transformation as the yeast plasmid preparation give extremely low plasmid yields.
31. We recommend doing a restriction enzyme digestion to confirm that the DNA isolated is that of the prey plasmid (choose a restriction enzyme that produces clearly different restrictions patterns in the bait and prey plasmids).
32. Cells can also be frozen and stored at –80°C as follows: Add 100 μL of 50% glycerol to each well of a 96-well V-bottom plate containing 100 μL of MG7α cells from an overnight culture. Mix with pipette tips and store at –80°C.
33. 96 R.E.A.L. for *E.coli* plasmid preparation is not necessary for yeast. However, the 96 R.E.A.L. system is advised for plasmid preparation for DNA sequencing (or any other method that produces high-quality plasmid DNA). As an alternative to the 96 R.E.A.L. kit, we recommend Whatman microplate unifilters (Cat. No. 7720-2830) for plasmid isolation in combination with a regular Birboim/alkaline lysis miniprep protocol. The protocol is basically the same, just change volumes: Solution I (200 μl), Solution II (200 μl), and solution III (800 μl). Recipes for these buffers are as follows *(22)*: Alkaline lysis solution I: 50 mM glucose, 25 mM Tris–HCl (pH 8.0), 10 mM EDTA (pH 8.0). Prepare Solution I from standard

stocks in batches of 100 mL. Autoclave for 15 min at 15 psi on liquid cycle and store at 4°C. Alkaline lysis solution II: 0.2 N NaOH (freshly diluted from 10 N stock), 1% (w/v) SDS. Prepare solution II fresh and use at room temperature. Alkaline lysis solution III: Mix 60 mL 5 M potassium acetate, 11.5 mL glacial acetic acid, and 28.5 mL H_2O. The resulting solution is 3 M with respect to potassium and 5 M with respect to acetate. Store the solution at 4°C.

34. We recommend running a gel and doing a restriction enzyme digestion to confirm that the isolated plasmid DNA corresponds to the prey plasmid. Choose a restriction enzyme that will give different patterns for the bait and prey plasmids.
35. In some occasions, in the yeast two hybrid system, false positives are obtained due to proteins available in the library, that otherwise would be in different compartments, this does not apply to the peptide aptamer libraries. In other cases, true interactors are identified only in yeast, due to proteins that could serve as a bridge to facilitate interactions between bait and prey *(6)*. Interaction can be confirmed within the yeast system by switching yeast strains, or by swapping domains (AD vs DNA-BD).
36. A common next step in a peptide aptamer screen is to evaluate their effects on protein function. Due to the broad nature of biological systems that can be evaluated, the description of such protocols is beyond the scope of this book chapter. In general, a functional assay to test protein that is to be targeted must be available, as well as a transfection/ transformation protocol. Peptide aptamer sequences must be transferred to a proper expression cassette and titration of the amounts of DNA corresponding to peptide aptamers might be required. Whenever possible, it is strongly recommended the use of positive controls, such as known partners of the protein of interest with mutations impairing function but not binding.
37. Plasmid DNA isolated using 96 R.E.A.L. prep plasmid kit is a good quality DNA which is suitable for sequencing. PCR primers specific to the activation domain can also be used to amplify the desired sequence for peptide identification. If it is difficult to obtain large amounts of prey plasmid DNA, PCR amplification of the target area is a valid alternative. Although each sequencing lab is different, approximately 350ηg of plasmid DNA or 10ηg of PCR product is necessary for sequencing. Many sequencing labs use BigDye Terminators. DNA as a PCR product or plasmid DNA must be quantified before sequencing. The DNA can be quantified using a spectrophotometer or quantification can be approximated by agarose gel using quantification markers *(23, 24)*.

Acknowledgments

This work was supported by NRI-USDA grant 2006-35319-17212 to L.H.-B.

References

1. Hoppe-Seyler, F. and Butz, K. (2000) Peptide aptamers: powerful new tools for molecular medicine. *J. Mol. Med.* **78,** 426–430.
2. Crawford, M., Woodman, R. and Ferrigno, P.K. (2003) Peptide aptamers: tools for biology and drug discovery. *Brief Funct. Genomic. Proteomic.* **2,** 72–79.
3. Geyer, C.R., Colman-Lerner, A. and Brent, R. (1999) "Mutagenesis" by peptide aptamers identifies genetic network members and pathway connections. *Proc. Natl. Acad. Sci. U.S.A.* **96,** 8567–8572.
4. Colas, P., Cohen, B., Jessen, T., Grishina, I., McCoy, J. and Brent, R. (1996) Genetic selection of peptide aptamers that recognize and inhibit cyclin-dependent kinase 2. *Nature* **380,** 548–550.
5. Estojak, J., Brent, R. and Golemis, E.A. (1995) Correlation of two-hybrid affinity data with in vitro measurements. *Mol. Cell. Biol.* **15,** 5820–5829.
6. Bickle, M.B., Dusserre, E., Moncorge, O., Bottin, H. and Colas, P. (2006) Selection and characterization of large collections of peptide aptamers through optimized yeast two-hybrid procedures. *Nat. Protoc.* **1,** 1066–1091.
7. Butz, K., Denk, C., Ullmann, A., Scheffner, M. and Hoppe-Seyler, F. (2000) Induction of apoptosis in human papillomavirus-positive cancer cells by peptide aptamers targeting the viral E6 oncoprotein. *Proc. Natl. Acad. Sci. U.S.A.* **97,** 6693–6697.
8. Lu, Z., Murray, K.S., Van Cleave, V., LaVallie, E.R., Stahl, M.L. and McCoy, J.M. (1995) Expression of thioredoxin random peptide libraries on the Escherichia coli cell surface as functional fusions to flagellin: a system designed for exploring protein-protein interactions. *Biotechnology (N Y)* **13,** 366–372.
9. Golemis, E.A., Gyuris, J. and Brent, R. (1994) In: *Current Protocols in Molecular Biology* (Ausbel, F.M., Brent, R., Kingston, R., Moore, D., Seidman, J., Smith, J.A. and Struhl, K., Eds.), pp. 13.14.1–13.14.17, John Wiley & Sons, New York.
10. Stern, M., Jensen, R. and Herskowitz, I. (1984) Five SWI genes are required for expression of the HO gene in yeast. *J. Mol. Biol.* **178,** 853–868.
11. Lopez-Ochoa, L., Ramirez-Prado, J. and Hanley-Bowdoin, L. (2006) Peptide aptamers that bind to a geminivirus replication protein interfere with viral replication in plant cells. *J. Virol.* **80,** 5841–5853.
12. Golemis, E.A. and Brent, R. (1992) Fused protein domains inhibit DNA binding by LexA. *Mol. Cell. Biol.* **12,** 3006–3014.
13. Golemis, E., Gyuris, J., Brent, R., Ausubel, F.M., Brent, R., Kingston, R.E., Moore, D.D., Seidman, J.G., Smith, J.A., and Struhl, K. (1996) In: *Current Protocols in Molecular Biology,* (Ausubel, F.M., Brent, R., Kingston, R.E., Moore, D.D., Seidman, J.G., Smith, J.A., and Struhl, K., Eds.), pp 13.14.1–13.14.17, Green & Wiley, New York.
14. Schiestl, R.H. and Gietz, R.D. (1989) High efficiency transformation of intact yeast cells using single stranded nucleic acids as a carrier. *Curr. Genet.* **16,** 339–346.
15. Griffith, M. and Gietz, R.D. (2003) Escherichia coli endA deletion strain for use in two-hybrid shuttle vector selection. *Biotechniques* **35,** 272–274, 76, 78.
16. Gietz, R.D., Schiestl, R.H., Willems, A.R. and Woods, R.A. (1995) Studies on the transformation of intact yeast cells by the LiAc/SS-DNA/PEG procedure. *Yeast* **11,** 355–360.
17. Kolonin, M.G., Zhong, J. and Finley, R.L. (2000) Interaction mating methods in two-hybrid systems. *Methods Enzymol* **328,** 26–46.
18. Stajich, J.E., Block, D., Boulez, K., Brenner, S.E., Chervitz, S.A., Dagdigian, C., Fuellen, G., Gilbert, J.G., Korf, I., Lapp, H., Lehvaslaiho, H., Matsalla, C., Mungall, C.J., Osborne, B.I., Pocock, M.R., Schattner, P., Senger, M., Stein, L.D., Stupka, E., Wilkinson, M.D. and Birney, E. (2002) The Bioperl

toolkit: Perl modules for the life sciences. *Genome Res.* **12,** 1611–1618.

19. Geyer, C.R. and Brent, R. (2000) Selection of genetic agents from random peptide aptamer expression libraries. *Methods Enzymol.* **328,** 171–208.
20. Kolonin, M.G. and Finley, R.L., Jr. (1998) Targeting cyclin-dependent kinases in Drosophila with peptide aptamers. *Proc. Natl. Acad. Sci. U.S.A.* **95,** 14266–14271.
21. Struhl, K., Stinchcomb, D.T., Scherer, S. and Davis, R.W. (1979) High-frequency transformation of yeast: autonomous replication of hybrid DNA molecules. *Proc. Natl. Acad. Sci. U.S.A.* **76,** 1035–1039.
22. Birboim H.C. and Doly, J. (1979) A rapid alkaline extraction procedure for screening recombinant plasmid DNA. *Nucleic Acids Res.* **7,** 1513–1523.
23. Van Criekinge, W. and Beyaert, R. (1999) Yeast two-hybrid: state of the art. *Biol. Proc. Online* **2,** 1–38.
24. CLONTECH (1996) Matchmaker LexA Libraries (PT3040-1) User Manual.

Chapter 20

Microbodies™

Hans-Ulrich Schmoldt, Matin Daneschdar, Harald Kolmar, and Michael Blind

Abstract

Microbodies™ are novel pharmacophoric entities which are derived from naturally occurring cystine-knot microproteins. They provide extremely stable scaffolds that can be engineered to high-affinity binding proteins. A peptide-grafting approach yielded specific ligands for human thrombopoietin receptor (TPO-R). Thrombopoietin (TPO) is the primary regulator of platelet production and acts by dimerization of its cognate receptor. Chemical cross linking of two anti TPO-R Microbodies™ resulted in highly potent TPO mimetics which are promising candidates for the treatment of TPO deficiencies. The approach demonstrates the high potential of dimeric Microbodies™ as synthetic receptor agonists.

Key words: Cystine-knot protein, peptide agonist, pharmacophoric scaffold, thrombopoetin receptor, Barnase.

1. Introduction

Cystine-knot microproteins (also referred to as *knottins*) are found in various species, such as plants, spiders, cone snails or humans (*1*, *2*). Microproteins are very small structured proteins typically composed of around 35 amino acids that contain at least six disulfide bond forming cysteines and adopt an extremely stable knotted structure (cystine-knot). Although microproteins have a well conserved three-dimensional structure in common, the primary sequences are quite variable. Due to their small size, microproteins are accessible by chemical as well as by recombinant synthesis (*3*–*5*). Hence, they are well suited as molecular scaffolds for the incorporation of foreign peptide sequences in order to build novel functional binding molecules with high affinities and selectivities (*see* **Note 1**). Microbodies™ are modified cystine knot microproteins addressing a specific target that have been

Günter Mayer (ed.), *Nucleic Acid and Peptide Aptamers: Methods and Protocols, vol. 535*

DOI 10.1007/978-1-59745-557-2_20 Springerprotocols.com

generated either by a rational approach incorporating functional peptide sequences into the respective microprotein scaffold (*6, 7*) or by screening of combinatorial libraries, e.g. using bacterial surface display (*8, 9*) or mRNA display (*10*). Here, we illustrate the general strategy of functionalizing microproteins by loop grafting to obtain Microbodies^{TM} that display high-affinity binding to the thrombopoietin receptor (TPO-R) and have thrombopoietin (TPO) mimetic activity (*6*). TPO is a hematopoietic growth factor that acts as the primary regulator of thrombocytopoiesis (*11, 12*). TPO receptor, c-Mpl, which is expressed on early hematopoietic progenitors, megakaryocytes and platelets, is activated by ligand-mediated homodimerization (*13*). TPO mimetic Microbodies^{TM} were designed on the basis of the squash trypsin inhibitor EETI-II (*14*) and a minimized C-terminal fragment of the human Agouti-related protein [AGRP(87-120)] (*15*), respectively (*see* **Fig. 20.1**).

Loop 1 (EETI-II) and loop 4 [AGRP(87-120)] that are responsible for binding of the natural targets were exchanged against a functional peptide sequence originally found by Cwirla et al. (*16*) to design TPO mimetic Microbodies^{TM}. Additionally, a single lysine residue was inserted into the AGRP(87-120) sequence at its C-terminus for chemical cross-linking aimed at generating dimeric Microbodies^{TM} that are not only able to bind but also to activate the TPO receptor via ligand-mediated receptor dimerization. The EETI-II scaffold innately contains a unique lysine residue whose ϵ-amino group can be used for chemical crosslinking using bifunctional chemical crosslinkers (see below).

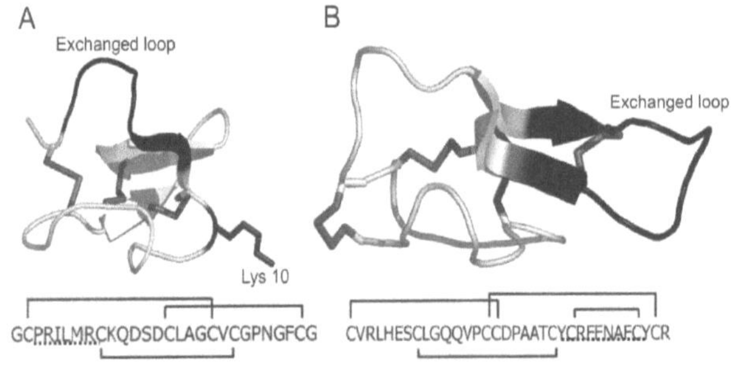

Fig. 20.1. Schematic representation of the three-dimensional structures of the Microbody^{TM} scaffolds EETI-II **(A)** and the minimized C-terminal fragment of AGRP(87–120) **(B)**. The disulfide bonds forming the knotted structure are indicated as *dark grey sticks*. β-strands are shown as *arrows*. Sequences and disulfide connectivities are indicated below. Exchanged loops are highlighted in *black* and marked with *dashed lines*, respectively. Lysine #10 of EETI-II which is used for chemical dimerization is shown as *sticks*; in the AGRP scaffold a single lysine residue was introduced in position #34 for dimerization replacing the arginine.

2. Materials

1. pBar100 plasmid (**5**) and helper plasmid pRep4 (Qiagen).
2. *E. coli* strain 71–18 [F' *lacI^q lacZ*ΔM15 *proA+B+* Δ*lac-proAB supE thi1*] (source B. Müller-Hill).
3. Oligonucleotide primers.
4. Restriction enzymes, Taq DNA polymerase, and T4 DNA ligase.
5. dYT medium (double yeast tryptone): 16 g bacto-tryptone, 10 g bacto-yeast extract, 5 g NaCl per liter.
6. TB (terrific broth): 12 g bacto-tryptone, 24 g bacto-yeast extract, 4 ml glycerol and 100 ml of a sterile solution of 0.17 M KH_2PO_4 and 0.72 M K_2HPO_4 per liter.
7. IPTG (isopropyl-β-D-thio-galactopyranoside).
8. Chromatography and reversed phase equipment.
9. SDS-PAGE (sodium dodecyl sulfate-polyacrylamide gel electrophoresis) and agarose gel equipment.
10. Chloramphenicol and kanamycin.
11. Disuccinimidyl suberate (DSS, Pierce).
12. Glacial acetic acid.
13. Device for sterile filtration of culture supernatant.
14. Fermentor or shake flasks for protein production.
15. SP-Sepharose XL cation-exchange resin (GE Healthcare).
16. Amberchrom CG-300 M reversed phase resin (Tosoh Bioscience).
17. Freeze-dry device.
18. Cyanogen bromide solution (5 M) (Fluka).
19. Semipreparative RP-HPLC column (C18)
20. M-07e cells (e.g. DSMZ, German Collection of Microorganisms and Cell Cultures).
21. Recombinant human TPO (rhuTPO, Cell Systems).
22. CellTiter kit (Promega).
23. 96-Well cell culture plates (Corning).
24. M-07e cell culture medium: RPMI1640 + Glutamax, 20% FCS, 1% Pen-Strep, 10 mg/l gentamicin, 20 μg/l rhuTPO (for starvation of the cells: culture medium with 1% FCS and without rhuTPO)
25. Microscope.
26. ELISA reader (e.g. Tecan Ultra).

3. Methods

The methods described below outline (1) construction of the Microbody^{TM} expression plasmids, (2) expression of the Microbodies^{TM} as Barnase' fusion proteins, (3) purification of the Barnase' fusion proteins, (4) cyanogen bromide cleavage and isolation of free Microbodies^{TM}, (5) chemical dimerization and (6) functional characterization of the Microbody^{TM} dimers.

3.1. Construction of Microbody^{TM} Expression Plasmids

The construction of the Microbody^{TM} expression plasmids is described in the following sections. It includes (a) the description of the Barnase' expression system, (b) the construction of the respective Microbody^{TM} genes and (c) cloning.

3.1.1. Barnase' Expression System

Due to their small size Microbodies^{TM} can be produced either using solid-phase peptide synthesis or by recombinant expression in *E. coli* (*see* **Note 2**). The herein applied Barnase' expression system relies on the genetic fusion of the respective Microbody^{TM} to an enzymatically inactivated mutant of an extracellular RNase (Barnase') from *Bacillus amyloliquefaciens* (5). In the plasmid pBar100 (*see* **Fig. 20.2A**) Barnase'-Microbody^{TM} fusions are

A

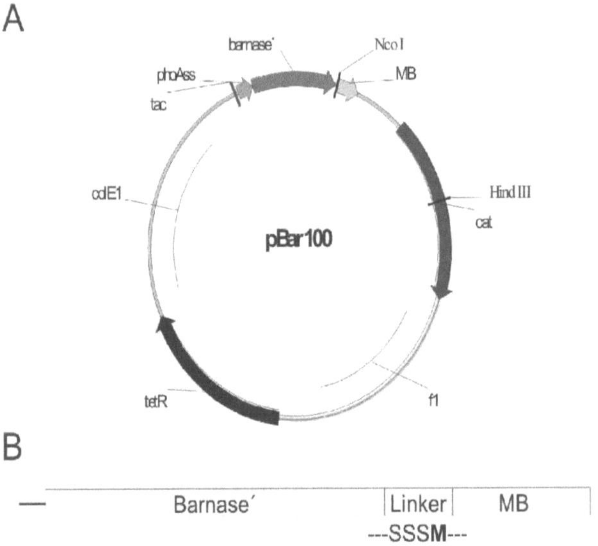

B

Fig. 20.2. **(A)** Schematic representation of the plasmid pBar100 harbouring the barnase' gene encoding for an enzymatically inactive H102A barnase variant; f1, replication origin; phoAss, alkaline phosphatase periplasmic signal sequence; cat, chloramphenicol resistance marker; tac, *tac* promoter sequence; tetR, tetracycline repressor encoding gene; colE1, colE1 replication origin; and MB, Microbody^{TM} encoding sequence. Restriction sites for cloning are indicated. **(B)** Schematic representation of the Barnase' fusion protein with a single methionine residue for chemical cleavage. MB, Microbody^{TM}.

located downstream of a phoA periplasmic leader sequence under the transcriptional control of a synthetic *tac* promoter. The fusion protein contains a short linker sequence with a single methionine residue (SSSM) for selective cleavage with cyanogen bromide at the junction of Barnase' and the fused Microbody™ (*see* **Fig 20.2B**). The fusion protein is expressed upon induction with IPTG (isopropyl-β-D-thio-galactopyranoside), directed to the periplasm and the culture supernatant where folding can occur in the oxidative milieu and from where it can be purified using conventional chromatography techniques (*5*).

3.1.2. Construction of Microbody™ Genes

The construction of the Barnase-Microbody™ expression sequences was performed according to standard procedures (*17*) and is not described here in detail due to space limitations (*see* **Note 3**). The genes encoding both Microbodies™ (named ETTP-2 and AGTP-4) were produced by PCR from templates pBar100-EETI-II M7I and pBar-100-AGRṔ, respectively. A specific oligonucleotide was used in the case of ETTP-2 that introduces the TPO mimetic sequence in loop 1 of the EETI scaffold whereas AGTP-4 was assembled in a two-step PCR reaction. Half of the *cat* gene was amplified for cloning reasons in both cases. The sequences of the corresponding Microbodies™ are indicated in **Table 20.1**.

3.1.3. Cloning

Cloning of Microbody™ genes was done according to standard procedures (*17*). Purified PCR products, consisting of the ETTP-2 or AGTP-4 Microbody™ gene, respectively, and half of the *cat* gene, were cleaved with restriction enzymes *Nco*I and *Hind*III and ligated into similarly digested pBar100 plasmids. The ligation product was transformed into *E. coli* 71–18 cells by standard methods (*17*). The cells were then plated on dYT plates containing 25 μg/ml chloramphenicol and incubated overnight at 37°C. Plasmid DNA of single clones was isolated and tested for presence of the insert by restriction enzyme digestion. The correct plasmid sequence was finally verified by DNA sequencing.

Table 20.1
Sequences of TPO mimetic Microbodies™ ETTP-2 and AGTP-4

Microbody™	Peptide sequence
ETTP-2	GCIEGPTLRQWLAARACKQDSDCLAGCVCGPNGFCGS
AGTP-4	CVRLHESCLGQQVPCCDPAATCGGTALAIEGPTLRQWLAARACKGS

Grafted peptide sequences and single lysine residues for dimerization are highlighted in bold.

3.2. Expression of Barnase' Fusion Proteins

The expression of ETTP-2 and AGTP-4 Barnase' fusion proteins involves transformation of *E. coli* strain 71–18 with the respective pBar100 plasmid and helper plasmid pRep4 which provides an additional amount of *lac* repressor avoiding possible toxic effects of protein expression during early growing phase and fermentation as well as IPTG-dependent induction of the expression cultures.

1. Transform *E. coli* 71–18 cells with the pBar100-ETTP-2 or the pBar100-AGTP-4 plasmid or any other Microbody^{TM} expression vector and pRep4 according to standard procedures (*17*).
2. Plate on dYT agar plates containing 25 μg/ml chloramphenicol and 37.5 μg/ml kanamycin.
3. Incubate overnight at 37°C.
4. Pick a single colony for inoculation of 50 ml dYT containing 25 μg/ml chloramphenicol and 37.5 μg/ml kanamycin.
5. Incubate overnight at 37°C in an incubation shaker.
6. Use the overnight culture for inoculation of 5 l TB medium containing 25 μg/ml chloramphenicol and 37.5 μg/ml kanamycin.
7. Incubate at 30°C with agitation until an OD_{600} of 3–5 (fermentor) or 1 (shake flasks) is reached.
8. Induce the cells with IPTG (1 mM final concentration) and incubate overnight at 30°C.

3.3. Purification of the Barnase' Fusion Proteins

The Barnase' fusion proteins are directed to the periplasm of *E. coli*, where oxidative conditions allow correct disulfide bond formation of the Microbodies^{TM}. The purification of the fusion proteins starts with an acidification step resulting in the complete release of the fusions from the periplasmic space into the culture medium from where they are subsequently purified using cation-exchange and reversed phase (RP) chromatography (*see* **Note 5**). An overview of the purification steps and the release of the Microbody^{TM} from the Barnase' fusion partner by CnBr cleavage is shown in **Fig. 20.3**.

3.3.1. Acidification Step

1. Cool the expression culture down to 4°C.
2. Add 55 ml of glacial acidic acid per liter of culture under continuous stirring.
3. Continue stirring for another 15 min.
4. Centrifuge at 4,000 × *g* for 30 min at 4°C and use the supernatant.
5. Remove insoluble particles from the culture supernatant by sterile filtration (0.2 μm).

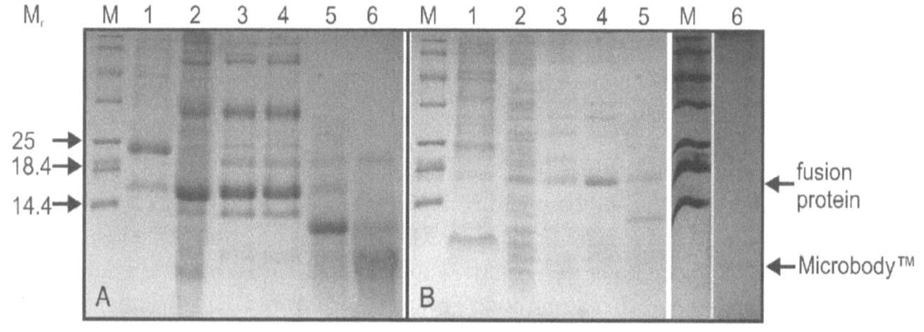

Fig. 20.3. Microbody™ expression and purification of (**A**) ETTP-2 and (**B**) AGTP-4. (1) Supernatant of bacterial liquid culture prior to acidification; (2) supernatant of acidified bacterial liquid culture (**Section 3.3.1**); (3) pooled fractions after cation exchange chromatography (**Section 3.3.2**); (4) pooled fractions after RP-chromatography (**Section 3.3.3**); (5) after cyanogen bromide cleavage of the fusion proteins (**Section 3.4**); (6) isolated Microbody™ after RP-HPLC (**Section 3.4**); *M*: molecular weight marker.

3.3.2. Cation-Exchange Chromatography

1. Dilute filtered supernatant 1:5 with H_2O.
2. Apply the diluted supernatant to a cation-exchange column (for example a XK26 column: 2.6 × 20 cm, 100 ml bed volume, containing SP-Sepharose XL).
3. Elute with a linear gradient ranging from 0 to 0.5 M NaCl in 25 mM Na-acetate buffer pH 5.0.
4. Analyze fractions using SDS-PAGE (15% gel) and pool fusion protein containing fractions.

3.3.3. RP-Chromatography

1. Apply the pooled elution fractions from the cation-exchange chromatography to a RP column (for example a XK26 column containing Amberchrom CG-300 M, 2.6 × 20 cm, 100 ml bed volume).
2. Wash with H_2O/0.1% (v/v) trifluoroacetic acid (TFA).
3. Elute with a linear gradient from 5 (v/v) to 60% acetonitrile/ 0.1% (v/v) TFA for 60 min at a flow rate of 7 ml/min.
4. Analyze fractions using SDS-PAGE (15% gel) and pool fusion protein containing fractions.
5. Freeze-dry the pooled elution fractions.
6. Determine the amount of lyophilized fusion protein by weighing (*see* **Note 4**).

3.4. Cyanogen Bromide Cleavage and Isolation of Free Microbodies™

Described below is the cyanogen bromide mediated chemical cleavage of fusion proteins (*see* **Note 6**) and subsequent isolation of free Microbodies™ using RP-chromatography (*see* **Fig. 20.3**).

1. Dissolve the respective lyophilized fusion protein in 20 μl 0.2 M HCl/8 M urea or alternatively in 0.2 M HCl/6 M guanidinium hydrochloride per milligram of protein.

2. Add in a fume hood 0.6 μl 5 M cyanogen bromide solution per milligram of fusion protein. Careful handling of the volatile and highly toxic CNBr in a fume hood is mandatory!
3. Incubate in a closed teflon vessel overnight at room temperature.
4. Dilute the sample 1:10 with H_2O/0.1% (v/v) TFA and apply to semi-preparative RP-HPLC for isolation of the free Microbodies^{TM}. A YMC J'sphere ODS-H80 column at a flow rate of 10 ml/min and a linear gradient from 10 to 80% eluent B (90% (v/v) acetonitrile/0.1% (v/v) TFA) can be used. The microprotein containing fraction commonly elutes at around 50% eluent B. Collect the flow through and the 0–15% acetonitrile fractions that might contain unreacted CNBr and dispose them as toxic waste appropriately.
5. Pool and subsequently freeze-dry the Microbody^{TM} containing fractions. Determine the amount of purified Microbody^{TM} by weighing.
6. Check the purity of the microprotein by analytical RP-HPLC using a C18 column (e.g. Phenomenex Synergi 4u Hydro-RP, 80 Å; 250 × 4.60 mm) and a gradient from 10 to 80% acetonitrile in 30 min.
7. Analysis can be done by ESI mass spectroscopy for verifying the correct mass of the Microbody^{TM}.

3.5. Chemical Dimerization

Activation of the TPO receptor relies on ligand-mediated homodimerization. Hence, two receptor binding sites on the agonistic ligand molecule are required. To achieve this, the TPO mimetic Microbodies^{TM} were designed to contain a single lysine residue for chemical cross-linking.

1. Dissolve the Microbody^{TM} at 2–4 mg/ml in dimethylformamide/dimethylsulfoxide (1:1) containg 1% triethylamine.
2. Add 1 eq. of disuccinimidyl suberate (DSS) to 2 eq. of protein.
3. Incubate overnight at room temperature.
4. Separate the dimer from the remaining monomers by RP-HPLC using analytical or semi-preparative columns as described in **Section 3.4.7**.
5. Analyze successful dimerization using Tris–Tricine SDS-PAGE (*18*).

3.6. Functional Characterization of the Microbody^{TM} Dimers

The dimeric Microbodies^{TM} can be assayed for functional activity by examining their ability to stimulate proliferation of the TPO-dependent megakaryoblastic cell line M-07e (*6*). Dimeric Microbodies^{TM} ETTP-2d and AGTP-4d are able to effectively induce a proliferative response (*see* **Fig. 20.4**). Induction of cell proliferation is measured as follows:

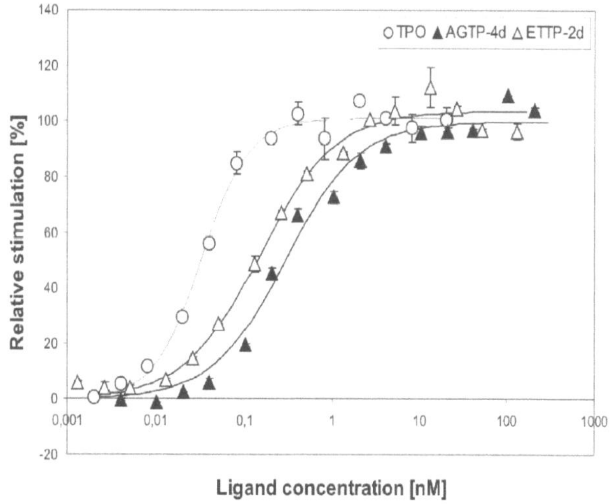

Fig. 20.4. Functional characterization of TPO mimetic dimeric Microbodies™ ETTP-2d and AGTP-4d compared to the natural cytokine TPO. M-07e cells were grown in the presence of the indicated concentrations of TPO or dimeric Microbodies™. Proliferative response was determined photometrically after incubation at 37°C with 5% CO_2 for 72 h. The chart shows relative stimulation of cell proliferation with the highest value of TPO-dependent stimulation set to 100%. Mean values and standard deviations were calculated from duplicate measurements.

1. Grow M-07e cells in RPMI supplemented with rhuTPO at 37°C and 5% CO_2. Wash the M-07e cells twice in M-07e cell culture medium devoid of rhuTPO.
2. Determine the cell concentration via counting the cells of an aliquot using a Neubauer counting chamber.
3. Seed the cells in a 96-well cell culture plate at 2×10^4 cells in 100 μl per well in medium devoid of rhuTPO.
4. Let the cells starve for 2 h at 37°C and 5% CO_2.
5. Add in duplicate or triplicate various concentrations of rhuTPO or the respective dimeric Microbody™ (0–1,000 nM, *see* **Fig. 20.4**).
6. Incubate for 72 h at 37°C and 5% CO_2.
7. Add 25 μl of CellTiter reagent and incubate for 4–6 h at 37°C and 5% CO_2.
8. Measure the absorbance at 492 nm using an ELISA reader and plot the optical density against ligand concentration. Plot the relative stimulation via ligand concentration by setting the highest value of TPO dependent stimulation to 100% (*see* **Fig. 20.4**)

4. Notes

1. A comprehensive overview of the knottin structural family (also referred to as the "Inhibitor Cystine Knot (ICK) family" or "cystine-knot microproteins") comprising sequence and structural databases is available at http://knottin.com (*19*).
2. For the expression of small, disulfide-rich proteins in *E. coli* cells alternative protocols have been published (*8*, *20–23*). In most attempts carrier proteins are used that direct the fusions into insoluble inclusion bodies requiring solubilization in denaturing agents, refolding and oxidation of the cysteines. This oxidation was also applied when Microbodies™ were chemically synthesized.
3. Due to their small size of around 100 nucleotides, variant microprotein genes are readily accessible through PCR amplification using a synthetic oligonucleotide of the length of approximately 60 nucleotides as template in fmol amounts that is extended using PCR primers which also introduce appropriate restriction sites for cloning. Alternatively, fully synthetic genes can be ordered from different providers and cloned directly into the vector of choice.
4. The average yield of Barnase' fusion proteins using the described system in small to medium scale fermentation devices (1 l/5 l fermentor vessels) is approximately 22 mg per liter of *E. coli* culture (*5*).
5. The RP chromatography is mainly performed to remove colored contaminating material which is not visible in a Coomassie stained SDS gel. This purification step can be omitted if the fusion protein containing fractions from cation exchange chromatography are colorless.
6. Selective cleavage of proteins at pre-defined sites allowing the removal of tags and carrier proteins can be in principle achieved by using proteolytic enzymes like thrombin, TEV protease or enterokinase or by using chemicals. Cyanogen bromide (CNBr) selectively cleaves a peptide bond at methionine residues while forming a homoserine lactone. Thus CNBr can be a relatively low-cost and specific alternative to proteases provided that the protein of interest contains no internal methionine residues. Frequently, formic acid is used as reaction solvent for cleavage with CNBr instead of urea or guanidinium hydrochloride. Formic acid is not recommended when working with disulfide-rich proteins because it tends to reduce the disulfide bonds (*24*).

Acknowledgments

The authors thank Dr. Sebastian Krause and Dr. Alexander Wentzel for substantial contribution in this project.

References

1. Craik, D.J., Daly, N.L. and Waine, C. (2001) The cystine knot motif in toxins and implications for drug design. *Toxicon* **39**, 1, 43–60.
2. Pallaghy, P.K., Nielsen, K.J., Craik, D.J. and Norton, R.S. (1994) A common structural motif incorporating a cystine knot and a triple-stranded beta-sheet in toxic and inhibitory polypeptides. *Protein Sci.* **3**, 1833–1839.
3. Avrutina, O., Schmoldt, H.-U., Kolmar, H. and Diederichsen, U. (2004) Fmoc assisted synthesis of a 29-residue cystine-knot trypsin inhibitor containing a guaninyl amino acid at the P1 po-sition. *Eur. J. Org. Chem.* **23**, 4931–4935.
4. Avrutina, O., Schmoldt, H.-U., Gabrijelcic-Geiger, D., Le Nguyen, D., Sommerhoff, C.P., Diederichsen, U. and Kolmar, H. (2005) Trypsin inhibition by macrocyclic and open chain variants of the squash inhibitor MCoTI-II. *Biol. Chem.* **386**, 1301–1306.
5. Schmoldt, H.-U., Wentzel, A., Becker, S. and Kolmar, H. (2004) A fusion protein system for the recombinant production of short disulfide bond rich cystine-knot peptides using barnase as a purification handle. *Protein Expr. Purif.* **39**, 82–89.
6. Krause, S., Schmoldt, H.-U., Wentzel, A., Ballmaier, M., Friedrich, K. and Kolmar, H. (2007) Grafting of thrombopoietin-mimetic peptides into cystine knot miniproteins yields high-affinity thrombopoietin antagonists and agonists. *FEBS J.* **274**, 86–95.
7. Reiss, S., Sieber, M., Oberle, V., Wentzel, A., Spangenberg, P, Claus, R., Kolmar, H. and Lösche, W. (2006) Inhibition of platelet aggregation by grafting RGD and KGD sequences on the structural scaffold of small disulfide-rich proteins. *Platelets* **17**, 3, 153–157.
8. Wentzel, A., Christmann, A., Krätzner, R. and Kolmar, H. (1999) Sequence requirements of the GPNG β-turn of the *Ecballium elaterium* trypsin inhibitor II explored by combinatorial library screening. *J. Biol. Chem.* **274**, 21037–21043.
9. Wentzel, A., Christmann, A., Adams, T. and Kolmar, H. (2001) Display of passenger proteins on the surface of Escherichia coli K-12 by the enterohemorrhagic *E. coli* intimin EaeA. *J. Bacteriol.* **183**, 7273–7284.
10. Baggio, R., Burgstaller, P., Hale, S.P., Putney, A.R., Lane, M., Lipovsek, D., Wright, M.C., Roberts, R.W., Liu, R., Szostak, J.W. and Wagner, R.W. (2002) Identification of epitope-like consensus motifs using mRNA display. *J. Mol. Recognit.* **15**, 3, 126–134.
11. Kaushansky, K., Lok, S., Holly, R.D., Broudy, V.C., Lin, N., Bailey, M.C., Forstrom, J.W., Buddle, M.M., Oort, P.J., Hagen, F.S. et al. (1994) Promotion of megakaryocyte progenitor expansion and differentiation by the c-Mpl ligand thrombopoietin. *Nature* **369**, 568–571.
12. de Sauvage, F.J., Hass, P.E., Spencer, S.D., Malloy, B.E., Gurney, A.L., Spencer, S.A., Darbonne, W.C., Henzel, W.J., Wong, S.C., Kuang, W.J. et al. (1994) Stimulation of megakaryocytopoiesis and thrombopoiesis by the c-Mpl ligand. *Nature* **369**, 533–538.
13. Methia, N., Louache, F., Vainchenker, W. and Wendling, F. (1993) Oligodeoxynucleotides antisense to the protooncogene c-mpl specifically inhibit *in vitro* megakaryocytopoiesis. *Blood* **82**, 1395–1401.
14. Heitz, A., Chiche, L., Le-Nguyen, D. and Castro, B. (1989) 1H, 2D NMR and distance ge-ometry study of the folding of *Ecballium elaterium* trypsin inhibitor, a member of the squash inhibitors family. *Biochemistry* **28**, 2392–2398.
15. Jackson, P.J., McNulty, J.C., Yang, Y.K., Thompson, D.A., Chai, B., Gantz, I., Barsh, G.S. and Millhauser, G.L. (2002) Design, pharmacology, and NMR structure of a minimized cystine knot with

agouti-related protein activity. *Biochemistry* **41**, 7565–7572.

16. Cwirla, S.E., Balasubramanian, P., Duffin, D.J., Wagstrom, C.R., Gates, C.M., Singer, S.C., Davis, A.M., Tansik, R.L., Mattheakis, L.C., Boytos, C.M. et al. (1997) Peptide agonist of the thrombopoietin receptor as potent as the natural cytokine. *Science* **276**, 1696–1699.
17. Sambrook, J., Fritsch, E.F. and Maniatis, T. (1989) *Molecular Cloning, A Labora-tory-Manual*, Second ed. Cold Spring Harbor, New York: Cold Spring Harbor Laboratory Press.
18. Schägger, H. and von Jagow, G. (1987) Tricine-sodium dodecyl sulfate-polyacrylamide gel electrophoresis fort he separation of proteins in the range from 1 to 100 kDa. *Anal. Biochem.* **166**, 2, 368–379.
19. Gelly, J.C., Gracy, J., Kaas, Q., Le-Nguyen, D., Heitz, A. and Chiche, L. (2004) The KNOTTIN website and database: a new information system dedicated to the knottin scaffold. *Nucleic Acids Res.* **32**, 1, 156–159.
20. Döbeli, H., Andres, H., Breyer, N., Draeger, N., Sizmann, D., Zuber, M.T., Wein-ert, B. and Wipf, B. (1998) Recombinant fusion proteins for the industrial production of disulfide bridge containing peptides: purification, oxidation without concatamer formation, and selective cleavage. *Protein Expr. Purif.* **12**, 404–414.
21. Fairlie, W.D., Uboldi, A.D., De Souza, D.P., Hemmings, G.J., Nicola, N.A. and Baca, M. (2002) A fusion protein system for the recombinant production of short disulfide-containing pep-tides. *Protein Expr. Purif.* **26**, 171–178.
22. Chen, X.M., Qian, Y.W., Chi, C.W., Gan, K.D., Zhang, M.F. and Chen, C.Q. (1992) Chemical synthesis, molecular cloning, and expression of the gene coding for the Trichosanthes trypsin inhibitor – a squash family inhibitor. *J. Biochem.* **112**, 45–51.
23. Kojima, S., Miyoshi, K. and Miura, K. (1996) Synthesis of a squash-type protease inhibitor by gene engineering and effects of replacements of conserved hydrophobic amino acid residues on its inhibitory activity. *Protein Eng.* **9**, 1241–1246.
24. Villa, S., De Fazio, G. and Canosi, U. (1989) Cyanogen bromide cleavage at methionine residues of polypeptides containing disulfide bonds. *Anal. Biochem.* **177**, 161–164.

Chapter 21

Peptide Aptamers for Small Molecule Drug Discovery

Carine Bardou, Christophe Borie, Marc Bickle, Brian B. Rudkin, and Pierre Colas

Abstract

Peptide aptamers have primarily been used as research tools to manipulate protein function and study regulatory networks. However, they also find multiple applications in therapeutic research, from target identification and validation to drug discovery. Because of their unbiased combinatorial nature, peptide aptamers interrogate the biological significance of numerous molecular surfaces on target proteins. Their use enables the identification and validation of some of these surfaces as interesting therapeutic targets to pursue. Peptide aptamers can subsequently be used to guide the discovery of small molecule drugs specific for these molecular surfaces.

Here, we present a high-throughput screening assay that identifies small molecules that displace interactions between proteins and their cognate peptide aptamers. AptaScreen is a duplex yeast two-hybrid assay featuring two luciferase reporter genes. It can be performed in 96- or 384-well plates and can be fully automated.

Key words: Peptide aptamers, yeast two-hybrid, high-throughput screening, luminescence, drug discovery.

1. Introduction

1.1. Peptide Aptamers in Therapeutic Research

Peptide aptamers were initially developed as research tools to manipulate protein function *(1)* and they are mostly used for this purpose by an increasing number of academic laboratories *(2)*. However, as the technology matured, it became obvious that peptide aptamers could find multiple applications in therapeutic research *(3, 4)*. Their use enables a high-confidence validation of therapeutic targets, since they introduce perturbations that are similar to those caused by most therapeutic

Carine Bardou and Christophe Borie contributed equally to this work

Günter Mayer (ed.), *Nucleic Acid and Peptide Aptamers: Methods and Protocols, vol. 535*

DOI 10.1007/978-1-59745-557-2_21 Springerprotocols.com

molecules and that differ from those caused by more popular target validation approaches that act upon protein expression levels *(5)*. When selected for their ability to confer given phenotypes to a cell population, peptide aptamers can be used to identify therapeutic targets *(6, 7)*. Recent advances in peptide chemistry and delivery pave the way to the use of peptide aptamers as biotherapeutic molecules against intracellular targets *(8)*. Finally, peptide aptamers can guide the discovery of small molecule drugs *(3)*. Because they are unbiased combinatorial molecules, peptide aptamers decorate numerous polymorphic protein surfaces, including *druggable* sites. When the binding sites of bioactive aptamers are mapped on their targets (by structural studies and/or yeast two-hybrid assays with target mutants *(9)*), the validated therapeutic targets are not only the protein targets themselves but, more precisely, specific molecular surfaces on these proteins *(10)*. If these surfaces are deemed *druggable* (i.e. amenable to small molecules), two complementary approaches can be undertaken to find drug candidates. First, virtual screening or target-based pharmacophore searches can be performed against these surfaces. Second, peptide aptamer displacement screening assays can identify small molecule hits that bind to these surfaces. This chapter focuses on a high-throughput aptamer displacement screening assay that we have recently developed and that has delivered small molecule hits against different protein targets.

1.2. AptaScreen

AptaScreen is a high-throughput displacement assay that identifies small molecules that disrupt interactions between target proteins and their cognate peptide aptamers. It rests on the hypothesis that peptide aptamers and small molecules binding to the same molecular surface on a target protein should trigger the same biological effects. AptaScreen is a dual luminescence yeast two-hybrid assay amenable to the 384-well plate format *(11)*. This duplex assay enables the quantification, in a single well, of two target–aptamer interaction phenotypes, presented by two yeast strains that contain either the firefly (*luc*) or the *Renilla* (*ruc*) luciferase reporters. This duplex scheme presents two major advantages. First, it doubles the screening throughput since two distinct target–aptamer interactions are screened at once. Second, it circumvents a major problem associated with cellular screening assays. Because the desired readout is a decreased yeast two-hybrid interaction phenotype (here, a decreased luciferase activity), it is important to be able to identify and discard immediately the numerous small molecules that are toxic to yeast and that cause a general, non-specific inhibition of transcription and translation. Consequently, only those molecules that inhibit

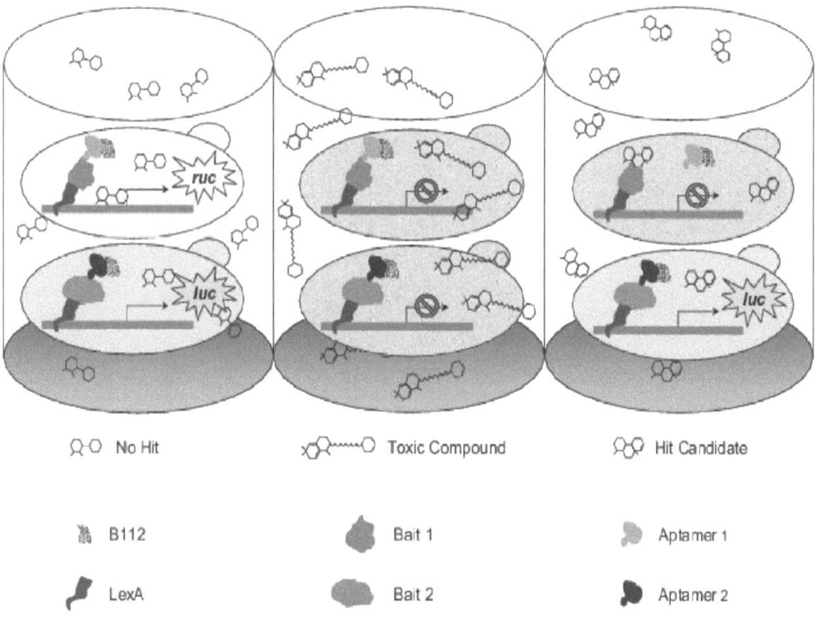

Fig. 21.1. Principle of the AptaScreen assay. The molecule delivered in the leftmost well does not affect either the *luc* or *ruc* signals and is thus considered as negative. The molecule delivered in the central well inhibits both *luc* and *ruc* signals and is thus flagged as toxic (or as interfering with the assay) and not further considered. The molecule delivered in the rightmost well selectively inhibits the *ruc* signal and is thus retained as a hit candidate.

one of the two luciferase reporters are considered as hit candidates, whereas those molecules that inhibit both reporters are flagged as toxic or interfering with the assay and discarded from further analysis (*see* **Fig. 21.1**).

Since yeast permeability to small molecules is always a concern in yeast-based screening assays, we have set up AptaScreen using an *erg6 Saccharomyces cerevisiae* mutant, which shows an enhanced permeability *(12)*. Our procedures can be applied to wild-type strains or to strains bearing mutations on other genes, such as those involved in drug efflux.

Although we have performed AptaScreen solely using a LexA-based yeast two-hybrid system and our own strains and plasmids, we believe that our procedures can be easily adapted to alternative yeast two-hybrid systems, such as the commonly used GAL4 system. However, an important feature of the LexA-based system resides in the galactose-inducible expression of the preys (here, peptide aptamers), which are placed under the control of the GAL promoter. AptaScreen takes full advantage of this feature. Because small molecules and galactose are added simultaneously, the small molecules are given a chance to interact with the target proteins

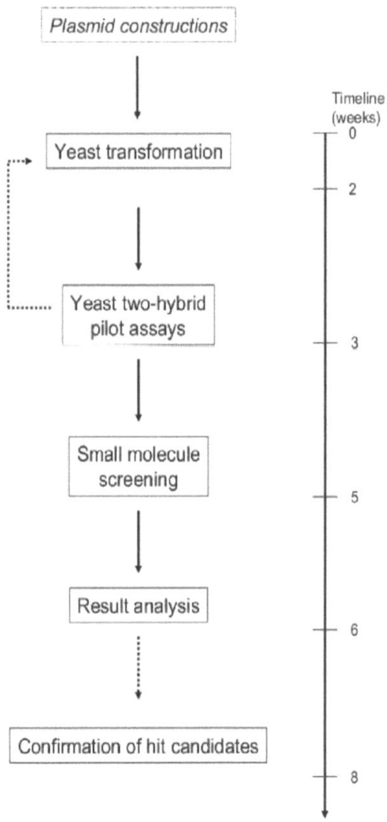

Fig. 21.2. Outline and timeline of the different procedures. Plasmid construction steps and hit confirmation experiments based on alternative biochemical interaction assays are not addressed in this chapter. *Dotted lines* indicate optional workflows that depend on the results obtained in the previous steps. The timeline is insensitive to the size of the screening, except for the screening step itself, which may last shorter or longer than 2 weeks, depending on the number of molecules to be screened and on whether a manual or automated procedure is used. As an indication, our automated platform allows us to screen up to 100,000 molecules in 2 weeks.

(the baits) in absence of their interacting aptamers. It is probable that this experimental scheme affords enhanced hit rates.

The outline and timeline of the different procedures described in this chapter are shown in **Fig. 21.2**. We routinely use a fully automated robotic platform and the 384-well format which enables a maximum throughput of 15,000 screened compounds /day. Here, we provide experimental protocols that can be performed either manually (using the 96-well plate format) or robotically (using the 96- or 384-well plate format).

2. Materials

2.1. Yeast Transformation

2.1.1. Strains and Plasmids

(*see* Note 1).

1. Yeast strain (*erg6*, *leu2*, *ura3*, *his3*)
2. Reporter plasmids bearing a 2 μ replication origin, a *URA3* marker and a *8lexAop::luc* or a *8lexAop::ruc* reporter gene, coding for the firefly or the *Renilla reniformis* luciferase, respectively.
3. Bait plasmids bearing a 2 μ or a *CEN/ARS* replication origin, a *HIS3* marker and that direct the constitutive expression of LexA–protein fusions under the control of an *ADH* promoter.
4. Prey plasmids bearing a 2 μ or a *CEN/ARS* replication origin, a *LEU2* marker and that direct the inducible expression of B112-peptide aptamer fusions under the control of the *GAL1* promoter.

2.1.2. Solutions and Buffers

1. TL solution: 100 mM LiAc, pH 7.2, 10 mM Tris–HCl, pH 7.5.
2. TL/PEG solution: 40% PEG 4000 in TL solution.
3. Salmon testes DNA (Sigma, St. Louis, MO).
4. Yeast synthetic solid and liquid media. SD: synthetic medium containing 2% glucose.
5. Yeast YPD rich medium.
6. Solution of 20% glucose (autoclaved, stored at 4°C).
7. Dimethyl sulfoxide (DMSO).
8. Solution of glycerol (87%, autoclaved, stored at room temperature).

2.2. Yeast Two-Hybrid Pilot Assays; Screening and Confirmation Assays

2.2.1. Solutions and Buffers

1. Solutions of 20% of glucose, 20% galactose, and 10% raffinose (autoclaved and stored at 4°C).
2. Solution of glycerol (87%, autoclaved, stored at room temperature).
3. Yeast synthetic liquid media. SR: synthetic medium containing 1% raffinose.
4. DMSO 10% (always make fresh – do not store).
5. Dual-Glo luciferase assay system (Promega, Madison, WI).

2.2.2. Multi-well Plates and Plate Reader

1. 96-well plates, white, half area, flat bottom (Greiner, Dutscher 675075).
2. 384-well plates, white, flat bottom (Greiner, Dutscher 781075).
3. Plate reader with luminescence module.

2.2.3. Automation Devices (Optional)

The minimal robotic platform that can be used to implement AptaScreen and reach a high-throughput includes:

1. A non-thermostated plate hotel.
2. A thermostated plate hotel (30°C).
3. A dispenser (with at least two channels and a minimal precision of 2 μL).
4. A pipetting head (96 or 384 pins, with a minimal precision of 2 μL).
5. A robotic arm.

3. Methods

3.1. Yeast Transformation

Transformation efficiencies in *erg6* strains are significantly lower than that observed in wild-type strains. Therefore, only one plasmid should be transformed at a time. We suggest introducing the reporter plasmids first and to freeze the transformants, which can then be used for several subsequent projects. We also suggest introducing the prey plasmids prior to the bait plasmids. Since bait expression is constitutive and since many baits show at least some toxicity for yeast, it is better to keep the bait expression period as short as possible. This can be achieved by performing the screening shortly after having introduced the bait plasmids or, better, by freezing the transformants.

1. Inoculate the yeast strain into 4 mL YPD or appropriate selective medium containing 2% glucose (*see* **Note 2**).
2. Incubate shaking at 230 rpm at 30°C overnight.
3. Inoculate the preculture into $n \times 2$ mL of the same medium (n, being the number of transformations to perform), aiming at a starting OD_{600} of approximately 0.3.
4. Incubate shaking at 230 rpm at 30°C to reach an OD_{600} comprised between 0.6 and 0.8.
5. Pellet the yeast 3 min at 4,600 rcf.
6. Discard the supernatant and resuspend the pellet into 1 mL H_2O.
7. Transfer into Eppendorf tubes.
8. Pellet the yeast 5 min at 3,500 rcf at 4°C.
9. Discard the supernatant and resuspend the pellet into 1 mL cold TL solution.
10. Pellet the yeast 5 min at 3,500 rcf at 4°C.

11. Discard the supernatant and resuspend the pellet into $n \times 50$ μL cold TL solution (*n*, being the number of transformations to perform). Keep on ice.
12. Dispense 50 μL of yeast in as many Eppendorf tubes as transformations to perform.
13. To each tube, add in the following order: 2.5 μL salmon testes DNA, 1–3 μg plasmid, 300 μL TL/PEG
14. Mix thoroughly by inverting the tubes several times (*see* **Note 3**).
15. Incubate 30 min at 30°C. Mix thoroughly once or twice during the incubation, by inverting the tubes several times.
16. Add 36 μL DMSO to each tube and mix by inverting.
17. Incubate 15 min at 42°C in a water bath or a heating block.
18. Add 1 mL H_2O to each tube.
19. Pellet the yeast 3 min at 3,500 rcf.
20. Discard the supernatant and resuspend into 100 μL H_2O using a pipette (*see* **Note 3**).
21. Plate on appropriate SD solid medium (*see* **Note 2**).
22. Grow the transformants for 2–3 days at 30°C (*see* **Note 4**).
23. To freeze transformants containing the reporter plasmids or all three plasmids, grow cultures in SD-U or SD-ULH, respectively, to reach an OD_{600} of approximately 3.5. Add glycerol (20% final concentration), mix well, and aliquot in cryotubes. Store at –80°C.

3.2. Yeast Two-Hybrid Pilot Assays

The goal of these pilot assays is to determine whether an interaction between a target protein and a peptide aptamer produces a yeast two-hybrid interaction phenotype that is sufficiently strong and above background to enable a screening campaign. Background is defined as the interaction phenotype observed when co-expressing a bait and a non-interacting peptide aptamer. Background corresponds to the well-known phenomenon of transcriptional auto-activation, which is frequently observed to various extents with many baits. Another purpose of these pilot assays is to determine whether the two interactions that will be used in the screening produce comparable yeast two-hybrid phenotypes (i.e. comparable *luc* and *ruc* values). We do not advise to match a pair of target–peptide aptamer interactions that produce a strong and a weak interaction phenotype.

An easy way to optimize signal to background ratios and to adjust two-hybrid interaction phenotypes is to try different combinations between 2 μ and *CEN/ARS*- bait and prey plasmids, using *luc* and *ruc*. Usually, varying the copy numbers of the plasmids significantly impacts the strength of the interaction phenotypes. Also, we have always observed that for a given interaction, *luc* signals are stronger than *ruc* signals.

These pilot assays entail two experiments. First, the yeast two-hybrid *luc* and *ruc* phenotypes are monitored individually for each transformed strain, over an 8 h time course, with a time point every hour. Then, a pair of strains is selected and combined. *luc* and *ruc* phenotypes are measured in single wells (duplex assay), at the optimal time defined in the kinetic experiment.

1. Pick a few transformed colonies and resuspend into 4 mL of SD-ULH.
2. Incubate shaking at 230 rpm at 30°C for 24 h to grow precultures.
3. Inoculate 0.5–1 mL of precultures into 3 mL of SR-ULH medium containing 1% glycerol (*see* **Note 5**).
4. Incubate shaking at 230 rpm at 30°C for 24 h.
5. Measure the OD_{600} of the cultures and adjust at 0.75 using the culture medium.
6. Prepare four times fewer 96 half-well white plates as strains to be tested (including the negative control involving a non-interacting peptide aptamer for each bait). The experiment is performed in duplicate. Column(s) 1 correspond(s) to negative controls, where aptamer expression is not induced. In column 1 of each plate, dispense 2.5 μL DMSO 10% (*see* **Note 6**) and 2.5 μL H_2O. Add 20 μL of yeast suspension (2 wells/ strain). In column 2 of each plate, dispense 2.5 μL DMSO 10% (*see* **Note 6**) and 2.5 μL galactose 20%. Add 20 μL of yeast suspension (2 wells/ strain). Column(s) 2 represent(s) the 8 h time point of the kinetics. Incubate the plates with lids on at 30°C without agitation for 8 h. Every hour, fill the subsequent column(s) as described for column(s) 2. The last column(s) to be filled (column(s) 10) correspond(s) to the 0 h time point. (*see* **Note 7**).
7. Prepare the Dual-Glo luciferase assay reagents as follows:

   R1 reagent: add 100 mL of Dual-Glo luciferase buffer into the bottle containing the lyophilized substrate. Mix gently to solubilize. This solution can be aliquoted, stored at −20°C, and thawed no more than twice. Prepare 2 mL /plate.

   R2 reagent: just before use, take the required volume of Dual Glo Stop&Glo buffer (2 mL /plate) from the stock stored at 4 °C. Add 1% (v/v) of Dual-Glo Stop&Glo substrate, taken from the stock stored at −20°C. Mix gently. (*see* **Note 8**).
8. At the end of the 8 h time course, add 20 μL of R1 reagent to all wells and incubate the plates with lids on, at 30°C for 1 h in the dark.

9. Measure the firefly *luc* luminescence using an appropriate plate luminometer.
10. In those plates containing yeast transformed with a *Renilla ruc* reporter plasmid, add 20 μL of R2 reagent to all wells and incubate plates with lids on at 30°C for 1 h in the dark.
11. Measure the *Renilla ruc* luminescence.
12. For each plate, calculate the mean luminescence values of the "uninduced" control and of the test wells for each time point, and plot the ratios over time. Analyze the kinetics of the different two-hybrid signals. Determine the time points and the combinations of bait and prey plasmids (in case different bait and prey plasmids were tested) that give robust *luc* and *ruc* signals and that minimize the difference between the signals of each interaction. Examples of typical results are given in **Fig. 21.3**.
13. To test the selected pair of strains in the conditions of a small-molecule screening assay and to determine a *Z'* value for the screening assay, prepare the two yeast suspensions as described in **Steps 1–5** and mix equal volumes of both suspensions.
14. Dispense 2.5 μL of 10% DMSO in each well of a 96- or 384-well plate (chose the format that will be used in the screening).
15. Dispense 2.5 μL of 20% galactose in one half of the plate (A1 to H6 or A1 to P12 for 96- or 384-well plates, respectively) and 2.5 μL of H_2O in another half.
16. Dispense 20 μL of the yeast suspension in all wells.
17. Incubate the plate at 30°C in the dark, for the optimal time determined in **Step 12** (*see* **Note 9**), with or without the lid (depending on what will be done during the screening).
18. Prepare the Dual-Glo luciferase assay reagents as described in **Section 3.2, Step** 7. The required volumes of R1 and R2 are 8 or 2 mL/plate for 384- or 96-well plates, respectively (*see* **Note 8**).
19. Dispense 20 μL of R1 reagent to all wells and incubate the plate (with or without the lid) at 30°C for 1 h in the dark.
20. Measure the firefly *luc* luminescence using an appropriate plate luminometer.
21. Dispense 20 μL of R2 reagent to all wells and incubate the plate (with or without the lid) at 30°C for 1 h in the dark.
22. Measure the *Renilla ruc* luminescence.
23. Determine the mean and the standard deviation of *luc* and *ruc* values of both halves of the plate (with and without

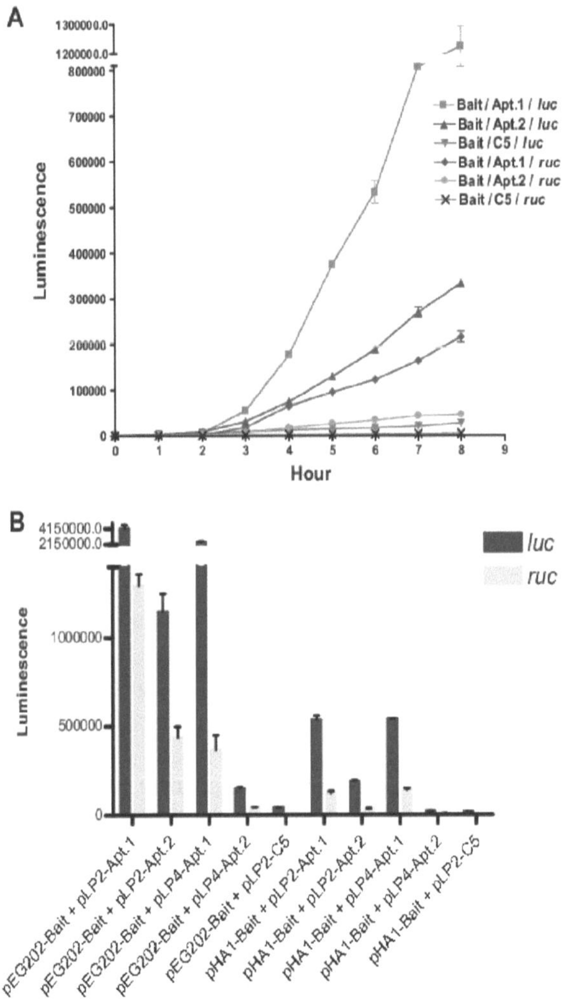

Fig. 21.3. Example of typical results from pilot assays. In this experiment, the two interactions to be tested involve a single bait protein ("Bait") and two peptide aptamers (Apt. 1 and 2) that bind two different molecular surfaces of the bait. (**A**) Kinetic experiment. For clarity, we only show the results obtained with the CEN/ARS bait plasmid (pHA1) and the 2 μ prey plasmid (pLP2). Apt. 1 shows a much higher apparent binding affinity for the bait than Apt. 2. As always observed for a given pair of interacting proteins, *luc* two-hybrid interaction phenotypes are significantly stronger than those obtained using the *ruc* reporter. Here, the transcriptional auto-activation level of the bait is negligible (compare *luc* and *ruc* values obtained with control aptamer C5 to that obtained with peptide aptamers Apt. 1 and 2). (**B**) Two-hybrid *luc* and *ruc* interaction phenotypes obtained with different combinations of bait and prey plasmids at a single 6 h time point. As already mentioned, *luc* values are higher than *ruc* values for a given interaction. The strength of the interaction phenotypes obtained with both reporters can be adjusted by the use of CEN/ARS (pHA1, pLP4) or 2 μ (pEG202, pLP2) bait and prey plasmids. Remarkably, the strength of the interaction phenotypes is more sensitive to the choice of the bait than that of the prey plasmid. In this experiment, the following combination was chosen for an AptaScreen: pHA1-Bait/pLP4-Apt.1/*luc* and pEG202-Bait/pLP2-Apt.2/*ruc*.

galactose). Calculate the Z' value by applying the following formula:

$$Z' = 1 - [3(\sigma M + \sigma m)/(M - m)]$$

M: mean of signals with galactose
m: mean of signals without galactose
σM: standard deviation of M
σm: standard deviation of m
(*see* **Note 10**)

3.3. Small-Molecule Screening Assay – AptaScreen

1. Inoculate 50 mL of frozen transformed stocks into 150 mL of SD-ULH medium.
2. Incubate shaking at 230 rpm at 30°C for 24 h.
3. Inoculate 100 mL of these cultures into 300 mL of SR-ULH medium containing 1% glycerol (*see* **Note 11**).
4. Incubate shaking at 230 rpm at 30°C for 24 h.
5. Measure the OD_{600} of the cultures and adjust at 0.75 using the culture medium.
6. Pool an equal volume of both cultures and mix well. Aim at a total volume of 8 mL per 384- or 2 mL per 96-well plate to be screened, respectively (*see* **Note 12**). Keep the mix under constant stirring with a sterile magnetic stirrer.
7. Prepare the plates containing the molecules to be screened. Fill columns 23 and 24 (for 384-well plates) or columns 11 and 12 (for 96-well plates) with 10% DMSO. All remaining wells contain the molecules at a recommended concentration of 100 μM in 10% DMSO (*see* **Note 13**).
8. Stack the plates in the desired order.
9. Pipet 2.5 μL from each well of these plates and deliver into flat bottom white plates (at the bottom of each well).
10. Dispense 2.5 μL of 20% galactose in each well in columns 1–23 or 1–11, for 384- and 96-well plates, respectively (*see* **Note 14**).
11. Dispense 20 μL of yeast suspension in each well.
12. Incubate each plate at 30°C for the optimized time, determined in the pilot assays (*see* **Note 15**).
13. Prepare the Dual-Glo luciferase assay reagents as described in **Section 3.2, Step** 7. The required volumes of R1 and R2 are 8 or 2 mL / plate for 384- or 96-well plates, respectively (*see* **Note 8**).
14. Dispense 20 μL of R1 reagent in all wells and incubate each plate at 30°C for 1 h (*see* **Note 16**).
15. Measure the firefly *luc* luminescence using an appropriate plate luminometer.

16. Dispense 20 μL of R2 reagent in all wells and incubate each plate at 30°C for 1 h (*see* **Note 16**).
17. Measure the *Renilla ruc* luminescence.

3.4. Result Analysis

1. Determine the *Z'* value for each plate and each reporter gene, using the formula given in **Section 3.2, Step 23**. Here, *M* and *m* are the mean values of the wells in columns 23 (or 11) and 24 (or 12), respectively. Plates showing *Z'* values lower than 0.5 should be discarded from further analysis.
2. For each well, calculate the percentage of *luc* and *ruc* inhibition using the following formula:

$$\%\text{inhib} = 100 - (luc\ \text{value} - mluc/Mluc - mluc) \times 100$$

(same for *ruc*) (*see* **Note 17**).
3. Identify those molecules that produce a differential inhibition of *luc* or *ruc* signal. The threshold values should be adjusted according to the overall results of the screening and the desired selection stringency. An example of a typical screening result is given in **Fig. 21.4**.

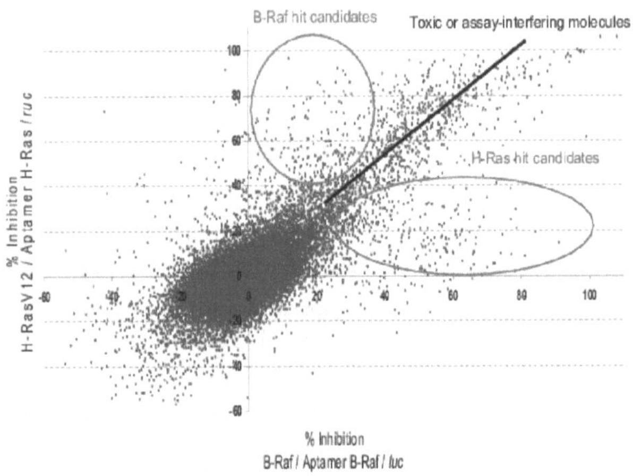

Fig. 21.4. Example of a typical screening result. Scatter plot of *luc* and *ruc* inhibition percentages obtained in a screening of 45,000 molecules against B-Raf/aptamer and H-Ras/aptamer interactions. The vast majority of the screened compounds do not significantly affect either signal (high density around the origin of both axes). A proportion of compounds enhance *luc* and/or *ruc* signals (negative percentages of inhibition, not exceeding 50%). A higher proportion of compounds inhibit *luc* and/or *ruc* signals. Areas comprising B-Raf and H-Ras hit candidates are delimited. Molecules lying in the vicinity of the diagonal axis show toxicity for yeast or interfere with the two-hybrid assay and are not further considered.

3.5. Confirmation of Hit Candidates

Every hit candidate identified by the result analysis must be confirmed first in the same conditions as the screening and then in a dose–response assay. The hit candidates are first pipetted from the plates used in the screening and delivered into 96- or 384-well plates (depending on the number of candidates). These candidates are tested as described in **Section 3.3**. The confirmed candidates are then tested in a dose–response assay. To this end, the molecules are pipetted from the undiluted stock plates (molecules at 1 mM in 100% DMSO) and subjected to serial dilutions. We suggest testing the following concentrations in duplicates: 50, 16.7, 5.5, 1.8, 0.6, 0 μM. Hence, eight hit candidates can be tested in a single 96-well plate (1 molecule /row). Preferably, hit candidates obtained with *luc* and *ruc* reporter genes should be confirmed in dose–response assays using *ruc* and *luc* reporter genes, respectively. Swapping the reporters will offer a more stringent confirmation by ensuring that the hit candidates do not interfere with the luciferases.

4. Notes

1. As already mentioned, alternative yeast two-hybrid systems can be used, with plasmids bearing different markers and transcriptional modules. In any case, we recommend the use of 2μ and *CEN/ARS* versions of bait and prey plasmids (*see* **Section 3.2**). The prey plasmids that are classically used in the LexA-based system and that bear a *TRP1* marker cannot be used in an *erg6* strain since *erg6 trp1* mutations form a synthetic lethal combination. Hence, we have constructed new prey plasmids that bear a *LEU2* marker.
2. According to the suggested sequence of transformations, the order of yeast (liquid or solid) media to be used is: YPD, SD-U, SD-UL, SD-ULH.
3. Do not vortex the tubes since PEG makes yeast fragile.
4. Usually, for the first two rounds of transformations, colonies are obtained 48 h after plating. For the third round of transformation, colonies are obtained 2–3 days after plating (depending on the bait plasmid).
5. It is also possible to inoculate cultures from frozen stocks of transformed yeast.
6. Although DMSO is not absolutely required in these pilot experiments, its use is recommended as it enables determination of the two-hybrid signals under the conditions of a small-molecule screening.

7. During the time course, yeast cultures are left at room temperature without agitation. Under these conditions, no significant growth is observed. Yeast must be thoroughly resuspended by vortexing before each time point.
8. Let R1 reagent thaw slowly at room temperature (do not hold in your hands or use a water bath). Both reagents should be protected from light and brought to room temperature before being added to yeast. When using an automated dispenser, the dead volume of the device should be added to the indicated volume of R1 and R2 reagents.
9. Usually, the optimal incubation time defined in **Section 3.2, Step 12** is 5 or 6 h.
10. The Z' value should be comprised between 0.5 and 1 (the higher, the better).
11. For those yeast transformants that grow well (and only for those), it is possible to inoculate directly the frozen stock into this 400 mL culture.
12. When using an automated dispenser, the dead volume of the device should be added to the indicated volumes of yeast suspension.
13. We routinely keep replicas of our chemical libraries at −20°C, in plates containing 2 μL of each molecule at 1 mM in 100% DMSO. We thaw the plates and dispense 18 μL of sterile water in each well to obtain molecules at 100 μM in 10% DMSO.
14. To aim at equal volumes in all wells, 2.5 μL H_2O should be added to the last column of each plate. However, we have observed that this step can be skipped without compromising the results, which is very advantageous when using a robotic procedure.
15. When using an automated platform, the exact incubation time is adjusted by the scheduling software in order to optimize the throughput, while ensuring that all plates are processed identically. To minimize the plate effect (*see* **Note 17**), it is advised to incubate the plates with lids on if the screening is performed manually or with a robot that can handle plate lids.
16. It will be impossible to perform 1 h incubations with some automated platforms, which will experience scheduling conflicts when screening large numbers of plates. In this case, the plates can be incubated 3–5 h at room temperature (after addition of each reagent), without compromising signal intensities. However, these longer incubation times will enhance the plate effect (*see* **Note 17**).

17. When using 384-well plates without the lids during incubation times, a plate effect will be observed, with stronger luminescence signals in the peripheral wells. This effect can be reduced by covering the plates and is, in any case, less pronounced with 96-well plates. A significant plate effect will affect the *M* and *m* values (calculated from the last two columns of the plates). When facing this problem, it is advised to calculate local mean values around each well and directly determine the ratio between the value measured in each well and the value of the local mean, without taking m values into account.

Acknowledgments

We are grateful to Jérôme Pansanel for the design and implementation of the screening result analysis method, and to Matthieu Schapira for his suggestions. We thank Eric Dusserre for his contribution to the first two-hybrid luminescence assays, Philippe Crouzeix and Guy-Nestor N'Gambai for their contributions to the screening result analysis method, Sandra Dollet for her contribution to the implementation of AptaScreen, and Hélène Bottin and Olivier Moncorgé for plasmid constructions. We are grateful to Marc Blondel for the gift of a *erg6* deletion strain, which we used to disrupt *erg6* in our own strain background. We are indebted to Christine Pernelle for her most useful advice and constant support. This work was supported by an OSEO / ANVAR innovation grant.

References

1. Colas, P., Cohen, B., Jessen, T., Grishina, I., McCoy, J. and Brent, R. (1996) Genetic selection of peptide aptamers that recognize and inhibit cyclin- dependent kinase 2. *Nature* **380**, 548–550.
2. Hoppe-Seyler, F., Crnkovic-Mertens, I., Tomai, E. and Butz, K. (2004) Peptide aptamers: specific inhibitors of protein function. *Curr. Mol. Med.* **4**, 529–538.
3. Baines, I.C. and Colas, P. (2006) Peptide aptamers as guides for small molecule drug discovery. *Drug Disc. Today* **11**, 334–341.
4. Colas, P. (2006) Peptide aptamers. In *Encyclopedic Reference of Genomics and Proteomics in Molecular Medicine* (Ganten, D. and Ruckpaul, K., eds.), Springer, Berlin Hedeilberg, pp. 1368–1372.
5. Abed, N., Bickle, M., Mari, B., Schapira, M., Sanjuan-Espana, R., Robbe Sermesant, K., et al. (2007) A comparative analysis of perturbations caused by a gene knockout, a dominant negative allele, and a set of peptide aptamers. *Mol. Cell. Prot.* **6**, 2110–2121.
6. Xu, X., Leo, C., Jang, Y., Chan, E., Padilla, D., Huang, B.C., et al. (2001) Dominant effector genetics in mammalian cells. *Nat. Genet.* **27**, 23–29.
7. de Chassey, B., Mikaelian, I., Mathieu, A-L., Bickle, M., Olivier, D., Nègre, D., et al. (2007) An antiproliferative genetic screening identifies a peptide aptamer that targets Calcineurin and upregulates its activity. *Mol. Cell. Prot.* **6**, 451–459.
8. Borghouts, C., Kunz, C. and Groner, B. (2005) Peptide aptamers: recent developments for cancer therapy. *Expert Opin. Biol. Ther.* **5**, 783–797.

9. Bickle, M., Dusserre, E., Moncorgé, O., Bottin, H. and Colas, P. (2006) Selection and characterization of large collections of peptide aptamers through optimized yeast two-hybrid procedures. *Nat. Protoc.* **1**, 1066–1091.
10. Pamonsinlapatham, P., Hadj-Slimane, R., Raynaud, F., Bickle, M., Corneloup, C., Barthelaix, A., et al. (2008) A RasGAP SH3 peptide aptamer inhibits RasGAP-Aurora interaction and induces caspase-independent tumor cell death, PLoS ONE 3, e2902.
11. Nieuwenhuijsen, B.W., Huang, Y., Wang, Y., Ramirez, F., Kalgaonkar, G. and Young, K.H. (2003) A dual luciferase multiplexed high-throughput screening platform for protein-protein interactions. *J. Biomol. Screen* **8**, 676–684.
12. Mukhopadhyay, K., Kohli, A. and Prasad, R. (2002) Drug susceptibilities of yeast cells are affected by membrane lipid composition. *Antimicrob. Agents Chemother.* **46**, 3695–3705.

Chapter 22

Synthesis and Application of Peptides as Drug Carriers

Robert Rennert, Ines Neundorf, and Annette G. Beck-Sickinger

Abstract

An efficient cellular drug delivery is a severe problem due to the charge, the hydrophilic character or the size of many therapeutic agents. High-drug doses, necessary to compensate the reduced bioavailability, often cause strong adverse effects. Synthetic drug delivery vectors will solve this problem, if limitations like low-cellular uptake efficiency or cytotoxicity can be overcome. Among these synthetic vectors, so-called cell-penetrating peptides (CPP) have proven their applicability as drug carriers. The ability to penetrate cellular membranes without the help of any receptor or transporter molecule was also found for derivatives of the native peptide hormone human calcitonin (hCT). We have shown that truncated hCT analogs with a branched peptide design and oligocationic side chain sequences – hCT(18-32)-k7 and hCT(9-32)-2br – are very interesting candidates as carrier peptides for drug delivery. Both peptides were found to efficiently shuttle covalently linked small molecules and non-covalently complexed DNA and RNA inside human embryonic kidney cells (HEK 293).

Key words: Cell-penetrating peptides, cell transfection, fluorescence microscopy, flow cytometry, drug delivery.

1. Introduction

Efficient drug delivery is a serious problem, both in the growing field of gene and stem cell therapy but also of "classical pharmaceuticals". The cellular administration of chemotherapeutics, therapeutic proteins, oligonucleotides or vector DNA is often hampered due to the semi-permeable character of the plasma membrane that prevents the entry of large and polar molecules inside the cell. Synthetic drug delivery vectors could solve this problem. During the last decade, a number of peptides – so-called cell-penetrating peptides *(1, 2)* – were found that are able to translocate themselves as well as covalently and non-covalently

Günter Mayer (ed.), *Nucleic Acid and Peptide Aptamers: Methods and Protocols, vol. 535*

DOI 10.1007/978-1-59745-557-2_22 Springerprotocols.com

linked cargoes. Recently, even human calcitonin (hCT), a native peptide hormone of 32 amino acids, was shown to possess cell-penetrating properties *(3, 4)*. Meanwhile, hCT-derived peptides were used to shuttle various cargoes into cells, e.g., chemotherapeutics *(5)*, proteins *(6)* or vector DNA *(7)*.

Since branched peptide designs and a high number of basic residues have been described to be of advantage for a peptidic drug carrier *(8, 9)*, we have synthesized and tested two branched, oligocationic derivatives based on N-terminally truncated hCT fragments, i.e., hCT(18-32)-k7 and hCT(9-32)-2br. Automated solid-phase peptide synthesis according to the Fmoc-strategy allow a rapid synthesis of the carrier peptides with yields in the milligram range and high purity. The branched peptide design is introduced by using a selective side chain protecting strategy for lysine residues. Initially, the resulting novel peptides should be tested to avoid any cytotoxic effects. For this purpose, a resazurin-based in vitro cell toxicology assay kit is commercially available that indicates the metabolic activity of the living cells by a fluorometrically detectable conversion of resazurin to the reduced resofurin. To examine the drug carrier potential of the novel peptides, the fluorophore 5(6)-carboxyfluorescein (CF) is covalently linked as small-model cargo, and the peptides are non-covalently complexed with either vector DNA encoding a fluorescent protein or with fluorescently labeled aptamer RNA. Fluorescence microscopy and flow cytometry proved the "high-cargo delivery" efficiencies of both novel carrier peptides.

2. Materials

2.1. Solid-Phase Peptide Synthesis

1. 30 mg (resin loading 0.5 μmol/g) 4-(2′,4′-dimethoxyphenyl-Fmoc-aminomethyl)phenoxy (Rink amide resin; NovaBiochem, Läufelfingen, Switzerland) for peptide amide synthesis are prepared per reaction vessel (2 ml polypropylene syringe from BD – Heidelberg, Germany – with a Sinterpor fluid filter from Angst+Pfister AG – Mörfelden, Germany).
2. As solvent *N,N*-dimethylformamide (DMF) from Biosolve (Valkenswaard, Netherlands) is used. For the Fmoc-deprotection piperidine (Fluka, Taufkirchen, Germany) in DMF (40% v/v) is prepared.
3. N^{α}-Fmoc-protected amino acid building blocks from IRIS Biotech (Marktredwitz, Germany) and 1-hydroxybenzotriazole (HOBt) from NovaBiochem are dissolved to 0.5 M in DMF. The side chain protecting groups for the amino acids are: *tert*-butyl (*t*Bu) for Tyr and Thr; *tert*-butyloxy (*t*BuO) for

Asp and Glu; trityl (Trt) for Asn, Gln and His; 2,2,4,6,7-pentamethyldihydrobenzofuran-5-sulfonyl (Pbf) for Arg; either *tert*-butyloxycarbonyl (Boc) or 1-(4,4-dimethyl-2,6-dioxocyclohex-1-ylidene)ethyl (Dde) for Lys according to the synthesis strategy.

4. As solvent for the amino acid coupling, a *N,N'*-diisopropylcarbodiimide (DIC)/DMF solution (1.65 M) is prepared. DIC is purchased from Sigma-Aldrich (Taufkirchen, Germany).
5. The amino acid building block Fmoc-Lys(Dde)-OH is introduced to allow a branched peptide design. The ε-amino group of lysine is protected by the base-labile group Dde, and can be selectively removed by using freshly prepared hydrazine in DMF (2% v/v).
6. For the peptide labeling with 5(6)-carboxyfluorescein (CF) CF, *O*-(7-azabenzotriazol-1-yl)-1,1,3,3-tetramethyluronium hexafluorophosphate (HATU) and *N,N*-diisopropyl-ethylamine (DIEA) (all reagents from Fluka) are freshly dissolved in 500 μl DMF (each 1.5 eq. to the peptide), and used immediately. To protect the hydroxy group of CF of side reactions, a solution of trityl chloride (Fluka) and DIEA (both 4 eq. to the peptide) in dichloromethane (DCM; Biosolve) is used.
7. Trifluoroacetic acid (TFA, peptide synthesis grade) from Riedel-de Haën (Seelze, Germany), thioanisole and *p*-thiocresole from Fluka, and diethyl ether (Biosolve) are necessary for the peptide cleavage from the resin.
8. TFA (HPLC grade; Fluka) in aqua bidest. (0.1% v/v) and in acetonitrile (ACN; Merck, Darmstadt, Germany) (0.08% v/v) is prepared freshly as HPLC solvent A and B, respectively.

2.2. Cell Culture

1. Dulbecco's Modified Eagle Medium (DMEM)/ Ham's F12 supplemented with 15% heat-inactivated fetal bovine serum (FBS) (PAA, Cölbe, Germany).
2. Phosphate-Buffered Saline (PBS) and a solution from trypsin/ethylenediamine tetraacetic acid (EDTA) (0.05% / 0.02% in PBS) was from PAA.
3. Cells are seeded on 96-well or 24-well plates, on coverslips (12 mm) in 24-well plates (all of them from TPP, Trasadingen, Switzerland), or in 8-well chamber slides (ibidi, Martinsried, Germany).

2.3. Cell Viability Assay

1. Stock solutions (1 mM) of the unlabeled or CF-labeled carrier peptides are prepared with RNase- and DNase-free aqua dest. (Invitrogen, Carlsbad, Canada), stored at –20°C, and diluted in DMEM/ Ham's F12 with FBS (PAA) as required.

2. Ethyl alcohol (70%) from Merck (Darmstadt, Germany) serves as positive control.
3. Resazurin-based in vitro toxicology assay kit solution (Sigma-Aldrich) is freshly diluted in serum-free Dulbecco's Modified Eagle Medium/Ham's F12 (1:5, v/v), pre-heated at 37°C, and added to the cells immediately.

2.4. Peptide-Mediated Cell Transfection with Vector DNA

1. Carrier peptide stock solutions (1 mM) in RNase- and DNase-free aqua dest. (Invitrogen), and stored at –20°C.
2. Vector DNA pEGFP-N1 (4,733 bp) encoding the enhanced green fluorescent protein (eGFP) is from Clontech (Mountain View, Canada). Maxi preparations were dissolved in RNase- and DNase-free aqua dest. at 1 μg/μl, and stored at –20°C.
3. Chloroquine (Fluka) is stored in aliquots as 2 mM stock solution in RNase- and DNase-free aqua dest. at –20°C.
4. The commercial transfection reagent Lipofectamine™ 2000 from Invitrogen is used as positive control.
5. Dulbecco's Modified Eagle Medium (DMEM)/Ham's F12 without and supplemented with 15% heat-inactivated FBS (PAA), as well as serum-free OptiMEM (Gibco) are used.

2.5. Peptide-Mediated Cell Transfection with Aptamer RNA

1. Carrier peptide stock solutions (1 mM) in RNase- and DNase-free aqua dest. (Invitrogen), and stored at –20°C.
2. The unspecific 3′-terminal Cy3-labeled aptamer RNA (101 bases, 270 μM RNA concentration, 220 μM Cy3 concentration) with the sequence 5′-GGAGCUCAGCCUUCACUGCCGGAGUGCA UUGUUGGCUGGGUGAUCGUGGUGGAGACCUUA CCCCUCUACGGGGGGGUGGAGGGCACCACGGUCG GAUCCAC-3′ was kindly given by Professor Dr. U. Hahn (Department for Biochemistry and Molecular Biology, University of Hamburg, Germany).
3. RNase- and DNase-free aqua dest. (Invitrogen), serum-free OptiMEM (Gibco) and PBS (PAA) containing 1% glucose are used.

2.6. Fluorescence Imaging of Peptide–Cargo Uptake

1. Stock solutions (1 mM) of the CF-labeled carrier peptides were prepared in RNase- and DNase-free aqua dest. (Invitrogen), stored at –20°C, and diluted in OptiMEM (Gibco) as required.
2. Bisbenzimid H33342 (Invitrogen) for nuclear counterstaining is dissolved in aqua dest. (0.5 mg/ml), and stored at –20°C.
3. Serum-free Hank's Balanced Salt Solution (HBSS): 1.26 mM $CaCl_2$, 0.49 mM $MgCl_2$, 0.41 mM $MgSO_4$, 5.33 mM KCl, 0.44 mM KH_2PO_4, 4.17 mM $NaHCO_3$, 137.93 mM NaCl, 0.34 mM Na_2HPO_4, and 5.56 mM D-glucose dissolved in aqua bidest., stored at –20°C.

4. Quenching solution: trypan blue (0.65 mM) (Fluka) in sodium acetate buffer (0.1 M sodium acetate trihydrate + 0.1 M acetic acid, 19:31 v/v, pH 4.5), stored at –20°C.
5. Paraformaldehyde (PFA) (Sigma-Aldrich) is diluted in PBS (PAA) (4% v/v), and stored at –20°C.
6. Mounting medium: Fluoromount-G (SouthernBiotech, Birmingham, UK).

2.7. Flow Cytometric Quantification of Peptide–Cargo Uptake

1. Stock solutions (1 mM) of the CF-labeled carrier peptides were prepared in RNase- and DNase-free aqua dest. (Invitrogen), stored at –20°C, and diluted in OptiMEM (Gibco) as required.
2. Quenching solution: trypan blue (0.65 mM) in sodium acetate buffer (0.1 M sodium acetate trihydrate + 0.1 M acetic acid, 19:31 v/v, pH 4.5), stored at –20°C.
3. Solution of trypsin/ethylenediamine tetraacetic acid (EDTA) (0.05%/0.02% in PBS) from PAA.
4. Dulbecco's Modified Eagle Medium (DMEM)/Ham's F12 supplemented with 15% heat-inactivated FBS from PAA and serum-free OptiMEM (Gibco) is used.
5. Filtrated (pore-size 0.2 μm) serum-free Hanks' Balanced Salt Solution (HBSS).

3. Methods

The carrier peptides are synthesized according to the Fmoc-strategy by using an automated multiple solid-phase peptide synthesizer (Syro, MultiSynTech, Bochum, Germany). In principle, all following steps can be done manually as well. Since a Rink amid resin is used, the peptides are synthesized as C-terminal amides. Prior to each amino acid coupling step the base-labile N^{α}-protecting group Fmoc has to be cleaved off from the building blocks, in a first step from the Rink amid resin. The peptide sequence is completed by repetitive Fmoc deprotection and amino acid coupling steps. The design and sequences of hCT(18-32)-k7 and hCT(9-32)-2br are shown in **Table 22.1**.

The cellular uptake efficiency of any carrier peptide can be investigated easily by using fluorescence-based techniques. This requires a fluorescent peptide label, in our case 5(6)-carboxyfluorescein (CF) *(10)* (*see* **Note 1**). Furthermore, CF serves as small cargo to examine the applicability of any cell-penetrating peptide as peptidic carrier. CF is introduced at the N^{α}-amino group of the otherwise fully protected peptide, while it is still bound to the

Table 22.1

Amino acid sequences, calculated and experimental molecular masses of hCT(9-32)-2br and hCT(18-32)-k7, and the viability of HEK 293 cells after 24 h treatment with 150 μM peptide solution. No relevant cytotoxic effect of both peptides was observed

Peptide	Sequence[a]	$MW_{calc.}$ (Da)[b]	$MW_{exp.}$ (Da)	Cell viability (%)[c]
hCT(9-32)-2br	LGTYTQDFNKFHTFPQTAIGVGAP-NH_2 AFGVGPDEVKRKKKPAFGVGPDEVKRKKKP	5,886.8	5,886.9	> 95
hCT(18-32)-k7	KFHTFPQTAIGVGAP-NH_2 AFKRKKKPAKRKK	3,165.9	3,166.0	> 95

[a] The peptides were synthesized as peptide amides by using a Rink amid resin.
[b] All peptides were also N-terminally labeled with CF (MW + 356.4 Da).
[c] Cell viability was determined by using a resazurin-based in vitro toxicology assay (Sigma).

resin. For a schematic illustration of the synthesis of CF-hCT(18-32)-k7 *see* **Fig. 22.1**.

3.1. Solid-Phase Peptide Synthesis

1. For each peptide a polypropylene syringe is equipped with a fluid filter, filled with 30 mg Rink amid resin, and furnished with a magnetic stirrer.
2. Initially, the resin is swollen in 800 μl DMF for 10 min with stirring. Subsequently, the solution is sucked off.
3. For Fmoc cleavage, 400 μl piperidine in DMF (40% v/v) are added to the resin and incubated for 3 min with stirring. The solution is removed, and the deprotection step is repeated with 400 μl piperidine in DMF (20% v/v) for 10 min with stirring. Then the solution is discarded, and the resin is washed four times with 600 μl DMF.
4. The first (C-terminal) amino acid is coupled by incubation of the resin with 300 μl amino acid building block solution (0.5 M in DMF with 0.5 M HOBt) and 100 μl 1.65 M DIC in DMF for 40 min with stirring. After washing with 800 μl DMF, the coupling step is repeated as described. Finally, the resin is washed two times with 800 μl DMF.
5. To complete the peptide sequence, the Fmoc-protecting group of the respective N-terminal amino acid has to be removed and the following amino acid (ongoing to the peptide N-terminus) has to be coupled as mentioned above in **Section 3.1 Steps 3 and 4**.

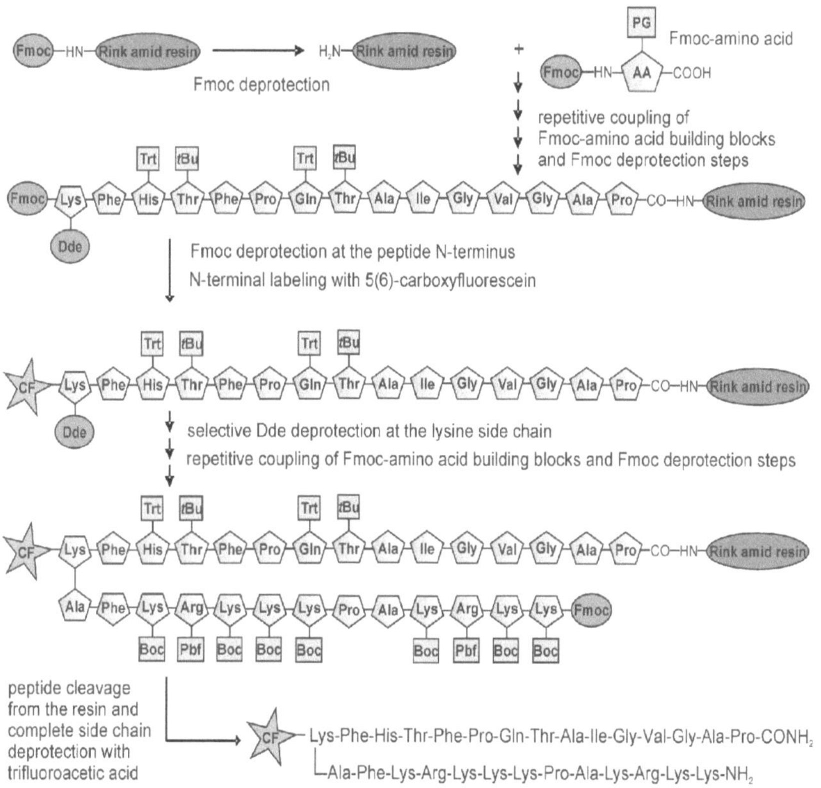

Fig. 22.1. Scheme of the solid-phase peptide synthesis (SPPS) of CF-hCT(18-32)-k7. The peptide amide is synthesized according to the Fmoc/*t* Bu strategy on the Rink amid resin. The sequence is built-on from the C- to the N-terminus by using N-terminally Fmoc-protected amino acid building blocks. Reactive side chains are further masked by acid-labile protecting groups (*squares*). To elongate the resin and the peptide sequence, respectively, the N-terminal base-labile Fmoc group has to be removed with piperidine. The following amino acid building block is coupled after activation by HOBt/DIC. These two steps are carried out repetitive until the sequence is completed. The free N-terminus can be labeled with 5(6)-carboxyfluorescein (CF). To avoid any side reactions, the CF hydroxy group should be protected with the acid-labile trityl group. The branched design of hCT(18-32)-k7 and hCT(9-32)-2br is achieved by introducing a Dde-side chain protected lysine. By using hydrazine, this Dde-protecting group can be selectively removed from the ε-amino group. Subsequently, the oligocationic side chain sequence is added as described above. If the synthesis is finished, the peptide is cleaved from the resin with trifluoroacetic acid, removing all acid-labile side chain protecting groups simultaneously. Finally, the peptide is analyzed by using RP-HPLC and MALDI-ToF mass spectrometry, and if necessary purified by using preparative RP-HPLC.

6. The CF label is N-terminally introduced by incubating the resin with 500 μl of the labeling solution (DMF containing CF, HATU and DIEA) for 2 h at room temperature under shaking. Subsequently, to protect the CF group against any side reactions, the resin is shaken under trityl chloride solution (in DCM with DIEA) for 16 h at room temperature. Finally, the resin is washed four times with each: DCM, MeOH, and diethyl ether, and is dried.

7. For the selective deprotection of a Dde-protected lysine residue, the fully protected, resin-bound peptide is incubated 10 × 10 min with 1 ml freshly prepared hydrazine in DMF. After each of the ten steps the resin is washed with 2 × 1 ml hydrazine solution. After the first and the tenth step, the removed hydrazine solution (3 ml) is collected, and its absorption is measured at 301 nm against a reference of fresh hydrazine in DMF. The Dde deprotection was successful if the absorption of the tenth fraction is <0.1. Otherwise, some more 10 min steps have to be carried out.
8. The complete peptide amide is cleaved from the resin with TFA/thioanisole/thiocresole (900:50:50 μl) within 3 h at room temperature, removing all acid-labile protecting groups (*see* **Note 2**). The peptide is 20 min precipitated from and subsequently washed at least five times with ice cold diethyl ether (*see* **Note 3**), then collected by centrifugation and lyophilized from 2–3 ml water/*tert*-butyl alcohol (3:1 v/v) (*see* **Note 4**).
9. To analyze the synthesized peptide by analytical RP-HPLC on a Vydac RP18-column (4.6 × 250 mm; 5 mm/300 Å), a linear gradient of 10–60% solvent B in A over 30 min with a flow rate of 0.6 ml min^{-1} is used. Peptide identification is performed by MALDI-ToF mass spectrometry (Voyager RP, Perspective Biosystems) by using α-cyano-4-hydroxycinnamic acid as matrix.
10. If the peptide purity is insufficient (< 90%), purification will be achieved by preparative RP-HPLC on a RP18-column (Waters, 5 μm, 25 × 300 mm) by using an appropriate linear gradient of solvent B in A over 50 min and a flow rate of 15 ml min^{-1}. Identification is performed by MALDI-ToF mass spectrometry.
11. Lyophilized peptides are stable at 4°C for at least some months.

3.2. Cell Culture

1. Human embryonic kidney cells HEK 293 are cultured in 75 cm^2 flasks at 37°C and 5% CO_2 in a humidified atmosphere.
2. At confluency, the cells are washed with PBS, detached with trypsin/EDTA (30 s) and resuspended in DMEM/ Ham's F12 with FBS.
3. The cells are distributed to 8-well chamber slides, 12 mm coverslips (in 24-well plates), 96-well or 24-well plates as required. In case of 8-well chamber slides and 96-well plates 25,000 cells per well will provide subconfluent cultures (~70%) after 48 h. In 24-well plates 150,000 cells per well will be subconfluent after 48 h (*see* **Note 5**).

3.3. Cell Viability Assay

1. To check the impact of the peptide on the cell viability, the carrier peptide is diluted to appropriate concentrations (1–150 μM) in standard cell culture medium. For each data point at least a triplicate of wells should be prepared. The resulting peptide solutions are used to replace the medium from sub-confluent HEK 293 cells in 96-well plates. Subsequently, the cells are incubated for 24 h under standard growth conditions.
2. After 24 h the cells are used for the fluorometric resazurin-based cell viability assay (*see* **Note 6**). As negative control untreated cells are investigated. Cells, incubated for 10 min after carefully replacement of the incubation solution by 70% alcohol, serve as positive control.
3. Subsequently, a freshly prepared resazurin solution (1:5 v/v in serum-free DMEM/ Ham's F12) is added (1:1 v/v) to the cells. Then the cells are incubated for 2 h under standard growth conditions. Finally, the metabolic conversion of resazurin to the reduced resofurin by the living cells is measured fluorometrically at 595 nm (after excitation at 550 nm) by using a multiwell plate-reader (Spectrafluor plus, Tecan, Crailsheim, Germany). The resulting data are normalized to the untreated cells, which are set 100% viable. EC_{50} values of the peptide–cargo impact on the cells can be calculated with software like GraphPad Prism.

The viability of HEK 293 cells after 24 h incubation with hCT(18-32)-k7 and hCT(9-32)-2br is shown in **Table 22.1**. No cytotoxic effects of both branched peptides were observed up to concentrations of 150 μM.

3.4. Peptide-Mediated Cell Transfection with Vector DNA

1. HEK 293 cells are transfected with vector DNA in 96-well plates at ~70% confluence.
2. The vector DNA pEGFP-N1 (0.5 μg) is complexed with the carrier peptide (1 mM solution) at a peptide–plasmid charge ratio of 30:1. Complexation is done in aqueous solution for 60 min at 37°C in sterile eppendorf tubes.
3. Meanwhile, the cells are pre-incubated for 60 min with serum-free OptiMEM, including 125 μM chloroquine (*see* **Note 7**).
4. After peptide–plasmid complexation, OptiMEM and chloroquine are added to the transfection sample to a final volume of 50 μl and a chloroquine concentration of 125 μM. As positive control cells are transfected with 0.5 μg vector DNA using 0.5 μl Lipofectamine™ 2000 according to the manufacturer's guideline. As negative control untreated cells and cells treated with pEGFP-N1 and chloroquine are investigated (*see* **Note 8**).
5. Subsequently, the pre-incubation medium is replaced by the transfection sample, followed by incubation for 4 h at standard growth conditions.

6. After that, the transfection medium is replaced by fresh cell culture medium, and the cells are cultured for 24 h.
7. The success of the carrier peptide-mediated cell transfection is detected after 24 h by monitoring the eGFP expression (eGFP$_{exc.}$: 488 nm, eGFP$_{em.}$: 509 nm) by using fluorescence microscopy. In **Fig. 22.2B** and **C** HEK 293 cells are shown that have been transfected by using hCT(18-32)-k7 and hCT(9-32)-2br.

3.5. Peptide-Mediated Cell Transfection with Aptamer RNA

1. HEK 293 cells in 24-well plates are transfected with aptamer RNA at ~70% confluence. Therefore, Cy3-labeled aptamer RNA is non-covalently complexed with the carrier peptide. The final concentrations at the cells are 1 μM aptamer RNA and 100 μM peptide in 300 μl OptiMEM.
2. Initially, the aptamer RNA has to be de- and renatured. For this purpose, the required volume of RNA stock solution (1.4 μl) is diluted with 8.6 μl of RNase- and DNase-free aqua dest., heated up to 70°C for 5 min, and cooled down to room temperature (~ 15 min).

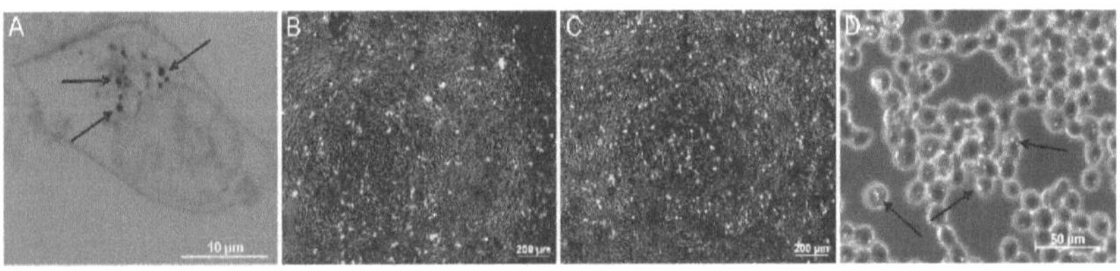

Fig. 22.2. Fluorescence microscopic images of cellular uptake in and transfection of HEK 293 cells by using CF-labeled and unlabeled hCT(18-32)-k7 and hCT(9-32)-2br. (**A**) The cells were incubated with 25 μM of CF-hCT(18-32)-k7 for 60 min. Ten minutes prior the end of the incubation bisbenzimid H33342 is added to the peptide solution for nucleus counterstaining. After the incubation solution is removed, external CF fluorescence is quenched by a trypan blue quenching step. The image exemplary shows the typical punctuated, cytoplasmatic fluorescence pattern of the hCT-derived carrier peptides (*arrows*), indicating an endocytotic mode of cell internalization *(14, 11, 12)*. Images (**B**) and (**C**) illustrates the peptide-mediated HEK 293 transfection with plasmid DNA encoding for eGFP. About 0.5 μg plasmid were complexed with hCT(18-32)-k7 (**B**) or hCT(9-32)-2br (**C**) in a 30:1 peptide–plasmid charge ratio for 60 min at 37°C. After that, pre-heated OptiMEM and chloroquine (CQ) were added to a final concentration of 125 μM CQ in 50 μl sample. Meanwhile, the cells were pre-incubated with OptiMEM containing 125 μM CQ. This pre-incubation medium was removed by the transfection mixture, the cells were incubated under standard growth conditions, then the transfection mixture was removed by fresh cell culture medium. The transfection rates and expression levels of eGFP were investigated after 24 h by using fluorescence microscopy and flow cytometry (data not shown). (**D**) HEK 293 cells were transfected with Cy3-labeled aptamer RNA by using hCT(9-32)-2br. Peptide and aptamer RNA (final concentrations in 300 μl: 100 and 1 μM, respectively) were complexed for 60 min at 37°C. Then pre-heated OptiMEM was added to a final volume of 300 μl, and the cells were incubated with this mixture for 60 min at 37°C. After that, the transfection mixture was removed, the cells were washed with PBS (containing 1% glucose) and investigated by using fluorescence microscopy. As shown in image **Fig. 2D** the aptamer RNA was internalized with good efficiency by using hCT(9-32)-2br (*arrows*). These results demonstrate the high potential of branched hCT-derived peptides as drug carriers. The scale bars are 10 μm (**A**), 200 μm (**B** and **C**) and 50 μm (**D**).

3. Then, the required volume of carrier peptide stock solution (30 µl of 1 mM) is added. After good mixing by pipetting the non-covalent peptide–aptamer complexation is carried out at 37°C for 1 h.
4. Subsequently, 260 µl pre-heated serum-free OptiMEM are added to reach a final volume of 300 µl (*see* **Note 8**).
5. Cells are rinsed with serum-free OptiMEM, and then incubated with the transfection mixture for 1 h under standard growth conditions.
6. After the incubation, the transfection mixture is removed, the cells are rinsed and covered with PBS (with 1% glucose), and the aptamer RNA internalization is investigated by fluorescence microscopy ($Cy3_{exc.}$: 550 nm, $Cy3_{em.}$: 565 nm). In **Fig. 22.2D**, the transfection with aptamer RNA is shown by using hCT(9-32)-2br.

3.6. Fluorescence Imaging of Peptide Uptake

1. Subconfluent cells are investigated for fluorescence microscopic imaging of the peptide uptake. To examine living, unfixed cells, 8-well chamber slides are used. Cells in 24-well plates on coverslips are used for subsequent fixation. In general, the investigation of living, unfixed cells should be preferred, since even with mild fixation procedures the occurrence of artifacts (delocalized peptide fluorescence) cannot be excluded *(13)*. However, in case of time consuming imaging procedures or to detect one very precise time point, cell fixation might be indispensable (*see* **Note 9**).
2. The CF-labeled carrier peptide (1 mM stock solution) is diluted in pre-heated OptiMEM to appropriate concentrations in the range of 1–50 µM. Subsequently, the cell medium is replaced by the prepared peptide solutions. Incubation at standard growth conditions should be investigated until at least 60 min. For the visualization of the nucleus a counterstaining with H33342 (5 µg/m*l* medium) 10 min prior the end of the peptide incubation is possible.
3. To discriminate between internalized and membrane surface-bound peptide fluorescence, a quenching step is necessary. Therefore, after the careful removal of the peptide–cargo solution the extracellular fluorescence is quenched by a 1 min washing step with acidic trypan blue buffer (in sodium acetate, pH 4.5) (*see* **Note 10**). Then, the trypan blue buffer is removed, and the cells are rinsed one time with ice cold OptiMEM.
4. For cell fixation, quenched and washed cells on 12 mm coverslips are incubated with 4% PFA/PBS for 30 min. Subsequently, the cells are washed three times for 10 min with HBSS and embedded on microscope slides with

Fluoromount-G. The resulting slides can be stored at 4°C for several weeks without loss of quality.

5. When investigating living cells, after trypan blue quenching and rinsing with OptiMEM the cells are immediately inspected under HBSS.
6. CF and H33342 are visualized with appropriate microscope filter sets ($CF_{exc.}$: 494 nm, $CF_{em.}$: 518 nm; $H33342_{exc.}$: 365 nm, $H33342_{em.}$: 418 nm). The internalization of CF-hCT(18-32)-k7 is exemplary illustrated in **Fig. 22.2A**.

3.7. Flow Cytometric Quantification of Peptide–Cargo Uptake

1. Flow cytometry is an appropriate technique for time-saving quantifications of the cellular uptake efficiency of a carrier peptide and its cargo, respectively. In principle, the incubation of subconfluent cells in the 96-well scale is sufficient, and should be carried out as described above and with CF-labeled carrier peptides.
2. Following the incubation with the peptide–cargo complex or the expression of the fluorescent protein, the cells have to be harvested from the 96-well plate.
3. If the uptake of CF-labeled peptides is investigated, a trypan blue quenching step will be necessary as described above. The quenched cells are washed one time with OptiMEM.
4. If the peptide-mediated DNA delivery (expression level and transfection efficiency) is investigated, the trypan blue quenching will be not necessary, and the cells are just washed with OptiMEM.
5. Subsequently, the cells are detached from the bottom by incubation with pre-heated (37°C) trypsin/EDTA for 30 s. Then, the catalytic activity of trypsin is stopped by addition of DMEM/ Ham's F12 supplemented with 15% FBS.
6. After the removal of this solution the cells are resuspended in ice cold, filtrated HBSS and analyzed by a flow cytometer.
7. For reproducible results at least 10,000 cells should be measured per sample. The flow cytometric data can be evaluated by using software like FlowJo.

Figure 22.3 shows the quantification of the cellular uptake of CF-hCT(18-32)-k7 as measured by using flow cytometry. The internalization of this carrier peptide is time- (**Fig. 22.3A**) and concentration-dependent (**Fig. 22.3B**). Furthermore, the observation that the uptake is markedly decreased at 4°C (**Fig. 22.3C**) supports the assumption of an energy-dependent internalization pathway of this peptide, i.e., endocytosis. Since a tenfold excess of unlabeled hCT(18-32)-k7 does not compete with the CF-labeled peptide (no decreased uptake as shown in **Fig. 22.3D**), the

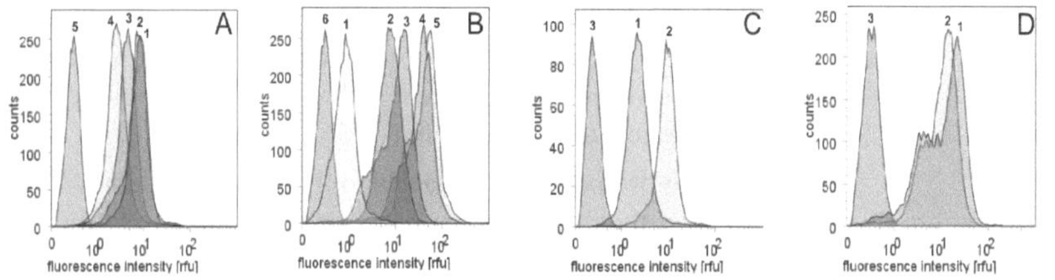

	Sample Name	Mean, FL 1
1	120 min	9,02
2	60 min	8,31
3	30 min	5,20
4	5 min	3,31
5	untreated cells	0,31

	Sample Name	Mean, FL 1
1	1 µM	1,08
2	5 µM	8,07
3	20 µM	13,52
4	50 µM	37,88
5	100 µM	49,11
6	untreated cells	0,32

	Sample Name	Mean, FL 1
1	4 °C	3,76
2	37 °C	10,20
3	untreated cells	0,26

	Sample Name	Mean, FL 1
1	+ 10 eq. competitor	15,63
2	20 µM	12,39
3	untreated cells	0,35

Fig. 22.3 Flow cytometric quantification of the internalization of CF-hCT(18-32)-k7 into HEK 293 cells. Subconfluent cells were incubated in 96-well plates with peptide solutions in OptiMEM. The influence of the incubation time (**A**), the peptide concentration (**B**), the incubation temperature (**C**) and an eventual competition by an excess of unlabeled peptide (**D**) was examined. After the peptide incubation a trypan blue quenching step was carried out to exclude surface-bound CF fluorescence. (**A**) 20 μM CF-hCT(18-32)-k7 were incubated at 37°C for up to 120 min. It is shown that the peptide uptake increases time-dependent. Already after 5 min the fluorescence intensity inside the cells is tenfold increased compared to untreated cells. This indicates a quite rapid internalization mechanism. After 60 min a maximum of peptide internalization is reached. (**B**) Various concentrations of the peptide were incubated at 37°C for 60 min. The concentration dependence of the peptide uptake into the cells is obvious. Good internalization rates are possible already with 5 μM CF-hCT(18-32)-k7. Thus, this peptide is a very efficient cell-penetrating peptide. (**C**) The temperature-dependence of the internalization was investigated by using 20 μM peptide for 60 min at 37 and 4°C. The markedly decrease at 4°C is a hint for an energy-dependent internalization pathway, since ATP-dependent mechanisms are inhibited at low temperatures. This finding supports the assumption of an endocytotic mode of action also for hCT(18-32)-k7. (**D**) As was shown by using 20 μM CF-labeled and 200 μM unlabeled hCT(18-32)-k7 at 37°C for 60 min no competition was observed. This indicates an unspecific internalization pathway, i.e., apparently no specific receptors or transporters are involved in the peptide internalization.

involvement of specific receptors or transporters in the uptake mechanism of CF-hCT(18-32)-k7 is unlikely.

4. Notes

1. The choice of the fluorescence label should not only depend on synthetic considerations but also on the microscopic and spectroscopic equipment that is available to excite and measure the fluorescence. We have found 5(6)-carboxyfluorescein to be excellent for the on-resin labeling at the N-terminus or N^{ε} of lysine residues. However, numerous other fluorescent marker compounds are available.
2. The use of the scavengers generates an unpleasant smell in the laboratory. Therefore, this procedure has to be done under an excellent hood.
3. Incomplete removal of the scavengers by washing with ice cold diethyl ether can be easily detected by the characteristic smell of the scavengers.

4. Instead of dissolving the peptide in water/*tert*-butyl alcohol (3:1 v/v), hydrophilic, well water-soluble peptides can be dissolved in pure aqua dest. and highly hydrophobic peptides in water/*tert*-butyl alcohol (1:3 v/v) as well.
5. This protocol can be adapted for many other cell lines. However, the detaching procedure has to be optimized for each cell line. In dependence of the investigated cell line, in some cases PBS/EDTA should be used instead of trypsin/EDTA. Also appropriate cell densities for the passage in various scales have to be tested for each cell line.
6. In our hands, the resazurin-based in vitro toxicity assay convinced with good reproducibility and ease of handling. Since this kit is not cytotoxic or cell-disrupting per se, the investigated cells could be cultured for further experiments even after the viability test.
7. The anti-malaria drug chloroquine was shown to enhance the gene expression after endocytotic cell transfections in several studies *(14, 15)*. However, before using chloroquine in transfection experiments the effect of various concentrations on the cell viability should be tested.
8. To avoid the disruption of the non-covalent peptide–DNA complexes, mixing should be done very carefully by pipetting.
9. If a cell fixation is necessary, a mild fixation protocol should be used in any case, e.g., 30 min with 4% PFA. To exclude any fluorescence delocalization due to the fixation, the fluorescence distribution should be compared in unfixed and fixed cells in advance.
10. Washing with trypan blue was shown to efficiently quench extracellular fluorescein fluorescence *(12, 16)*. To improve the fluorescence quenching, we dissolved the trypan blue in an acidic buffer (~ pH 4.5), since 5(6)-carboxyfluorescein is quenched by acidic pH. Trypan blue is toxic, therefore while preparing the stock solution breathing protection has to be used.

Acknowledgments

The authors would like to thank Doris Haines, Regina Reppich and Kristin Löbner for technical assistance in peptide synthesis and analysis. We thank Prof. Dr. Ulrich Hahn for providing the aptamer RNA. This work was supported by EU grant QoL-2001-01451.

References

1. Gupta, B., Levchenko, T.S. and Torchilin, V.P. (2005) Intracellular delivery of large molecules and small particles by cell-penetrating proteins and peptides. *Adv. Drug Deliv. Rev.* **57**, 637–651.
2. Mae, M. and Langel, U. (2006) Cell-penetrating peptides as vectors for peptide, protein and oligonucleotide delivery. *Curr. Opin. Pharmacol.* **6**, 509–514.
3. Schmidt, M.C., Rothen-Rutishauser, B., Rist, B., Beck-Sickinger, A., Wunderli-Allenspach, H., Rubas, W., Sadee, W. and Merkle, H.P. (1998) Translocation of human calcitonin in respiratory nasal epithelium is associated with self-assembly in lipid membrane. *Biochemistry* **37**, 16582–16590.
4. Trehin, R., Krauss, U., Muff, R., Meinecke, M., Beck-Sickinger, A.G. and Merkle, H.P. (2004) Cellular internalization of human calcitonin derived peptides in MDCK monolayers: a comparative study with Tat(47-57) and penetratin(43-58). *Pharm. Res.* **21**, 33–42.
5. Krauss, U., Kratz, F. and Beck-Sickinger, A.G. (2003) Novel daunorubicin-carrier peptide conjugates derived from human calcitonin segments. *J. Mol. Recognit.* **16**, 280–287.
6. Machova, Z., Muehle, C., Krauss, U., Trehin, R., Koch, A., Merkle, H.P. and Beck-Sickinger, A.G. (2002) Cellular internalization of enhanced green fluorescent protein ligated to a human calcitonin-based carrier peptide. *Chembiochem* **3**, 672–677.
7. Krauss, U., Mueller, M., Stahl, M. and Beck-Sickinger, A.G. (2004) *In vitro* gene delivery by a novel human calcitonin (hCT)-derived carrier peptide. *Bioorg. Med. Chem. Lett.* **14**, 51–54.
8. Futaki, S., Nakase, I., Suzuki, T., Youjun, Z. and Sugiura, Y. (2002) Translocation of branched-chain arginine peptides through cell membranes: flexibility in the spatial disposition of positive charges in membrane-permeable peptides. *Biochemistry* **41**, 7925–7930.
9. Futaki, S., Suzuki, T., Ohashi, W., Yagami, T., Tanaka, S., Ueda, K. and Sugiura, Y. (2001) Arginine-rich peptides. An abundant source of membrane-permeable peptides having potential as carriers for intracellular protein delivery. *J. Biol. Chem.* **276**, 5836–5840.
10. Fischer, R., Mader, O., Jung, G. and Brock, R. (2003) Extending the applicability of carboxyfluorescein in solid-phase synthesis. *Bioconjug. Chem.* **14**, 653–660.
11. Foerg, C., Ziegler, U., Fernandez-Carneado, J., Giralt, E., Rennert, R., Beck-Sickinger, A.G. and Merkle, H.P. (2005) Decoding the entry of two novel cell-penetrating peptides in HeLa cells: lipid raft-mediated endocytosis and endosomal escape. *Biochemistry* **44**, 72–81.
12. Rennert, R., Wespe, C., Beck-Sickinger, A.G. and Neundorf, I. (2006) Developing novel hCT derived cell-penetrating peptides with improved metabolic stability. *Biochim. Biophys. Acta* **1758**, 347–354.
13. Richard, J.P., Melikov, K., Vives, E., Ramos, C., Verbeure, B., Gait, M.J., Chernomordik, L.V. and Lebleu, B. (2003) Cell-penetrating peptides. A reevaluation of the mechanism of cellular uptake. *J. Biol. Chem.* **278**, 585–590.
14. Cheng, J., Zeidan, R., Mishra, S., Liu, A., Pun, S.H., Kulkarni, R.P., Jensen, G.S., Bellocq, N.C. and Davis, M.E. (2006) Structure-function correlation of chloroquine and analogues as transgene expression enhancers in nonviral gene delivery. *J. Med. Chem.* **49**, 6522–6531.
15. Erbacher, P., Roche, A.C., Monsigny, M. and Midoux, P. (1996) Putative role of chloroquine in gene transfer into a human hepatoma cell line by DNA/lactosylated polylysine complexes. *Exp. Cell. Res.* **225**, 186–194.
16. Hed, J., Hallden, G., Johansson, S.G. and Larsson, P. (1987) The use of fluorescence quenching in flow cytofluorometry to measure the attachment and ingestion phases in phagocytosis in peripheral blood without prior cell separation. *J. Immunol. Methods* **101**, 119–125.

Index

U

V

X

Y

MIX
Papier aus verantwortungsvollen Quellen
Paper from responsible sources
FSC® C105338

If you have any concerns about our products,
you can contact us on
ProductSafety@springernature.com

In case Publisher is established outside the EU,
the EU authorized representative is:
Springer Nature Customer Service Center GmbH
Europaplatz 3, 69115 Heidelberg, Germany

Printed by Libri Plureos GmbH
in Hamburg, Germany